Computer-Based Modeling of Novel Carbon Systems and Their Properties

CARBON MATERIALS: CHEMISTRY AND PHYSICS

A comprehensive book series which encompasses the complete coverage of carbon materials and carbon-rich molecules from elemental carbon dust in the interstellar medium, to the most specialized industrial applications of the elemental carbon and derivatives. A great emphasis is placed on the most advanced and promising applications ranging from electronics to medicinal chemistry. The aim is to offer the reader a book series which not only consists of self-sufficient reference works, but one which stimulates further research and enthusiasm.

Series Editors

Dr. Prof. Franco Cataldo
Director of Lupi Chemical
Research Institute
Via Casilina 1626/A
00133 Rome
Italy

Professor Paolo Milani
University of Milan
Department of Physics
Via Celoria, 26
20133, Milan
Italy

VOLUME 3:
COMPUTER-BASED MODELING OF NOVEL CARBON SYSTEMS AND THEIR PROPERTIES: BEYOND NANOTUBES

Volume Editors

Prof. Luciano Colombo
Department of Physics
University of Cagliari
Cittadella Univeritaria
I-09042 Monserrato (Ca)
Italy

Prof. Annalisa Fasolino
Theory of Condensed Matter
Institute for Molecules and Materials
Radboud University Nijmegen
Heyendaalseweg 135
6525 AJ Nijmegen
The Netherlands

For other titles published in this series, go to
http://www.springer.com/series/7825

Luciano Colombo • Annalisa Fasolino
Editors

Computer-Based Modeling of Novel Carbon Systems and Their Properties

Beyond Nanotubes

Springer

Editors
Prof. Luciano Colombo
Department of Physics
University of Cagliari
Monserrato, Italy
luciano.colombo@dsf.unica.it

Prof. Annalisa Fasolino
Theory of Condensed Matter
Institute for Molecules and Materials
Radboud University Nijmegen, Nijmegen
The Netherlands
a.fasolino@science.ru.nl

ISSN 1875-0745 e-ISSN 1875-0737
ISBN 978-94-007-3254-4 ISBN 978-1-4020-9718-8(eBook)
DOI 10.1007/978-1-4020-9718-8
Springer Dordrecht New York Heidelberg London

Printed on acid-free paper

Springer is part of Springer Science+Business Media (www.springer.com)

Preface

The number of observed or guessed atomic architectures formed by elemental carbon has unexpectedly increased in the last decades. In addition to graphite and diamond, a multiplicity of other structures – such as well ordered fullerenes and nanotubes, or less ordered structures like nanoporous and amorphous carbon – have been synthesized and studied. Similarly to what happens for diamond with respect to graphite, the new metastable phases (although basically sp^2-bonded) are definitely other than graphite (still to consider as the most stable phase) as for their physical and chemical features in a wide range of temperatures and pressures. Interestingly, the mechanical, structural, and electronic properties of these new forms of carbon are extremely different, ranging from soft to hard, from compact to open, from insulating to metallic.

Theory and simulations have largely contributed to understand and characterize the new carbon-based systems and led to the prediction of new ones. The multiplicity of possible arrangements, the diversity of behaviors, and the complex interplay between structures and properties poses a formidable challenge to theoretical and computational physicists. A number of different methods are therefore needed, ranging from model-potential molecular dynamics, to tight-binding calculations, to first-principles simulations. As a general feeling, it is also becoming evident that their combination into a unique theoretical and computational tool is actually needed to make new progress in this field.

This volume presents a unique survey of the theoretical modeling of all phases of carbon – other than single fullerene molecules or nanotubes – from natural crystalline forms found on earth and in meteorites to artificial (hypothetical) nanofoams. In addition, the present volume deals with the computational techniques used to understand and predict the structure and properties of such carbon systems, as well as reports about the present state of the art, including controversial aspects like the occurrence of magnetism, and presents open questions for the future. Although the main focus is on carbon-based systems, the computational challenges posed by their diverse structural, bonding, mechanical and electronic properties are relevant for all materials and make the present volume valuable for the whole community of computational condensed matter physics.

Each chapter is a self-contained authoritative exposition by scientists with an international reputation, sharing their knowledge and tricks of the trade. As a whole, the book provides a basis towards a unified theoretical description of carbon, the most fascinating element in Nature.

The first part of the volume is mostly devoted to the structural properties of several novel carbon structures.

Chapter 1 by Ghiringhelli and Meijer opens with a description of the phase diagram of carbon that includes its traditional phases, diamond, graphite and liquid carbon. Graphite and diamond melt into a liquid with a network-like microscopic structure at extremely high values of pressure and temperature. The phase diagram of carbon is experimentally poorly known and recent simulations give access to its determination.

The diamond structure can also become extremely stable at the nanoscale, as extensively discussed in Chapter 2 by Galli, where nanoparticles with diamond-like structures, formed in a wide variety of natural environments (around stars or in CVD phases) or under very high pressure and temperature conditions, are investigated.

Since the discovery of fullerenes and nanotubes, that are not treated in detail here, several other new forms of nanostructured carbon mostly based on the graphitic sp^2 bond configuration have been identified. Chapter 3 by Seifert, Kuc and Heine introduces the carbon nanofoams, hypothetical ordered graphitic structures that display interesting analogies with carbon nanotubes. Although not yet firmly established, these structures might have been possibly experimentally realized.

The second part of the volume is more focused on the physico-chemical properties of several exotic carbon architectures.

Chapter 4 by Carlsson introduces a more general type of structure based on sp^2 bonding called nanoporous carbon, a class of materials that have been obtained by a number of experimental methods. Nanoporous carbon and nanofoams promise to have specific cathalitic action and great potential for chemical applications.

Chapter 5 by Marks examines amorphous carbon, currently among the most interesting materials for applications. Computer simulations have largely contributed to the complex characterization of this disordered type of structure. The contribution provides a critical exam of the criteria for theory to be accurate enough to discern between different structural models.

Chapter 6 by Blase, Benedek and Bernasconi examines the exceptional mechanical properties of clathrate structures. The criteria for designing hard materials are examined, also in connection with electronic properties and possible occurrence of superconductivity.

Finally in Chapter 7 by Ganchenkova, Vehviläinen and Nieminen, the intriguing, but still controversial observation of ferromagnetism in nanostructured carbon, such as polymerised fullerenes and ion-irradiated graphitic materials is examined and the conditions for ferromagnetic ordering defect-related magnetism are explored.

Luciano Colombo
Annalisa Fasolino

Contents

Chapter 1
Liquid Carbon: Freezing Line and Structure Near Freezing

Luca M. Ghiringhelli[1] and Evert Jan Meijer[2]

Abstract This chapter deals with the phase diagram of carbon with emphasis on the liquid phase occurring in extreme conditions of temperature and pressure. After presenting a critical review of the experimental results and still unresolved issues, the authors discuss the possibility of modeling carbon by use of empirical potentials. Also the techniques to evaluate numerically the free energy of each phase are presented in detail. The second part of the chapter discusses in detail the structure of the liquid in different ranges of pressure, the pressure–density equations of state at different temperatures and the possibility of a liquid–liquid phase transition.

1.1 Introduction

Carbon exhibits a rich variety of solid structures such as the familiar crystalline graphite and diamond state, or amorphous states such as glassy carbon or carbon black. More recently additional (metastable) phases have been found or predicted. These include carbynes [1,2] and M-carbon [3]. In addition to the bulk phases there are various other recently discovered structures including fullerenes [4], nanotubes [5], and graphene [6].

Knowledge of the phase diagram of carbon is of crucial importance for a better understanding of a wide variety of physical phenomena and properties of carbon-based materials. For example, the phase diagram determines the carbon content of the interior of the Earth and other planets and it determines the optimal conditions for the manufacturing of synthetic diamonds. Knowledge of the graphite melting line is relevant for the formation mechanism of low dimensional layered structures that are important in fundamental and technological applications.

[1]Fritz Haber Institute of the Max Planck Society, Faradayweg 4-6, D-14195, Berlin, Germany
e-mail: luca@fhi-berlin.mpg.de

[2]Van't Hoff Institute for Molecular Sciences & Amsterdam Center for Multiscale Modeling, Universiteit van Amsterdam, Nieuwe Achtergracht 166, NL-1018 WV, Amsterdam, The Netherlands
e-mail: e.j.meijer@uva.nl

L. Colombo and A. Fasolino (eds.), *Computer-Based Modeling of Novel Carbon Systems and Their Properties*, Carbon Materials: Chemistry and Physics 3,
DOI 10.1007/978-1-4020-9718-8_1,

The phase diagram of carbon has been intensively studied, both experimentally and by theoretical and numerical methods [1, 7–11, 11–26]. These studies covered pressures (P) and temperatures (T) ranging up to 100 GPa and 10,000 K. Until recently the knowledge of the carbon phase diagram was still fragmented because experiments under these conditions are difficult if not outright impossible, whereas quantitative theoretical and numerical predictions were hampered by the fact that the existing atomistic models for carbon had serious flaws that made them unsuited for quantitative predictions. Recent advances in modeling of carbon allows for the calculation of the phase diagram of carbon and the structure of the liquid with unprecedented accuracy. This provided significant progress in the understanding of the behavior of carbon under extreme conditions [26–30]. Knowledge of the location of the melting line and the liquid structure near melting is an important requisite in understanding the (non)-existence of a liquid–liquid phase transition and the process of homogeneous nucleation of the liquid into graphite or diamond. The latter issue is key to answering the question of the existence of diamond in planet interiors.

In the present contribution we provide a review of some of the recent computational studies of the phase diagram and the liquid structure of carbon. First, we summarize some important characteristics of the carbon phase diagram. Then we discuss the modeling of the inter-atomic interaction with a focus on a particular class of bond-order potentials, the so-called LCBOP models. Subsequently we discuss in detail the computational methods to determine the thermodynamic stability of the diamond, graphite, and liquid phase together with a presentation of the calculated phase diagram for a well tested version of the LCBOP model. These results are discussed in a context of experimental and other computational studies. Finally, we address the structure of liquid carbon.

1.2 Carbon Phase Diagram: Some Important Characteristics

In the pressure and temperature range up to 100 GPa and 10,000 K, the well established thermodynamic stable phases of carbon are the crystalline graphite (G) and diamond (D) phase at lower temperatures, and a liquid (L) phase at higher temperatures (Fig. 1.4). The graphite-diamond coexistence line has been relatively well characterized up to 2,400 K [7, 15]. For the graphite melting line a large amount of experimental data are available [1, 8, 9, 13, 14, 16]. The experiments have in common that the melting temperature shows little variation with pressure, and that most of the measured graphite melting $P - T$ lines [8, 9, 16] show a maximum around $P = 6$ GPa. However, the nature of the maximum is not well established. The experimental estimates for the melting temperatures show a large spread. It appears that the estimated melting temperature depends significantly on the heating rate of the sample [13, 14], yielding values from 3,700 to 5,000 K below 0.01 GPa. In a recent comprehensive review of graphite melting [31] the melting temperature is proposed to be in the range of 4,600–5,000 K. The precise nature of the maximum in the melting curve is important, because a discontinuous change of slope of the melting curve at this point would imply the existence of a liquid–liquid phase transition (LLPT) line, branching off from the graphite melting curve. This will be discussed in more detail in Section 1.5.

Early shock wave experiments [11] provided evidence for the existence of diamond at $P = 140$ GPa at a temperature beyond the temperature of the Graphite–Diamond–Liquid triple point (T $\sim$ 4,000 K), implying that the carbon diamond melting line has a positive slope in the $P - T$ diagram. Recent experimental data have provided evidence [32] that diamond is stable up to at least 800 GPa.

1.3 Modeling Carbon

Realistic modeling of the carbon phase diagram involving the liquid, graphite, and diamond phase requires an accurate description of the inter-atomic interactions, combined with a precise evaluation of the relative stability of the involved phases. This requires the evaluation of the free energy of state points in all phases involved. Presently, density-functional theory (DFT) based ab initio MD simulations would provide the best possible approach. However, the computational cost associated with DFT calculations renders such an approach unfeasible, in particular when combined with free-energy calculations.

A viable alternative is to model the inter-atomic interactions by a functional description, whose parameters are (partly) fitted to a selected database. Such a functional description (also referred to as empirical, semi-empirical or classical potentials) serves several purposes, ranging from the modeling of minimum energy structures for surface reconstructions, grain boundaries or related defects, to the description of the liquid structure and thermodynamic stability.

According to Brenner [33], an analytic potential needs to be:

- *Flexible* The function should be flexible enough to accommodate the inclusion of a relatively wide range of structures in a fitting database.
- *Accurate* The potential function must be able to accurately reproduce quantities such energies, bond lengths, elastic constant, and related properties entering a fitting database.
- *Transferable* The functional form of the potential should be able to reproduce related properties that are *not* included in the fitting database. In practice the potential should be able to give a good description of the energy landscape for any possible realistic configuration characterized by the set of atomic positions $\{\mathbf{r}_i\}$.
- *Computationally Efficient* The function should be of such a form that it is tractable for a desired calculation, given the available computing resources.

Focusing on carbon, there is a noble lineage of increasingly successful parameterizations of the empirical potentials, from the Tersoff [34] potential, via the Brenner potentials [35] and their modifications [36–39] up to Los-Fasolino LCBOP [27,29]. The crucial characteristic of these potentials is that they account for the "bond order" : the potentials formalize and parameterize the idea that for covalently bonded systems an increasing number of bonds per atom modifies (typically decreases) the bond energy *per bond*. With increasing complexity, many-body contributions due to angular correlations [34], conjugation effects [35] and torsional interactions [36] were included. These short ranged potentials proved to describe increasingly well hydrocarbon molecules (when hydrogen is included) and the diamond phase. However, more disordered structures were less well described. After an earlier attempt to introduce longer range (pair) correlations [38], the seamless inclusion of non-bonded interactions led to the semi-empirical long range bond order potential (LCBOPI) by Los and Fasolino [27] that is partly based on ab initio data. The inclusion of conjugation dependent torsional interactions (LCBOPI$^+$) (Ref. [29] in Appendix A, and Ref. [40] in chapter 6.7) added the necessary flexibility to describe in a proper way the transformation between diamond and graphite and the structure of the liquid phase. LCBOPI$^+$ also accounts properly for the inter-planar interactions in graphite. In a subsequent development, the introduction of middle range interactions (i.e. allowing for smooth bond breaking and forming), together with a revision of the definition of the torsional angle, yielded LCBOPII [29], which performs as well as density functional in the liquid phase (wherever it was tested) and opens the way for an accurate description of surfaces and their reconstructions.

The LCBOP family is the first empirical potential that is capable of providing an accurate description of the graphite, diamond, and liquid phase. This makes the LCBOP's uniquely suited to predict the carbon phase diagram and the properties of liquid carbon.

1.4 The Graphite–Diamond–Liquid Phase Diagram of LCBOPI$^+$

1.4.1 Computational Methods

The properties of the liquid, graphite, and diamond phases were determined by Monte Carlo (MC) simulations. Coexistence lines were determined by locating points in the $P - T$ diagram with equal chemical potential for the two phases involved. To this purpose, we first determined the chemical potential for the liquid, graphite, and diamond at an initial state point ($P = 10$ GPa, $T = 4{,}000$ K). Subsequently, the liquid/graphite, liquid/diamond, and graphite/diamond coexistence pressures at $T = 4{,}000$ K were located. In turn, these coexistence points served as the starting point for the determination of the graphite melting, diamond melting, and graphite/diamond coexistence lines, obtained integrating the Clausius–Clapeyron equation (this procedure is also known as Gibbs–Duhem integration):

$$\frac{\mathrm{d}T}{\mathrm{d}P} = \frac{T\Delta v}{\Delta h} \tag{1.1}$$

where Δv is the difference in specific volume, and Δh the difference in molar enthalpy between the two phases (calculated as $h = u + Pv$, being u the potential energy per particle).

The first point is in turn accomplished in two sub-steps. Firstly a Helmholtz free energy (F) at a given volume (V) and temperature (T) can be calculated via thermodynamic integration. In a canonical system, coexistence between phases can be found via the Helmholtz double tangent construction, after F is evaluated at other V and T, by integrating its gradient, a quantity that can be measured in a MC simulation (see e.g. Ref. [21]). As an alternative [41], the one we chose, one can transform F into the chemical potential μ (coinciding with the specific Gibbs free energy in a one-component system), the latter as a function of P and T, knowing accurately enough the equation of state of each phase. Coexistence at a given T is found at that P where μ for the different phases cross.

For all phases, the free energies at the initial state point $F^{\maltese}$ was determined by transforming the systems into a reference system F^{ref} of known free energy, using $U_\lambda = (1-\lambda)U^{\maltese} + \lambda U^{\text{ref}}$. Here, $U^{\maltese}$ and U^{ref} denote the potential energy function of the LCBOPI$^+$ and of the reference system, respectively. The transformation is controlled by varying the parameter λ continuously from 0 to 1. The free-energy change upon the transformation was determined by thermodynamic integration:

$$
\begin{aligned}
F^{\maltese} &= F^{\text{ref}} + \Delta F^{\text{ref}\rightarrow\maltese} \\
&= F^{\text{ref}} + \int_{\lambda=0}^{\lambda=1} d\lambda \left\langle \frac{\partial U_\lambda}{\partial \lambda} \right\rangle_\lambda \\
&= F^{\text{ref}} + \int_0^1 \mathrm{d}\lambda \left\langle U^{\text{ref}} - U^{\maltese} \right\rangle_\lambda
\end{aligned}
\tag{1.2}
$$

The symbol $\langle ... \rangle_\lambda$ denotes the ensemble average with the potential U_λ.

For the liquid phase the reference system was taken to be a Lennard–Jones 12-6 (LJ) system, described by the well known interaction energy:

$$
U^{LJ} = 4\varepsilon \left(\left(\frac{\sigma}{r}\right)^{12} - \left(\frac{\sigma}{r}\right)^{6} \right)
$$

The reference free energy (F^{ref}) of the liquid is:

$$
F^{\text{ref}} = F^{\text{LJ}} = F^{\text{id}} + F^{\text{ex}}_{\text{LJ}} \tag{1.3}
$$

The ideal-gas contribution is:

$$
\frac{\beta F^{\text{id}}}{N} = 3\ln\Lambda + \ln\rho - 1
$$

where N is the number of particles in the box, $\Lambda = h/\sqrt{2\pi m\, k_B T}$ is the de Broglie wavelength, m is the mass of one atom, and ρ is the number density. The LJ liquid excess free energy ($F^{\text{ex}}_{\text{LJ}}$) has been accurately parameterized [42] by means of (NVT) MC and MD simulations.

The LJ σ parameter was determined by matching the first peak of the radial distribution functions ($g(r)$) of the LCBOPI$^+$ and LJ liquid at the same position, ensuring optimal similarity between the structure of the two liquids. The LJ ε parameter was chosen such that, at the selected $T = 4{,}000$ K, the LJ liquid was above the critical temperature: this is done in order to avoid possible unwanted transitions, since the thermodynamic integration method works under the hypothesis that no boundary between phases is ever crossed on varying λ. On the other hand, the liquid should not be too far from the critical temperature: in fact, the $g(r)$ given by the LCBOPI$^+$ (see Section 1.5) has pronounced secondary peaks beyond the first coordination shell. Thus, a rather structured LJ liquid had to be preferred for the coupling. The requirements are matched by putting the LJ liquid in proximity of the critical temperature.

For the solid phases the Einstein crystal, whose free energy is analytically known, was taken as reference system [43]. For the Einstein solid, U^E is:

$$U^E = \frac{\alpha}{2} \sum_{i=1}^{N} (\mathbf{r}_i - \mathbf{r}_{i,0})^2$$

where the $\mathbf{r}_{i,0}$ are the equilibrium (i.e. at $T = 0$ K) lattice positions of the particles. In the Einstein solid, the fixed equilibrium lattice positions are referred to an absolute frame, so that if a particle is moved, then the crystal as a whole cannot. When $\lambda \sim 0$ (i.e. the system is on the LCBOPI$^+$ side) the center of mass of the system (CoM) is free to drift: if L is the box size, the CoM mean square displacement $\langle r^2 \rangle_{CoM}$ becomes of the order of L^2. Should this happen, the integral of Eq. 1.2 becomes sharply peaked for small values of λ. In fact, the particles are allowed to drift far away from their absolute equilibrium lattice positions, since the coupling with the Einstein solid is mild, but in Eq. 1.2 appears the energy $U^{\maltese} = U^E$, that can become uncontrollably large. In order to circumvent this problem for small λ, a physically well founded remedy is to perform a simulation under the constraint that the CoM of the solid is fixed [43–45], so that $\langle r^2 \rangle_{CoM}$ is of the order of $\langle r^2 \rangle_0$, the mean square displacement of a particle from its lattice site in a real (i.e. interacting) crystal. This constraint calls for a slight modification of Eq. 1.2. We label with $E(CM)$ the Einstein solid with fixed center of mass, $\maltese(CM)$ the LCBOPI$^+$ system with fixed center of mass, so that [43–45]:

$$\begin{aligned} F^{\maltese} &= F^{\mathrm{E(CM)}} + \Delta F^{\mathrm{E(CM)} \rightarrow \maltese(CM)} + \Delta F^{\maltese\mathrm{(CM)} \rightarrow \maltese} \\ &= F^{\mathrm{E(CM)}} + \int_0^1 \mathrm{d}\lambda \left\langle U^{\mathrm{ref}} - U^{\maltese} \right\rangle_\lambda + \Delta F^{\maltese\mathrm{(CM)} \rightarrow \maltese} \end{aligned} \quad (1.4)$$

Specifically:

$$\frac{\beta F^{\mathrm{E(CM)}}}{N} = 3\ln\Lambda - \frac{3}{2}\ln\left(\frac{2\pi}{\beta\alpha}\right) - \frac{3}{2N}\left(\ln\left(\frac{\alpha\beta}{2\pi}\right) + \ln N\right) \quad (1.5)$$

The last term on the right hand side represents the (finite size) correction for the fixing of the CoM. Note its dependency on $1/N$, which consistently makes the correction vanish in the thermodynamic limit.

$$\frac{\beta \Delta F^{\maltese(\mathrm{CM})\to\maltese}}{N} = -\frac{1}{N}\ln\frac{V}{N_{ws}} \tag{1.6}$$

where N_{ws} is the number of Wigner–Seitz cells in the simulation box. If n_{ws} is the number of atoms per Wigner–Seitz cell, $N_{ws} = N/n_{ws}$. Note that also this term, a purely finite size effect, vanishes in the thermodynamic limit.

In reporting the results (in Section 1.4.2) we will group differently the terms of the previous three equations: it is indeed natural to group the terms proportional to $\frac{1}{N}$, so that:

$$\frac{\beta F^{\mathrm{E}}}{N} = 3\ln\Lambda - \frac{3}{2}\ln\left(\frac{2\pi}{\beta\alpha}\right) \tag{1.7}$$

$$\frac{\beta \Delta F^{\frac{1}{N}}}{N} = -\frac{1}{N}\left(\frac{3}{2}\ln\left(N\frac{\alpha\beta}{2\pi}\right) + \ln\frac{V}{N_{ws}}\right) \tag{1.8}$$

The coupling of (hot) graphite to an Einstein crystal, whose average atomic positions are constrained to a fixed reference system, displayed a peculiar feature. Due to the softness of the interplanar interactions (0.07 k_BT at 4,000 K), graphite neighboring sheets are allowed to slide. Also this is a finite size effect: to correct for this we found necessary to attach any sheet to its CoM, independently from the others.[1]

The Einstein crystal spring constant, α, was determined by requiring that the mean-squared displacement from the equilibrium lattice positions is equal for the Einstein crystal and the carbon crystal:

$$\frac{3}{\beta\alpha} = \left\langle \frac{1}{N}\sum_{i=1}^{N}(\mathbf{r}_i - \mathbf{r}_{i,0})^2 \right\rangle$$

Therefore α was fixed by calculating the right hand side in a simulation with the LCBOPI$^+$.

[1] Equation 1.5 then becomes:

$$\frac{\beta F^{\mathrm{E(CM)}}}{N} = 3\ln\Lambda - \frac{3}{2}\ln\left(\frac{2\pi}{\beta\alpha}\right) - \frac{3N_s}{2N}\left(\ln\left(\frac{\alpha\beta}{2\pi}\right) + \ln NN_s\right)$$

where N_s is the number of sheets. Equation 1.6 becomes:

$$\frac{\beta \Delta F^{\maltese(\mathrm{CM})\to\maltese}}{N} = -\frac{N_s}{N}\ln\frac{V}{N_{ws}}$$

where, in $N_{ws} = N/n_{ws}$, one has to define the Wigner–Seitz cell within a graphite sheet; this leads to $n_{ws} = 2$. Equation 1.8 becomes:

$$\frac{\beta \Delta F^{\frac{1}{N}}}{N} = -\frac{N_s}{N}\left[\frac{3}{2}\ln\left(NN_s\frac{\alpha\beta}{2\pi}\right) + \ln\frac{V}{N_{ws}}\right].$$

Table 1.1 Parameters for the polynomial fitting of the 4,000 K isotherms of the three phases, according to: $P(\rho) = a + b\rho + c\rho^2$

	a [GPa]	b [GPa nm^3]	c [GPa nm^6]
Liquid	89.972	−1.9654	0.011 092
Diamond	74.809	−3.6307	0.019 102
Graphite	108.29	−2.2707	0.011 925

In order to estimate the chemical potential μ along the 4,000 K isotherm we integrated from the initial state point a fit, $P(\rho) = a + b\rho + c\rho^2$, through simulated (P, T) state points along the 4,000 K isotherm. Here, ρ is the number density, and a, b, and c are fit parameters (see Table 1.1). This yields for the chemical potential [41]:

$$\beta\mu(\rho) = \frac{\beta F^{\maltese}}{N} + \beta\left[\frac{a}{\rho^{\maltese}} + b\ln\frac{\rho}{\rho^{\maltese}} + b + c\left(2\rho - \rho^{\maltese}\right)\right] \tag{1.9}$$

Here, $\rho^{\maltese}$ denotes the number density at the initial state point, N the number of particles, and $\beta = 1/k_{\mathrm{B}}T$, with k_{B} the Boltzmann constant. Details on this equation are given in Appendix A.

1.4.2 *The LCBOPI$^+$ Phase Diagram*

For calculating the three $F^{\maltese}$ we performed independent Monte Carlo (MC) simulations for three phases. Three samples of 216 particle of the three systems were prepared, the solids in their lattice positions, and the liquid in a simple cubic arrangement. The three phases were equilibrated with NPT MC simulations at the chosen $T = 4{,}000$ K and at $P = 10$ GPa.[2] The integer N = 216 permits the atoms to be arranged both in a defect-free diamond and cubic lattice, aligned with the sides of a cubic cell, while bonding perfectly across its faces to periodic-image atoms. The same requirements are fulfilled for 216 atoms in a defect-free graphite lattice, arranged in three sheets, but in a rectangular periodically replicated cell, with resulting edge-size ratios 1:1.5:1.7. The first, in-plane, ratio (1:1.5) is defined by the lattice geometry (hexagons), while the interplanar ratio (1:1.7) is pressure dependent. In fact, the rescaling of the box was allowed to be independent on the three axes for the equilibration of the solid phases, while kept intrinsically isotropic for the melting of the cubic crystal and the subsequent equilibration of the liquid phase. The equilibrium densities $\rho^{\maltese}$, expressed in 10^3 kg/m^3, were 3.425 for diamond, 2.597 for graphite, and 2.421 for the liquid. Three configurations at the equilibrium volume were then chosen as starting points for the three thermodynamic integrations. The value of α

[2] For the correct application of the method it is not needed to have the three states at the same P. It is only required that the phases share a broad stable region in pressure at the chosen T.

was set to 453,000 and 39,700 kJ/(mol nm^2) for diamond and graphite, respectively. The parameters σ and ε for the LJ fluid were 0.127 nm and 31.84 kJ/mol.

The reference free energies $\beta F^{\mathrm{E}}/N$ were $-5.755\ k_B T$, and $-1.912\ k_B T$ for graphite and diamond, respectively, while the reference free energy for the liquid was found $(\beta/N)\left(F^{\mathrm{id}} + F^{\mathrm{ex}}_{\mathrm{LJ}}\right) = -10.863\ k_B T$. The integration in Eq. 1.2 yields for $\beta F^{\maltese}/N$ the values $-25.090 \pm 0.006\ k_B T$ (graphite), and $-24.583 \pm 0.002\ k_B T$ (diamond), $-25.137 \pm 0.002\ k_B T$(liquid).

The values of λ for the sampling were defined by a 10-point Gauss–Legendre integration scheme. The scheme avoids the sampling of the systems at the two boundary values of λ. A 10 point scheme assures exact result whenever the integrand function of Eq. 1.2 $\left(\langle U^{\mathrm{ref}} - U^{\mathrm{LCBOPI}^+}\rangle_\lambda\right)$ can be reasonably described with a polynomial up to order $2 * 10 + 1 = 21$. When $\lambda = 0, 1$ the system performs its random walk on the basis of only one of the two potentials, thus in principle is allowed to assume configurations completely avoided by the other potential, in such a way that the integrand of Eq. 1.2 could diverge. Should this be the case, the integration scheme would yield a poor estimate of the integral. We thus ascertained that the integrand never indeed diverged at $\lambda = 0, 1$. For the three phases, we run at each λ point an NVT MC simulation of 500,000 cycles.

In Fig. 1.1 $\langle U^{\mathrm{ref}} - U^{\mathrm{LCBOPI}^+}\rangle_\lambda$ versus λ is shown. The absence of spurious phase boundary crossings throughout the integration over λ was checked by looking at the

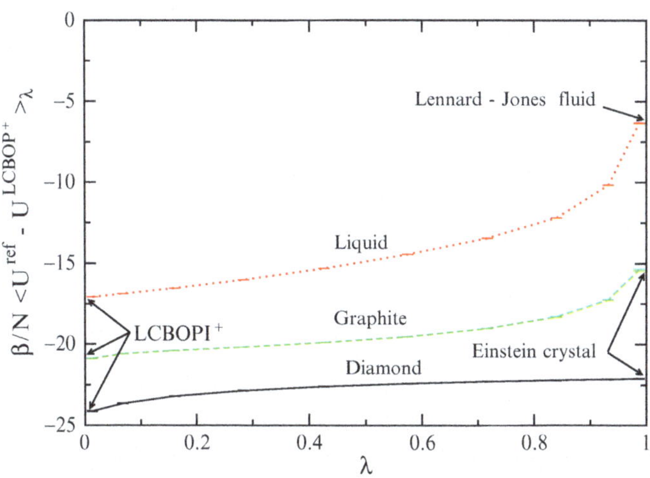

Fig. 1.1 Plots of the quantity $\beta/N\ \langle U^{\mathrm{ref}} - U^{\mathrm{LCBOPI}^+}\rangle_\lambda$ (see Eqs. 1.2 and 1.4) as a function of the coupling parameter λ for the liquid, graphite, and diamond phase. On the *left side* of the horizontal axis ($\lambda = 0$) is the pure LCBOPI$^+$. On the *right side* ($\lambda = 1$) is the reference system, i.e. the Lennard–Jones liquid for the liquid phase and two Einstein crystals (with different coupling constant) for graphite and diamond phase. The temperature is 4,000 K and the pressure is 10 GPa for the three phases, at $\lambda = 0$ (along the integration path the volume, rather than the pressure, is conserved). The simulated λ-points are marked by their *error bars*, that are almost reduced to a single dash at this scale

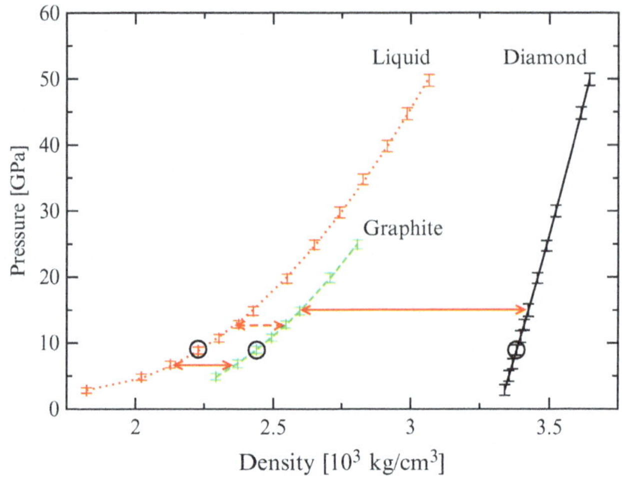

Fig. 1.2 Equations of state at 4,000 K for the liquid, graphite, and diamond phase. The *lines* are the quadratic polynomial fits to the data. The *circles* indicate the points, at 10 GPa, where the thermodynamic integration (Eq. 1.2) was performed. The *solid arrows* connect coexisting (stable) points, i.e. liquid/graphite and graphite/diamond. The *dashed arrow* links the liquid/diamond coexisting point, that is metastable relative to the graphite phase

distribution of $(U^{\mathrm{ref}} - U^{\mathrm{LCBOPI}^+})$.[3] Since the points were run in parallel in order to accumulate more statistics, only shorter independent simulations were performed by increasing and then decreasing λ, each new λ point starting from the final configuration of the previous. The absence of hysteresis in this process completely rules out phase boundary crossings. The isotherms for the three phases, calculated via NPT MC simulations together with their fit, are shown in Fig. 1.2.

The three μ curves (at $T = 4{,}000$ K), as given in Eq. 1.9, but expressed as functions of P, are shown in Fig. 1.3. The three curves, μ_{L}, μ_{G}, μ_{D}, as given in Eq. 1.9, intersect in pairs in three points (these points are shown as a solid triangle, square and diamond in Fig. 1.4). The intersections locate the graphite/liquid coexistence at 6.72 ± 0.60 GPa ($\mu_{\mathrm{GL}} = -24.21 \pm 0.10\ k_{\mathrm{B}}T$), and the graphite/diamond coexistence at 15.05 ± 0.30 GPa ($\mu_{\mathrm{GD}} = -23.01 \pm 0.03\ k_{\mathrm{B}}T$). The third intersection locates a diamond/liquid coexistence at $= 12.75 \pm 0.20$ GPa ($\mu_{\mathrm{DL}} = -23.24 \pm 0.03\ k_{\mathrm{B}}T$). Even though both diamond and the liquid are there metastable, this point can be taken as the starting one for the Clausius–Clapeyron integration of the diamond melting line. Starting from the three coexistence points at 4,000 K, the coexistence lines were traced–by integrating the Clausius–Clapeyron equation using the trapezoidal-rule predictor-corrector scheme [46]. The new value of the coexisting P at a given T was taken when two iterations differed less than 0.01 GPa, this being the size of the single uncertainty in the calculation of $\mathrm{d}P/\mathrm{d}T$. This normally took two to three iterations to be obtained.

[3] The distribution usually exhibits a bimodal shape in case of phase boundary crossing.

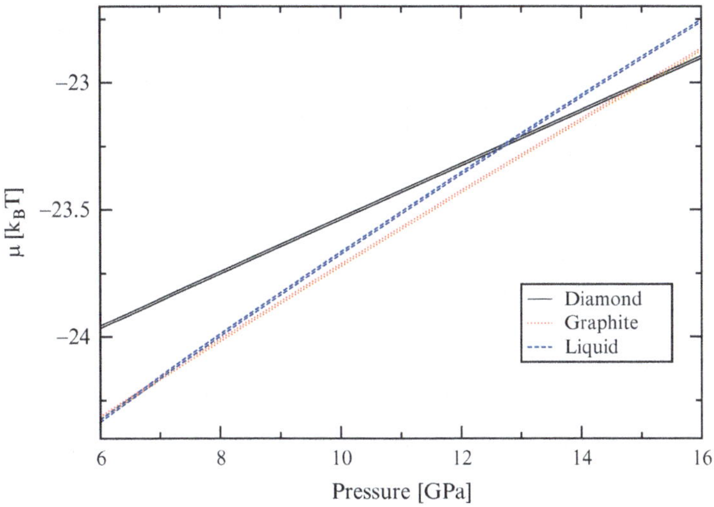

Fig. 1.3 Plot of the chemical potential μ for the three phases at $T = 4{,}000$ K; we plot *double lines*, which marks the boundaries of the numerical uncertainty of the calculated μ. These curves represent Eq. 1.9, but are expressed as a function of pressure using $P(\rho) = a + b\rho + c\rho^2$. The main source of error was the thermodynamic integration; the uncertainty in the equations of state was at least an order of magnitude less

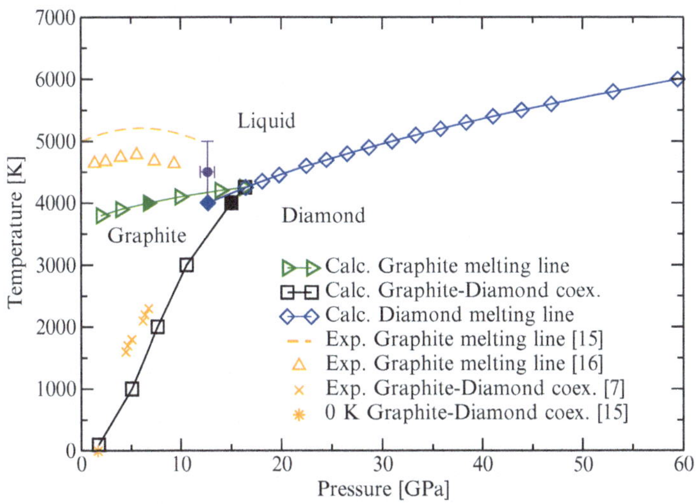

Fig. 1.4 Phase diagram of carbon up to 60 GPa. The *solid right triangle*, *square*, and *diamond* are the three coexistence points found by equating the chemical potentials at 4,000 K (see text). The *open right triangles*, *squares*, and *diamonds* are the calculated coexistence points, propagated via Gibbs–Duhem integration. The *solid circle* with *error bars* indicates the experimental estimate for the liquid/graphite/diamond triple point [15, 18, 20]. The *dashed line* is the experimental graphite *melting line* from Ref. [15]. The *up triangles* are graphite melting state points from Ref. [16]. The *crosses* represent experimental graphite/diamond coexistence from Ref. [7]. The *asterisk* represent the theoretical graphite/diamond coexistence at zero kelvin, as reported in Ref. [15]

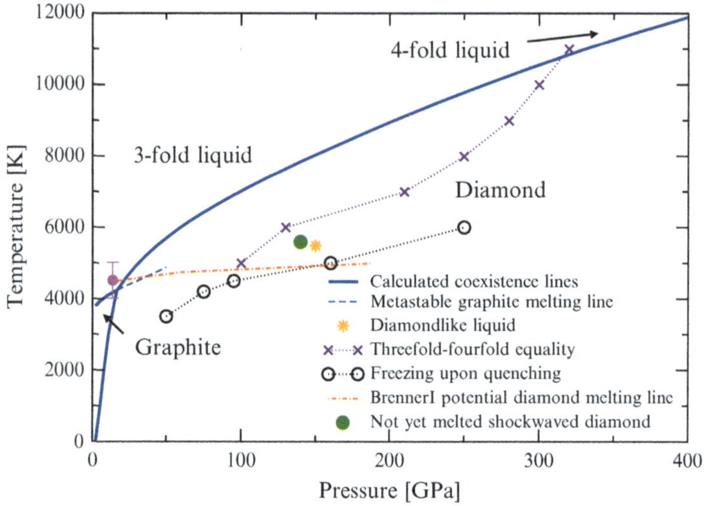

Fig. 1.5 Phase diagram of carbon at all calculated pressures. The thick *solid lines* are the calculated phase boundaries. The *dashed line* is the metastable prolongation of the graphite *melting line*, from Gibbs–Duhem integration; the line stops just before the simulated graphite became instable, displaying large density fluctuations. The *dashed-dotted line* departing from the experimental guess for the triple point (*solid circle* with *error bar* [15, 18, 20]) is the diamond *melting line* calculated in Ref. [21] with the BrennerI potential. The *solid circle* is the final point of the shock wave experiment of Ref. [11] at which diamond is not yet melted. Crosses mark the liquid with equal amount of three- and fourfold atoms; *circles* represent state points in which the sample freezes; in the region in between the two series is the "diamond-like liquid": the star is the point reported in Ref. [47]

Table 1.2 Pressure (P), temperature (T), solid and liquid densities (ρ) along the melting lines

P [GPa]	T [K]	ρ_G [10^3 kg/m^3]	ρ_L [10^3 kg/m^3]
Graphite melting line:			
2.00	3,800	2.134	1.759
6.70	4,000	2.354	2.098
16.4	4,250	2.623	2.414
Diamond melting line:			
16.4	4,250	3.427	2.414
25.5	4,750	3.470	2.607
43.9	5,500	3.558	2.870
59.4	6,000	3.629	3.043
99.4	7,000	3.783	3.264
148.1	8,000	3.960	3.485
263.2	10,000	4.286	3.868
408.1	12,000	4.593	4.236

The calculated phase diagram in the $P - T$ plane is shown in Fig. 1.4 for the low pressure region, and in Fig. 1.5 for the full range of pressures and temperatures considered. Table 1.2 lists the densities of selected points on the coexistence lines. The three coexistence lines meet in a triple point at 16.4 $\pm$ 0.7 GPa and 4,250 $\pm$ 10 K.

Table 1.3 Pressure (P), temperature (T), and melting enthalpy (Δh_m, calculated as the enthalpy of the liquid subtracted of the enthalpy of the underlying solid phase) along the melting lines

P [GPa]	T [K]	Δh_m [kJ/mol]
Graphite melting line:		
2.00	3,800	68.8
5.24	3,950	65.6
9.94	4,100	67.8
16.4	4,250	64.7
Diamond melting line:		
16.4	4,250	95.9
25.5	4,750	111.5
43.9	5,500	130.8
59.4	6,000	143.9
99.4	7,000	160.5
148.1	8,000	174.7
263.2	10,000	195.3
330.5	11,000	208.1
408.1	12,000	221.7

The graphite/diamond coexistence line agrees very well with the experimental data. In the region near the liquid/graphite/diamond triple point, that has not been directly probed in experiments, the graphite/diamond coexistence line bends to the right, departing from the usually assumed straight line. Analysis of our data shows this is mainly due to the fast reduction with increasing pressure of the interplanar distance in graphite at those premelting temperature. This causes an enhanced increase of the density in graphite, yielding a decrease of $\mathrm{d}T/\mathrm{d}P$.

Table 1.3 shows the melting enthalpy Δh_m for graphite and diamond. These are calculated as the difference in enthalpy between the solid and the melt at coexistence. Our calculated melting enthalpies of graphite are sensibly lower than the values around 110 kJ/mol reported in shock heating melting experiments in the past years [15, 16], nonetheless our values retain the feature of being rather constant along the graphite melting line. No experimental data are known about the melting enthalpies of diamond: we note that they increase monotonically with temperature (and pressure).

The calculated graphite melting line is monotonically increasing in a small temperature range around 4,000 K. In contrast to data inferred from experiments it shows no maximum and is at a somewhat lower temperature. In agreement with the experiments the coexistence temperature is only slowly varying with pressure. Inspection reveals that this behavior is due to (1) the limited variability of the melting enthalpy, and (2) a similar bulk modulus for liquid and graphite such that Δv is almost constant.

We have extended the calculation of the graphite melting line to the region in which both graphite and the liquid are metastable towards diamond, with the aim to look for a possible maximum in the line. The results are shown as a dashed line in Fig. 1.5. We stopped the Gibbs–Duhem integration at ~50 GPa, where the 216-particles graphite sample started showing huge volume fluctuations during the NPT sampling. The integration algorithm became instable, forbidding any further analysis. Looking at this metastable melting line, it is clear that its slope does not continue to decrease with increasing pressure as in the stable region; thus, the hypothesis of an hidden maximum appears to be rejected.

The slope of the diamond melting line is consistent with the only experimental point available [11] (see Fig. 1.5). When compared to the diamond melting line of the Brenner model [21], the LCBOPI$^+$ diamond melting line has a steeper slope yielding significantly higher temperatures for the diamond melting line.

1.4.3 Diamond Melting According to LCBOPII

We have not carried out an extensive numerical evaluation of the phase diagram as predicted by the LCBOPII. Yet, using direct free-energy difference calculations starting from the diamond melting line of the LCBOPI$^+$, we determined one point and the slope at that point of the diamond melting line for the LCBOPII. We sampled with the LCBOPI$^+$ a liquid and a diamond sample at the same phase point on the calculated diamond melting line and at intervals a virtual swapping between the two potentials, LCBOPI$^+$ and LCBOPII, was performed. This means that the energy of independent configurations during this run was evaluated also with the LCBOPII. We chose T_m = 6,000 K, which gave P_m = 59.44 GPa for the coexistence for the LCBOPI$^+$. We found for LCBOPII T_m = 5,505 K at the same pressure. The slope of the melting line was evaluated by means of the Clausius–Clapeyron equation. We found a slope of 28.04 K/GPa. We compare it to the very close value of 28.97 K/GPa as given by the LCBOPI$^+$ at the same pressure. Thus, by means of this single point evaluation, we found that LCBOPII has a melting line at lower temperature than the LCBOPI$^+$, but the slope should be similar between the two potentials.

1.4.4 Recent Developments

Recently, the melting curve of diamond in a range up to 2,000 GPa has been studied by ab initio MD simulations using density functional theory. Wang et al. [48] determined the relative stability of the diamond and liquid phase by evaluating the free energy of both phases. Correa et al. [26] determined the melting temperature using a "two phase" simulation method, where the system initially consists of a liquid and a diamond structure that are in contact. Subsequently the melting temperature is estimated by locating the temperature at which the system spontaneously evolves towards a liquid or a crystalline structure. In both ab initio MD studies it was found that the diamond melting curve shows a maximum; around 450 GPa [26] or 630 GPa [48].[4] Subsequent laser-shock experiments [49] provided data consistent

[4] The difference between these two values gives a hint on the uncertainties related to the two different methods used for calculating coexistence, given that the DF-MD set-up is quite similar in the two works.

with this observation, indicating a negative melting slope most probably in the region of 300–500 GPa. When comparing the LCBOPI$^+$ diamond melting curve, that monotonically increases with pressure, to the ab initio MD results of Refs. [26, 48] we see a significant deviation from 200 GPa onwards. This might be attributed to an incorrect description of the liquid structure at high compression. Indeed, LCBOPI$^+$ has not been validated against high density structures with coordination beyond four. These are typical configuration that might become more dominant in the pressure region beyond 200 GPa.

1.5 The Nature of Liquid Carbon: Absence of a *First-Order* Liquid–Liquid Phase Transition

In this section we re-examine the issue of the liquid–liquid phase transition (LLPT) for carbon. A short review of the relevant findings is given in Section 1.5.1. In Section 1.5.2 we describe the liquid at 6,000 K as predicted by several bond order potentials (see Fig. 1.9 for the complete list) and compare the results with density functional (DF) based molecular dynamics (MD) calculation. The temperature was chosen to agree with the DF-MD based analysis of the liquid Wu et al. [22] There the isotherm at 6,000 K was originally chosen for two reasons. Firstly it is expected to be far from coexistence in the density interval studied; this assumption is based on the phase diagram calculated by Glosli and Ree [21] using one of the Brenner bond order potential [35] (the authors do not specify which parameterization they use). Secondly, the 6,000 K isotherm is predicted by Glosli and Ree [20], with the BrennerI [35] bond order potential, to cross the liquid–liquid coexistence line. In Section 1.5.3 the analysis will be extended at all the relevant regions of the phase diagram, with the aim of ruling out the presence of a LLPT, at least within the scope of the LCBOP family. Intriguingly, signatures of a LLPT transition are nonetheless hinted at by LCBOPII for the strongly undercooled liquid, where the liquid would be anyhow dynamically arrested into a glass (see Section 1.5.2.3). A characteristic that carbon would share with water.

1.5.1 A Short History of Carbon LLPT

An Analysis of Experimental Data The possibility of a liquid–liquid phase transition (LLPT) in liquid carbon has been firstly investigated by Korsunskaya et al. [17], analyzing data on the graphite melting line proposed by Bundy [8] (those data showed a maximum melting temperature at 6.5 GPa). By fitting the data from Bundy into the original two levels model of Kittel [50] and postulating the existence of two liquids, Korsunskaya et al. found the critical temperature T_c of the LLPT. The model is fitted with three points on the graphite melting line, with the respective derivatives, and with the heat of melting at a selected pressure. The authors assume that:

1. Liquid and solid have different compressibilities.
2. The nature of liquid carbon is described univocally by the relative fraction of the two liquids.
3. Each of the two liquids presents a volume change on melting, heat of melting and entropy of melting that are independent of T, P, and the fraction s: the volume change upon melting for the liquid is a linear combination of the volume changes of the pure species (i.e. for $s = 0, 1$), while heat and entropy of melting combine according to the regular solution rules.
4. The overall entropy jump on melting is independent of T (that is equivalent to assuming the same heat capacity in the liquid and the solid).

The fitting procedure gives an estimate for the critical pressure of ~6.5 GPa and for the critical temperature of the searched transition at 3,770 K, i.e. below the melting temperature. The fitted value for the entropy of melting is the same for the two liquids, thus implying a vertical slope ($\mathrm{d}T/\mathrm{d}P$) of the coexistence line (in the metastable liquid region just below the critical temperature).

When the slope of two out of the three coexisting line meeting at a triple point is known, the slope of the third is also determined. On the basis of their results, the authors were thus able to calculate also the diamond melting line: they predicted it to have a negative slope. Note that the slope of the *graphite* melting line, and the slope of the diamond/graphite coexistence, as extracted from Bundy's data [7,8], together with the densities of the phases obtained by fitting to the two levels model *implied* (via Clausius–Clapeyron equation) a negative slope of the diamond melting line. Different values of the slopes of the graphite boundary lines, and of the densities of the phases can yield rather different slope of the diamond melting line, as we have shown in Section 1.4.

Consistently with the slope of the fitted graphite melting line, the low density liquid (LDL, $s = 0$) is less heavy, and the high density liquid (HDL, $s = 1$) is heavier than the coexisting graphite. The nature of the two liquids is predicted as follows: at low pressure graphite melts into a liquid of neutral particles, which interact predominantly through dispersion (London) forces. Upon increasing pressure[5] the liquid metallizes into a close packed liquid. No assumption is made on the local structure.

A Semi-empirical Equation of State The modern discussion on the LLPT for carbon, starts with the elaboration of a semi-empirical equation of state for carbon, valid also at high P and T, by van Thiel and Ree [18,51]. The equation of state is constructed on the basis of experimental data and electronic structure calculations. It is postulated the existence, in the graphite melt, of a mixture of a threefold (sp_2) and

[5] The transition in the stable liquid region is supercritical, thus continuous, but taking place in a short range of pressures around 6.5 GPa.

a fourfold (sp_4) liquid. The model of *pseudo-binary mixture* is assumed to describe the mixing of the two liquids [52]; the mixing energy J of the two liquids is written as: $\beta J = \left[A_0/\left(1 + (P/P_0)^{3/2}\right)\right] s(1-s)$ where $\beta = 1/(k_B T)$, A_0 and P_0 are fitting parameter, and s is the fraction of the sp_3 liquid. The value of these fitting parameters is essential to determine the possibility of the occurrence of a first order transition. Van Thiel and Ree show that fitting A_0 in order to obtain the graphite melting points of Bundy [8], the slope of the graphite melting line predicted by their model inverts its sign discontinuously in correspondence of the maximum, so that a first order LLPT arises. On the other hand, if they fit to the data from Ref. [9], the value of A_0 decreases so that the T_c of the LLPT drops below the melting line and the transition between the two liquids becomes continuous in the stable liquid region. As pointed out by Ponyatovsky [53] the expression for βJ proposed by van Thiel and Ree involves two ambiguities. Firstly, extrapolating the coexistence line between the two liquids at atmospheric pressure, the coexistence temperature would be $T \sim 3{,}700$ K: this would imply that the sp_3 liquid (and the glass) would be more stable than the sp_2 at room pressure up to very high temperatures, which is in contrast to the experimental data. Furthermore, J is proposed to have a linear dependence on T, so that, when $T \to 0$, also the mixing energy would tend to zero, i.e. at zero temperature the regular solution would become an ideal solution. This is extremely unusual.

Experimental Suggestions from the Graphite Melting Line Togaya [16] found a maximum in the melting line at $P_{max} = 5.6$ GPa. The author fitted the six experimental points with two straight lines: with positive slope at pressures lower than P_{max}, with negative slope at pressures higher than P_{max}. The discontinuous derivative of the melting curve at the maximum would imply there a triple point graphite/LDL/HDL, as a starting point of a LLPT coexistence line.

Prediction of a Short Range Bond Order Potential In Ref. [20] Glosli and Ree reported a complete study of a LLPT simulated with the Brenner bond order potential [35] in its version with torsional interactions [36]. The authors simulated in the canonical (NVT) ensemble several samples at increasing densities at eight different temperatures. By measuring the pressure, they show the familiar van der Waals loop denouncing mechanical instabilities at certain imposed densities. Using the Maxwell equal-area construction, the authors calculated the LLPT coexistence line, ending in a critical point at $T = 8{,}802$ K and $P = 10.56$ GPa. The lowest temperature coexistence point was calculated at $T = 5{,}500$ K and $P = 2.696$ GPa. The LDL/HDL coexistence line should meet the graphite melting line at its maximum, but unfortunately the BrennerI potential does not contain non bonded interactions, thus it cannot describe neither bulk graphite nor its melting line. To overcome this deficiency, the authors devised an ingenious perturbation method. Assuming constant slope of the negative sloped branch of the graphite melting line[6] and fixing the

[6] The authors adopted the graphite melting line measured by Togaya [16]. This melting line is reported in Fig. 1.4, together with our results. According to Glosli and Ree, from the maximum of that melting line would branch off the LLPT coexistence line.

graphite/diamond/HDL triple point at a value taken from the experimental literature, they give an estimate of the graphite/LDL/HDL triple point, at $T = 5{,}133$ K and $P = 1.88$ GPa. The LDL was found to be mainly twofold (sp) coordinated with a polymeric-like structure, while the HDL was found to be a network forming, mainly fourfold, (sp_3) liquid. Following the predictions of this bond order potential, the sp_2 coordinated atoms would be completely avoided in the liquid. The authors identified the reason in the presence of torsional interactions. In fact, the increase in density demands an increase in structures with higher coordination than the sp, which is entropically favored at low densities. Each bonds of the sp_3 structures can freely evolve around the bond axis, while bonds between sp_2 sites are constrained in a (almost) planar geometry by the torsional interactions: this implies a low entropy for a liquid dominated by sp_2 sites. This low entropy would eventually destabilize the sp_2 sites towards the sp_3. To prove this conjecture, the authors calculated two relevant isotherms in the original version of the potential, without torsional interactions, finding no sign of a LLPT. Since some torsional interactions are definitely needed to mimic the double bond reluctancy to twist, the authors concluded that the LLPT predicted by the Brenner bond order potential with torsion is more realistic than its absence when torsional interactions are switched off.

Tight binding calculations [54] showed no evidence of van der Waals loops at some of the temperatures analyzed in Ref. [20]. As Glosli and Ree note, the tight binding model used in [54] is strictly two-center, thus the torsional interactions *cannot* be described.

An Ab Initio Confutation of the LLPT In Ref. [22], Wu et al. reported on a series of NVT-CPMD simulations at 6,000 K from density $1.27 - 3.02 \times 10^3$ kg/m^3, in a range where the BrennerI potential showed the first order LLPT at the same T. No sign of a van der Waals loop was found: in contrast to the BrennerI results of the previous section, two approaching series starting from the lowest and the highest density, were found to meet smoothly at intermediate densities. Looking for the reasons of the failure of the BrennerI potential, the authors calculated, with the same density functional (DF) used in the CPMD simulations, the torsional energy of two model molecules. One, $(CH_3)_2CC(CH_3)_2$ (see Fig. 2 in Ref. [30] for a schematic representation), was chosen so that the bond between the two central atoms represents a double bond in a carbon network: two sp_2 sites are bonded each to two sp_3 sites; the peripheral hydrogens are needed to saturated the sp_3 atoms and are intended to have no effect on the central bond. The second molecule, $(CH_2)_2CC(CH_2)_2$ (see again Fig. 2 in Ref. [30] for a schematic representation) is a portion of a completely sp_2 coordinated network: in the bond order language, the central bond is conjugated. The two molecules were geometrically optimized in their planar configurations and then twisted around the central bond axis in steps of $\pi/12$. In each configuration the electronic wave function was optimized, without further relaxations, to give the total energy, that was compared to the planar configuration total energy. The difference is the torsional energy. The DF calculations found a surprising picture (see Section 6.7 in Ref. [40] or Fig. 3 in Ref. [30]): while the double bond torsional energy was only slightly overestimated by the BrennerI potential at intermediate angles, the DF torsional energy for the conjugated bond showed a completely different scenario compared to the classical prediction. It shows a maximum at $\pi/4$, while the planar and orthogonal configuration have basically the same energy. For the BrennerI potential, the torsional energy in this conjugated configuration is monotonically increasing with the torsion angle, just as for the double bond configuration. On average, considering that the conjugated configuration would be characteristic of a mainly sp_2 coordinated liquid, the torsional interactions are enormously overestimated by the classical potential. As a further proof, the authors tried to lower torsional energy of the conjugated bond in the classical potential, by tuning the proper parameter, and found a much less pronounced LLPT. Note that the functional form of the torsional interactions for the BrennerI potential *cannot* reproduce the DF data here mentioned. Wu et al. concluded: "Brenner potential significantly overestimates the torsional barrier of a chemical bond between two- and three-center-coordinated carbon atoms due to the inability of the potential to describe lone pair electrons"; and: "Brenner potential parameters derived from isolated hydrocarbon molecules and used in the literature to simulate various carbon systems may not be adequate to use for condensed phases, especially so in the presence of lone pair electrons". The bond-order (Brenner like) potential (LCBOPI$^+$) [29] incorporates a much more flexible and transferable definition of the torsional interactions. In Section 1.5.2 we will show that the conclusion of Wu et al. is not necessarily true for all BOPs; we will show how our definition of the torsional interaction is able to reproduce relevant features of liquid carbon, as described by density functional based molecular dynamics.

1.5.2 Properties of Liquid Carbon, According to Selected BOPs

1.5.2.1 Computational Methods

We performed Monte Carlo (MC) simulations of 128 particles in a cubic box with periodic boundary conditions with all the bond order potentials we chose for the comparison. We sampled at 6,000 K the constant volume (NVT) ensemble for densities smaller than 2.5×10^3 kg/m^3 and the constant pressure (NPT) ensemble for larger densities where the increase of pressure is steeper, with an overlap between the two regions to check for consistency. The pressure in the NVT simulation for the BOPs was calculated without virial evaluation, by means of virtual volume rescaling [55,56].

We performed constant volume DF-MD simulations using the Car–Parrinello method [57] as implemented in the CPMD package [58]. The system consisted of 128 atoms in a cubic box with periodic boundary conditions at nine densities and a temperature $T = 6{,}000$ K, imposed by means of a Nosé–Hoover [59] thermostat. We used the Becke [60] exchange and Perdew [61] correlation gradient corrected functional (BP) with a plane wave basis set cut off at 35 Ry and sampled the Brillouin zone only in the gamma point. BP gives a correct description of bulk diamond. Each state point was studied for 5 ps, starting from a sample equilibrated via the LCBOPI$^+$; only minor structural changes occurred in the first tenths of ps. Since liquid carbon is metallic, we imposed a thermostat for the electronic degrees of freedom in order to ensure a proper implementation of the Car–Parrinello scheme [62].

1.5.2.2 Radial Distribution Functions

In Fig. 1.6 we present the radial distribution functions ($g(r)$) obtained for the LCBOPII, LCBOPI$^+$, and DF-MD at four selected densities along the 6,000 K isotherm. Taking DF-MD as a reference, we see that the LCBOPII is a major improvement with respect to the LCBOPI$^+$. In particular the minimum between the first and second shell is now properly described. Here we should note that the radial distribution functions and the coordination fractions, at $\rho = 3.75 \times 10^3$ kg/m^3 were used as a test system in the development of the potential.

The figures also show that the LCBOPII reproduces the DF-MD values for the peak positions, and the height of the second and third peak. Only the fist-peak height is slightly overestimated by the LCBOPII, consistent with the fact that the LCBOPII showed larger values for the higher coordination numbers (Fig. 1.14).

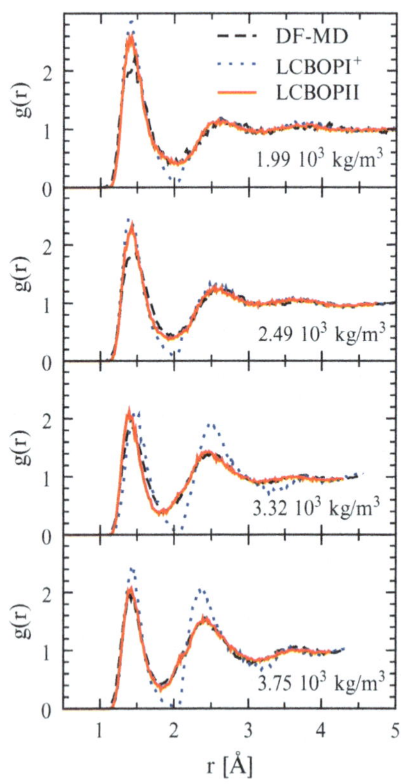

Fig. 1.6 Comparison of the radial distribution functions at 6,000 K and four selected densities between the LCBOPII (*solid lines*), the LCBOPI$^+$ (*dotted lines*), and the reference data taken from our own DF-MD simulations (*dashed lines*)

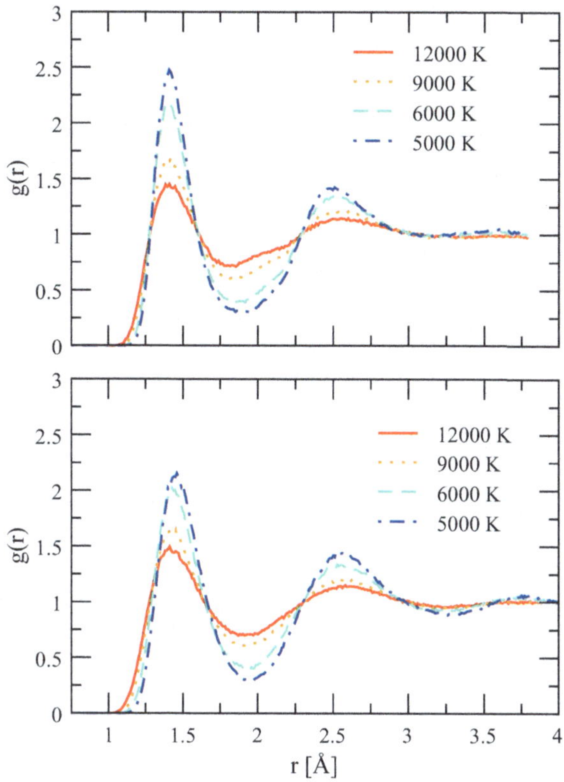

Fig. 1.7 Comparison of the radial distribution functions at 2.9×10^3 kg/m^3 and four temperatures between the LCBOPII (*top panel*) and reference data taken from Ref. [63] (*bottom panel*). *Solid lines* are at 12,000 K, *dotted lines* at 9,000 K, *dashed lines* at 6,000 K, and *dotted–dashed lines* are at 5,000 K

Figure 1.7 compare the LCBOPII radial distribution functions for a liquid at 2.9×10^3 kg/m^3 at four different temperatures with 64-atom DF-MD data from Ref. [63], calculated using the local density (LDA) functional. This figure makes clear that, up to 12,000 K, also the temperature dependence is well reproduced by the LCBOPII. The temperature dependence is typical for a liquid: the peak heights decrease with increasing temperature, while minima increase, indicating a gradual loss of structure. It is striking that all the curves cross at the same points at $g(r) = 1$. In fact, at ~0.165, ~0.230, and ~0.285 nm, the value of $g(r)$ is 1, regardless of the temperature.

Angular Distribution Function In Fig. 1.8 we show the comparison of the angular distribution functions between LCBOPII and DF-MD calculations at the same densities as for the radial distribution function (see Fig. 1.6). The overall agreement is impressive also in this case. The position of the main peak together with its broadness agrees at all densities. Compared to the LCBOPI$^+$, this is a significant

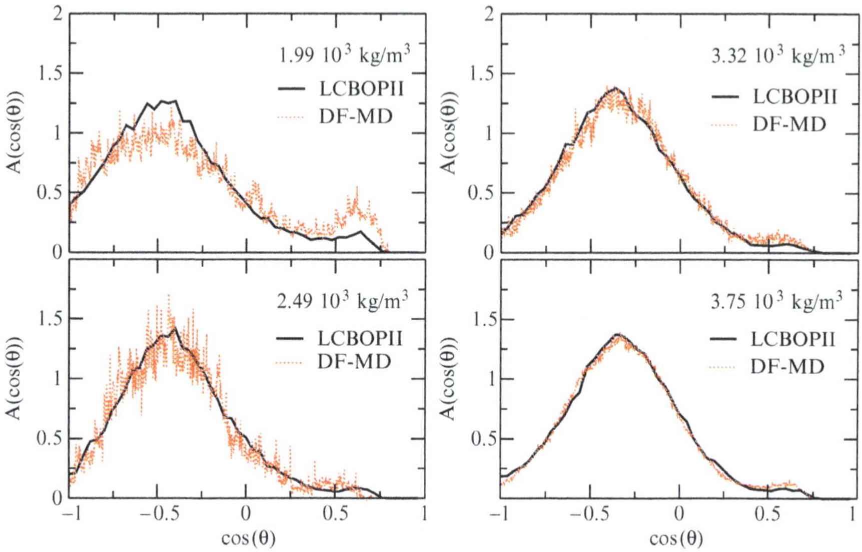

Fig. 1.8 Angular distribution functions at 1.99 (*top left panel*), 2.49 (*bottom left*), 3.32 (*top right*), and 3.75×10^3 kg/m^3 (*bottom right*), and $T = 6{,}000$ K; we compare DF-MD (*dotted lines*) with the LCBOPII (*solid lines*)

improvement as the results of Ref. [47] showed that there was only a qualitative agreement between the LCBOPI$^+$ and DF-MD, with the LCBOPI$^+$ underestimating the presence of sub 90 degrees structures. The improvement of the LCBOPII over the LCBOPI$^+$ should be attributed to the softening of the angular part in the potential yielding a lower energy for fourfold coordinated structures at small angles. As for the radial distribution functions, the major discrepancies are at lower densities: in this case the position and/or the broadness of the secondary peak at low angles is only qualitatively reproduced.

1.5.2.3 Equation of State

Figure 1.9 shows the equations of state of all the mentioned potentials. All the points were calculated by us, except for the AIREBOII potential [38] points in the bottom panel, that come from Ref. [39], and the DF-MD data of Wu et al., marked with PBE (see below). The data from Ref. [22] are calculated with a different functional (i.e. PBE, as labelled in Fig. 1.9) and a different cut-off energy (50 Ry). In view of the relatively low cut-off energy (35 Ry), we had to correct the pressures for the spurious contribution due to Pulay forces [64]. In the density range where we can compare with the results of Wu et al. [22], the pressures that we compute are some 15% lower than those reported by Wu et al. The difference in calculated pressures between the two DF-MD simulations should be attributed to the use of a different functional. Different plane-wave cutoffs and pseudopotentials should not contribute

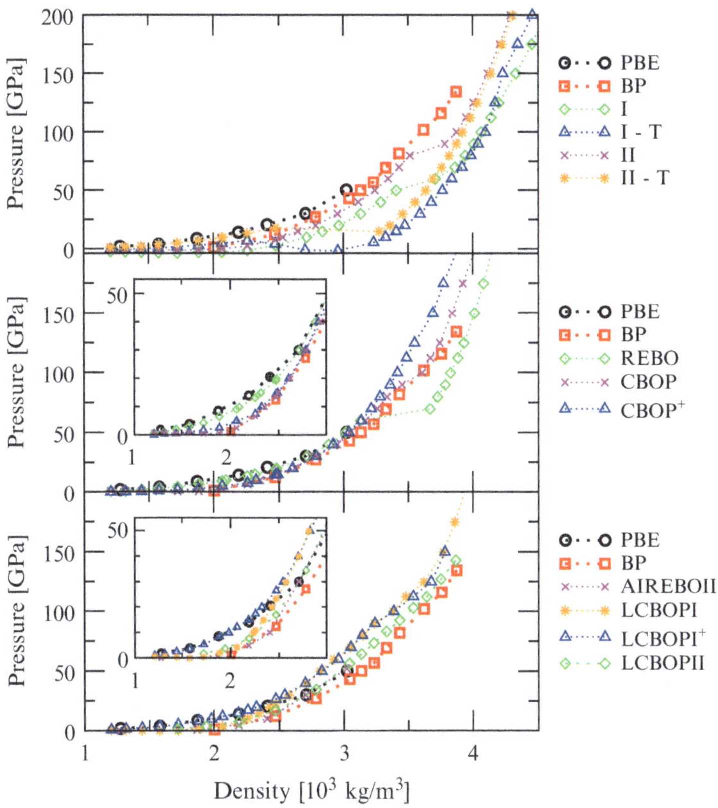

Fig. 1.9 Equations of state at $T = 6{,}000$ K for all the potentials tested. The abbreviation 'PBE' and 'BP' refer to two types of gradient corrected density functionals (see text), and are taken as reference data in these plots. The Brenner potentials are labelled 'I' and 'II' for their original version [35], i.e. without torsional interaction, and 'I-T' and 'II-T' for the version [36] with added torsional interactions. The REBO potential is presented in Ref. [37]: the data points come from our simulations. The CBOP is presented in Ref. [27]; with CBOP$^+$ is labelled the CBOP with modified angular interactions and added torsional interactions, in the same way as for the LCBOPI$^+$, but with appropriate parameters. LCBOPI [27], the LCBOPI$^+$, and the LCBOPII are presented in Refs. [29,40]. The AIREBOII potential is a slight modification of the AIREBO potential [38]; it is described in Ref. [39] (the data points shown here are taken from the same reference)

significantly to this discrepancy, when one considers that both setups yielded good binding energies. We have checked that our samples were indeed liquid: over the whole isotherm we have observed diffusive behavior in both the MC-LCBOPI$^+$ and the DF-MD simulations, the latter indicating a self-diffusion coefficient at least of order 10^{-5} cm^2/s.

The two series of DF-MD data are taken as reference, and for this reason repeated in each panel. The original Brenner potentials [35], BrennerI and BrennerII, with and without torsional interactions [36], are in the top panel. The center panel is for the subsequent generation of *short range potentials*: REBO [37] and CBOP [27].

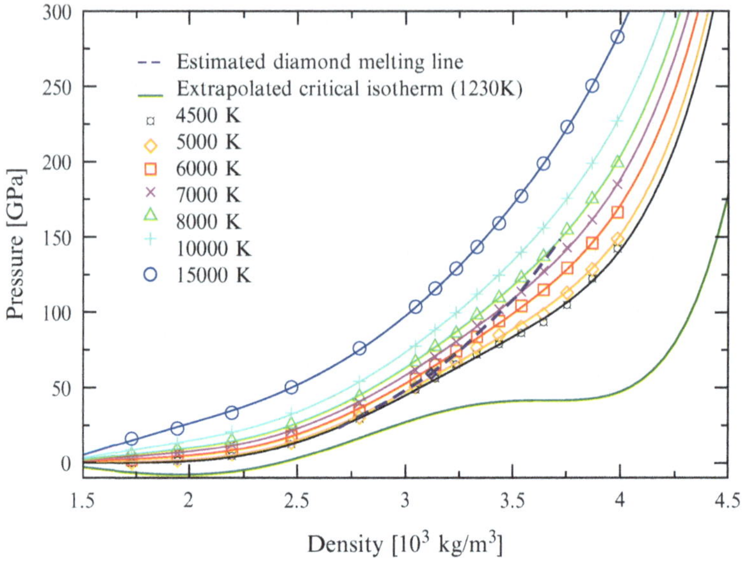

Fig. 1.10 Equations of state for the LCBOPII at seven different temperatures. *Error bars*, not shown, are within the symbol size. The relative error is around 1% at the highest pressure and increases up to 10% at the lowest (i.e. smaller than 5 GPa). The *thick line* at the bottom is the extrapolated critical isotherm, at 1,230 K. It has a horizontal inflection point at density 3.66×10^3 kg/m^3 and pressure 41.74 GPa. The *dashed line* is the estimated *diamond melting line*, started from the calculated point shown as a diamond (at 5,505 K, 59.4 GPa, and 3.12×10^3 kg/m^3, see previous section), and extrapolated assuming constant $\mathrm{d}T/\mathrm{d}P$

With *short range potential* we mean a potential which includes only covalent interactions. We label with CBOP$^+$ the CBOP with modified angular interactions and added torsional interactions, in the same way as for the LCBOPI$^+$. The bottom panel groups the potentials with non-bonded interactions: AIREBOII, LCBOPI and LCBOPI$^+$. The bottom panel also shows the equation of state as given by the LCBOPII [29].

In Fig. 1.10 we have plotted the pressure–density curves for 7 isotherms. Figure 1.11 shows the equation of state surface, fitted to data coming from 7 isotherms, each sampled at 15 densities. In Fig. 1.10 also the estimated coexistence line is plotted. In the stable region all curves show a regular monotonic increase of the slope of the curve. In the undercooled region we observe for the 4,500 and 5,000 K isotherms, in a small density region around 3.3×10^3 kg/m^3, a decrease of the slope. For the LCBOPI$^+$, this wiggling of the pressure–density curve was a pronounced feature at 6,000 K, and associated with a rapid switching of the dominant coordination from three- to fourfold. For the LCBOPII, the same coordination change occurs in the wiggling region around 3.3×10^3 kg/m^3 of the 4,500 and 5,000 K isotherms (see below).

A wiggle in the $\rho - P$ equation of state denounces a sudden decrease of the isothermal compressibility ($1/\rho \ \mathrm{d}P/\mathrm{d}\rho$). Thinking pressure as the leading

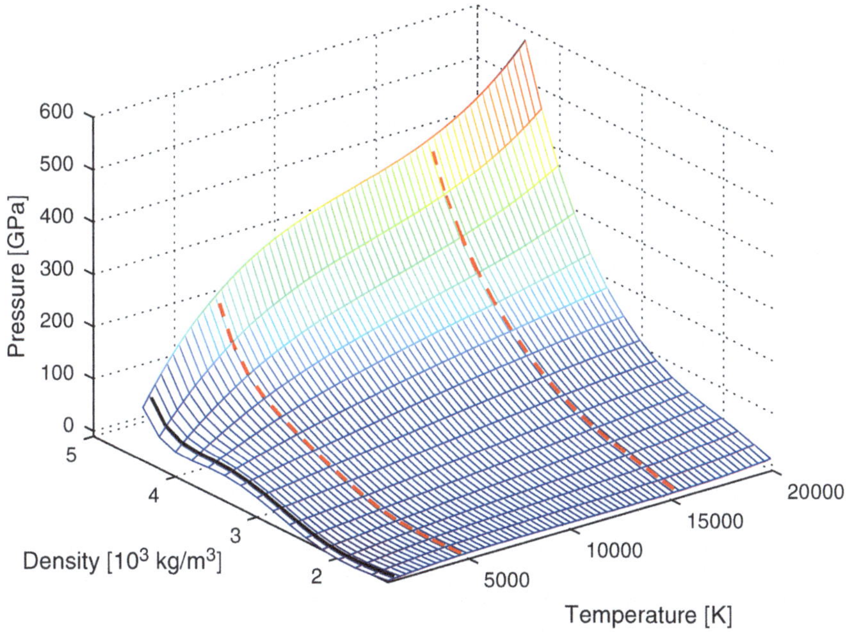

Fig. 1.11 Three dimensional density–temperature–pressure plot of the fitted equation of state for LCBOPII, $P = P(\rho, T)$. The *solid line* is the (deeply supercooled) critical isotherm (at 1,230 K), while the two *dashed lines* are the two extreme isotherms of the simulated state points (i.e. 4,500 and 15,000 K)

parameter, in this region the liquid reacts to an increase of pressure with an increase of density, higher than at the surrounding densities. This is understandable thinking that on the left and on the right of the wiggle, the liquid needs mainly to shorten covalent bonds to increase density. In the transition region, a bigger increase of density is readily achieved by receiving an extra neighbor. Given the shape of the equation of state at lower temperatures, nothing would prevent to speculate the existence of a liquid–liquid phase transition at even lower temperatures, if ever a first order one. We observed that samples at temperatures lower than ~4,000 K rather freezed, especially at densities higher than 3×10^3 kg/m^3. This makes the speculation impossible to prove with this sample size.

In Ref. [65], we proposed a 19 parameters fit for the Equation of State for LCBOPII. The polynomial fit reproduced the wiggling of the 4,500 and 5,000 K isotherms for the undercooled liquid. Extrapolating the fit at low temperatures, inaccessible in the liquid phase with our system size, the wiggles become more pronounced, yielding an inflection point with zero slope in the $P - \rho$ plane for the 1,230 K isotherm at 3.66×10^3 kg/m^3, and 41.74 GPa. This behavior is typical for the critical isotherm. At lower temperatures the isotherms of the fit function show a van der Waals loop, indicating a first-order phase transition associated with a density change. However, in our simulations the system freezes below 4,000 K, especially at densities higher than 3×10^3 kg/m^3. Hence it would be rather speculative to propose the presence of a liquid–liquid phase transition. Still it might be interesting to explore the undercooled liquid by considering (much) larger system size and perform careful annealing to see if a scenario similar to that of liquid water, with its speculated liquid–liquid phase transition hidden in the glass region [66–68] would appear.

1.5.2.4 Coordination

The coordination fractions are shown in Figs. 1.12 and 1.13; for a clearer comparison, these figures share the same labelling criterion for the two-, three-, and fourfold

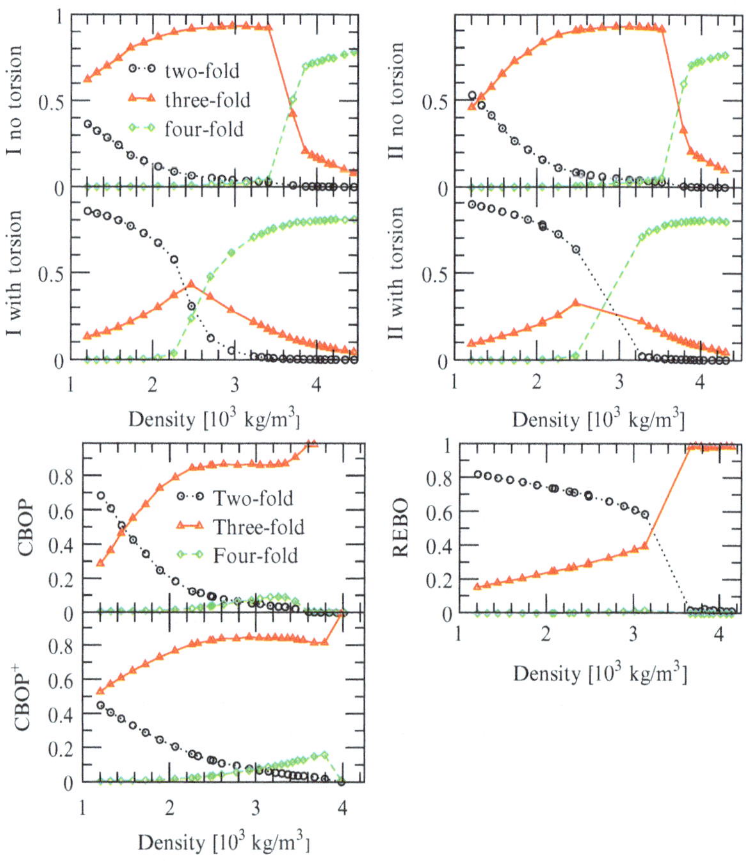

Fig. 1.12 *Top four panels:* coordination fractions at $T = 6,000$ K for the Brenner potentials of Ref. [35] (BrennerI and BrennerII, without torsional interactions) and Ref. [36] (BrennerI and BrennerII, with torsional interactions). *Bottom three panels:* coordination fractions at $T = 6,000$ K for more recent short range potentials [27,37], see text for the abbreviations. *Circles* are always for twofold, triangles for threefold, and diamonds for fourfold sites (atoms)

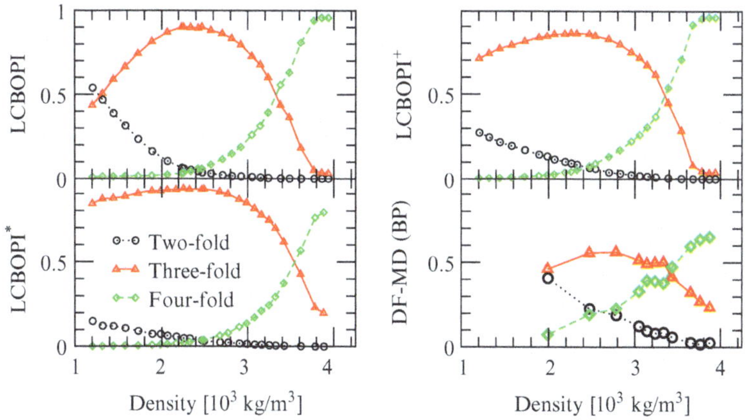

Fig. 1.13 Coordination fractions at $T = 6{,}000$ K for long range potentials (LCBOPI, LCBOPI*, and LCBOPI$^+$) before the introduction of LCBOPII (Refs. [27,40], see also text for the abbreviations). The density functional (BP) data are reported for comparison. The data for LCBOPI$^+$ and BP are shown in Fig. 1.14, too. *Circles* (*black*), *triangles* (*red*), and *diamonds* (*green*) indicate the fraction of two-, three-, and fourfold coordinated atoms, respectively

sites, and the same axes. For the AIREBOII potential no coordination data are available; 'LCBOPI*' (in Fig. 1.13) indicates an intermediate version of the LCBOPI$^+$, i.e. with the softened angular interactions at low coordination [40], but without torsional interactions.

Figure 1.14 shows the coordination fractions calculated with the LCBOPII, LCBOPI$^+$, and DF-MD along the 6,000 K isotherm. We see that, except for a slight overestimation of the threefold fractions in the low-density regime, the LCBOPII results reproduce the DF-MD data very well, both for the density dependence as for the absolute values. The LCBOPII improves the predictions of the LCBOPI$^+$: at densities up to $\sim 3.4 \times 10^3$ kg/m^3, the LCBOPII predicts less threefold and more two- and fourfold sites, thus getting closer to the DF-MD data. At higher densities, where the LCBOPI$^+$ overestimated the fourfold fraction, the coordination fractions predicted by the LCBOPII almost perfectly recover the DF-MD data. Fivefold coordinated atoms (not shown in Fig. 1.14) only appear in the high-density region. At 3.75×10^3 kg/m^3 the fraction for the LCBOPII is 0.1, slightly larger than DF-MD value of 0.07. Note that this is a remarkable achievement of the potential, as the structures used to develop the LCBOPII did not have fivefold coordination. We also note that in the high density range, short range BOPs hardly show coordination beyond three, while for the LCBOPI$^+$ the fivefold fraction remained negligible. The appearance of fivefold coordinated structures in LCBOPII calculations is due to the presence of the "middle range" part in the potential and to the softening of angular correlations. In fact, in older BOPs as well as in the LCBOPI, a small angle such as 60 degree had a significant energetic penalty, fitted to a 12-fold structure (i.e. an fcc lattice) that applied also for 5-fold sites. With the LCBOPII the penalty at small angle for this lower coordination has been reduced [29]. As already shown

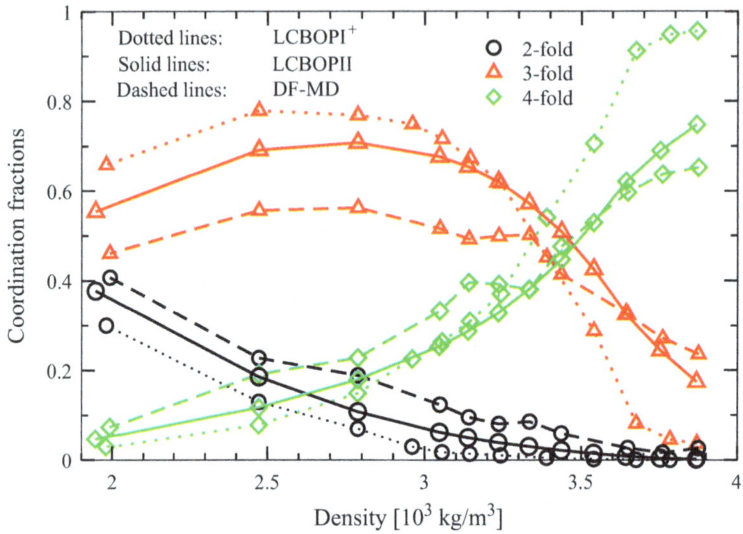

Fig. 1.14 Comparison of coordination fractions at 6,000 K between the LCBOPII (*solid lines*) and reference data coming from our own DF-MD (*dashed lines*). Data at density 2.79×10^3 kg/m^3 and higher are the same shown in Ref. [47]. *Circles* always represents twofold sites, *triangles* threefold, and *diamonds* fourfold. Fivefold sites are not shown but can be deduced by subtraction, since no atoms with a single bond (onefold) or sixfold sites were observed at these densities. The *error bars*, not shown, are within the symbol size, ~0.01 for the LCBOPII, and ~0.02 for the DF-MD points

in Ref. [47] the REBO potential yields a negligible fourfold fraction at all the densities: the threefold atoms replace the twofold upon increasing density, until the spurious LLPT at which all the atoms become threefold. The transition is also appears for CBOP, but the fraction of fourfold atoms raises until ~10% before the transition. We mention that the environment dependent potential (EDIP) introduced by Marks [69] is similar to the LCBOPII in reproducing the DF-MD coordination fractions at 5,000 K [70].

In Fig. 1.15 we show the average coordination fractions for the LCBOPII at seven temperatures. Onefold coordinated atoms appear only in a small amount (a few %) at the lowest densities, and are not shown. Considering the density dependence, we observe that, for all temperatures, the atoms are mainly two- and threefold coordinated in the low density region, with the twofold sites gradually replaced by three- and fourfold sites upon increasing density. At 6,000 K the two- and threefold fractions match around $\rho = 1.73 \times 10^3$ kg/m^3. Note that the maximum of the threefold fraction is at $\rho \sim 2.75 \times 10^3$ kg/m^3 for all temperatures. Beyond this density the dominant coordination is three- and fourfold. The threefold sites are replaced by fourfold sites over a relatively short density range around $\rho = 3.4 \times 10^3$ kg/m^3. The fivefold fraction only appears with a significant fraction in the high-density region and shows a marked temperature dependence. This also implies a stronger temperature dependence of the fourfold fraction in the high-density region.

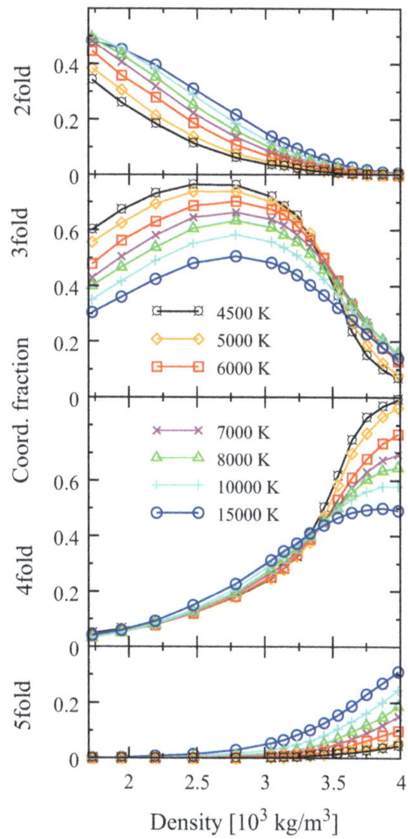

Fig. 1.15 Coordination fractions at seven temperatures according to the LCBOPII. The fraction of sixfold coordinated atoms were negligible at all simulated state points. Each panel shows a different coordination fraction interval but y-axes share the same unit, so that they can be directly compared. The *error bars*, not shown, would be smaller than the symbol size, typically around 0.01

1.5.2.5 Discussion

Looking at all the equations of state and coordination fractions, BrennerI and BrennerII with torsional interactions show the LLPT from a mainly twofold to a mainly fourfold liquid, as discussed in Section 1.5.1. We show that also for the version without torsional interactions, the potentials present an abrupt transition from a mainly threefold to a mainly fourfold liquid from density $3.4–3.8 \times 10^3$ kg/m^3. The REBO potential displays a quite good reproduction of the DF-MD *equation of state* at lower densities (up to 3.2×10^3 kg/m^3), but, also at low density, the structure of the liquid is far from the reference data (see Fig. 1.13, bottom right panel): twofold atoms are predominant where DF-MD predicts threefold to be. Furthermore, the liquid undergoes a LLPT at 3.2×10^3 kg/m^3. The high density liquid is revealed to be, upon visual inspection of the snapshots, a completely threefold graphite-like liquid consisting of almost defect-less sliding sheets that eventually get stuck upon further increasing of the pressure. We note that the transition occurred regularly upon compression for different, independent, samples. Note that the fourfold fraction remains always negligible. Also the CBOP and the CBOP$^+$ present this spurious transition, but the predominance of threefold sites in the lower density regime is correctly given by both these potentials. Note that the equation of state of CBOP is the one in better agreement with our DF-MD data. Nonetheless, as it is dramatically shown by the coordination fractions, a good equation of state is not sufficient to assess the accuracy of a potential.

With the introduction of non-bonded interactions, the fraction of fourfold becomes eventually non negligible at higher densities. For these potentials, the coordination fractions depict a similar scenario at high density, whereas the main differences are found at intermediate densities. The original potential, LCBOPI, predicted a too stiff variation of the threefold fraction, both when increasing, substituting the twofold predominance, and when decreasing, substituted by the fourfold sites. The introduction of softened angular correlation for low coordinated atoms, the case of LCBOPI*, increased the stability of the threefold sites versus two- and fourfold sites. The introduction of torsional interactions, i.e. LCBOPI$^+$, yielded a picture closer to the reference data. Anyway the agreement remains qualitative since only the trends and the fold-type of dominant coordination is reproduced, while the absolute values are always shifted towards overestimation of the dominant fraction.

DF-MD predicts a marked switching of dominant coordination from three to four around 3.4×10^3 kg/m^3. Judging from the isotherm of Fig. 1.9, the transition seems to be continuous with no sign of a van der Waals loop. These results are consistent with the tight binding MD simulations of Ref. [54]. In contrast, between 3.3 and 3.6×10^3 kg/m^3, where the switch of dominant coordination takes place, the MC results based on the LCBOPI$^+$ display large fluctuations in density at the imposed pressure of 100 GPa, resulting in a slight bending of the isotherm of Fig. 1.9.

Our results provide no evidence of a first-order transition but rather indicate a pronounced but continuous change of dominant coordination.

It is rather surprising that the LCBOPI$^+$ potential reproduces the transformation to a predominantly fourfold coordinated liquid, while the REBO potential does not. The latter is in fact a potential known to give rather accurate predictions in several environments [37]. The high temperature liquid is, anyway, the most demanding benchmark for a semiempirical potential, due to the presence of a large variety of local structures. It thus becomes crucial to have a reasonable estimate of the interaction energy for configurations that are very far from cluster and lattice structures in their minimum energy. Apparently, the isotropic long-ranged interactions play a crucial role in stabilizing the high density liquid. This behavior is rather puzzling as long-ranged interactions were predicted (see, e.g., Ref. [21]) to play a negligible role at these high densities. It has to be recognized that the Brenner original potentials, even without torsional interactions, present at a certain density the switch to the dominant fourfold coordination, but with a strongly first order LLPT, that is as spurious as the one predicted by the torsional interaction corrected versions.

Finally, we note that long-ranged interactions were introduced in Ref. [27] to describe threefold coordinated graphitic phases, and no attempt was made to make the long range interactions dependent on the local environment. Torsional interactions appear to be important, since, without them, the calculated pressures would be too high for high densities and too low at low densities. We conjecture that the combination of torsional interactions, and long range forces is required to give the best description of the liquid.

1.5.3 No Liquid–Liquid Phase Transition in the Stable Liquid Region

In Section 1.4 we reported the calculation of the phase diagram according to the LCBOPI$^+$. With the knowledge of the full melting lines, we can state that there is no LLPT in the stable liquid phase. One indication is the smoothness of the slopes of the melting lines. A further argument lies in the structure of the liquid near freezing. We determined several structural properties of the liquid at coexistence with the solid phases (the properties were calculated during the same simulations of the state points needed for the Gibbs–Duhem integration of the previous section). Figure 1.16 shows the coordination fraction in the liquid along the coexistence lines with graphite and diamond. Since the coexistence lines are unimodal, the graph can be shown indifferently with density, pressure, or temperature on the horizontal axis. We chose to plot the graph with a linear scale in density. The dashed line is the graphite/diamond/liquid triple point. On the left hand side of the triple point, the liquid coexists with graphite, while on the right hand side it coexists with diamond. The threefold and twofold coordination fractions remain rather constant when the

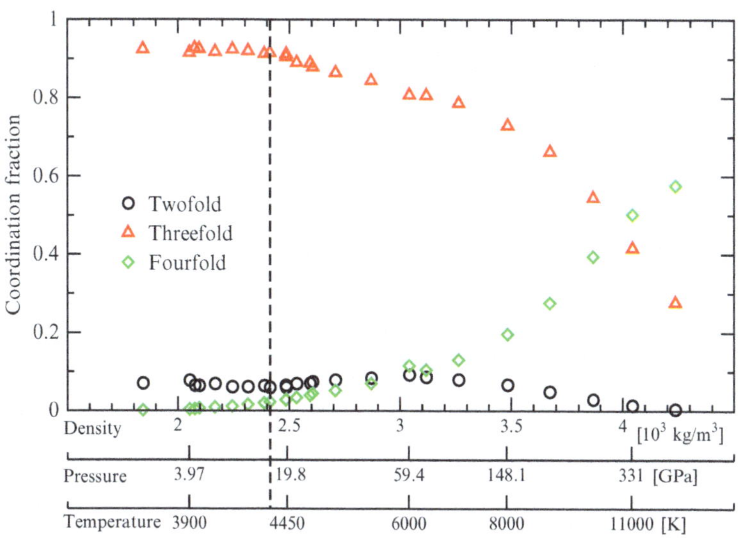

Fig. 1.16 Coordination fraction of the liquid along the *melting line(s)*, according to LCBOPI$^+$. The *melting lines* are unimodal, thus the fixing of ρ fixes also P and T. We arbitrarily chose the ρ scale to be linear. The *dashed line* marks the liquid/graphite/diamond triple point

liquid coexists with graphite, with the fourfold slightly increasing, to account for the increase in density. Along the diamond melting line, on the right hand side of the dashed line, the threefold coordinated atoms are gradually replaced by fourfold coordinated atoms. However, only at (3.9×10^3 kg/m^3, 300 GPa, and 10,500 K) the liquid has an equal fraction of threefold and fourfold coordinated atoms. The change of dominant coordination is rather smooth; moreover, it is fully reversible, without signs of hysteresis, in the region around the swapping of dominant coordination. These results contradict the generally assumed picture (see e.g. Ref. [18]) that diamond melts into a fourfold coordinated liquid.

The interrelation between three and fourfold sites, was further investigated by calculating the partial radial distribution functions ($g_{ij}(r)$) of the liquid at 300 GPa, and 10,500 K. Partial radial distribution functions are defined as the probability of finding a j-fold site at a distance r from a i-fold site[7]; the total radial distribution function g is recovered by: $g = \sum_i g_{ii} + 2\sum_{i \neq j} g_{ij}$. We show the results in Fig. 1.17; we focus on the three predominant curves, describing the pair correlations between threefold atoms (g_{33}), between fourfold atoms (g_{44}), and the cross pair correlation between three- and fourfold sites (g_{34}). Disregarding the rather pronounced minimum in correspondence of the dip around 2 Å of the g_{33} and the g_{34}

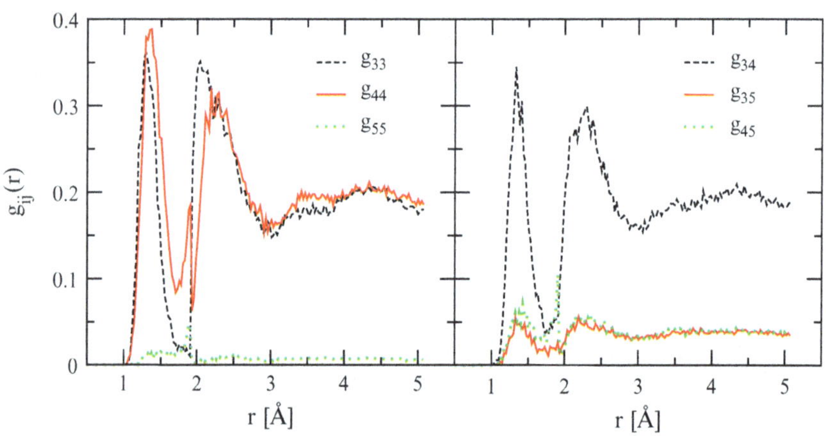

Fig. 1.17 Partial distribution functions g_{ij} of the LCBOPI$^+$ liquid at 10,500 K and ~ 300 GPa, where the liquid coexists with diamond and three- and fourfold atoms are equally present (i.e. where the *dotted line* with cross symbols in Fig. 1.5 cuts the *diamond melting line*). The left panel is for the diagonal contributions (i.e. for $i = j$), while the right panel is for the cross correlations (i.e. for $i \neq j$)

[7] The factor two multiplying the off-diagonal partial distribution functions ($g_{ij}(r)$, with $i \neq j$) is needed when those distributions are calculated according to the literature (e.g. Refs. [71, 72]). The algorithm calculating the $g_{ij}(r)$ browses the pairs of particles only once, as is commonly done for the total $g(r)$. If the algorithm browsed over all the neighbors of each particle, the factor two would clearly not be needed.

(see introduction to Ref. [29] for a discussion of this problem), the similarity of three curves at all distances r is striking. The two sites (i.e. threefold and fourfold atoms) are almost indistinguishable: if there was a tendency towards a phase transition, one would expect some segregation of the two structures. In contrast, looking at distances within the first neighbors shell, a threefold site seems to bond indifferently to a three- or a fourfold site, and viceversa. Furthermore, the partial structures up to the third, quite pronounced, peak at $\sim$4.5 Å, are almost the same for these three partial radial distribution functions.

We determined the properties of the metastable liquid in the stable diamond region. Figure 1.5 shows the liquid $P - T$ state points (crosses) that exhibit an equal number of three and fourfold coordinated atoms. It ranges from high-pressure high-temperature region where the liquid is thermodynamically stable down into the diamond region, where the liquid is metastable for the LCBOPI$^+$. The state point in which this line cuts the diamond melting line is the state point for which we plotted the partial radial distribution function g_{ij} in Fig. 1.17. In Fig. 1.5, the circles indicate state points in which the LCBOPI$^+$ liquid freezes in the simulation. Enclosed by the two set of points lies a liquid state point with a diamond-like structure liquid.[8] This suggests that a (meta)stable liquid with a dominantly fourfold coordination may only exist for pressures beyond $\approx$100 GPa and could imply that the freezing of liquid into a diamond structure might be severely hindered for a large range of pressures beyond the graphite/diamond/liquid triple point. In Ref. [47] it is also pointed out that at 6,000 K the equation of state shows a change of slope around the transition to the fourfold liquid. At even lower temperature this feature becomes more and more evident, but for temperatures lower than $\sim$4,500 K the liquid freezes into a mainly fourfold coordinated amorphous structure. This observation is consistent with quenching MD simulations [70, 73] to obtain the tetrahedral amorphous carbon. In those simulations a mainly threefold liquid freezes into an almost completely fourfold amorphous.

1.6 Summary and Outlook

We have demonstrated how a well designed semi-empirical potential (LCBOP) provides detailed insight in the properties of carbon at extreme conditions. Applying advanced molecular simulation techniques to the LCBOP model provides an accurate description of the melting of graphite and diamond and a detailed picture of the structure and thermodynamics of the liquid. A detailed analysis of the liquid structure over a wide range of pressures and temperature validates the most advance LCBOP model (LCBOPII) as competitive to state-of-the-art DFT models. Analysis of the liquid structure near the melting line reveals that the liquid undergoes

[8] A fourfold coordinated liquid with a rather pronounced diamond-like structure in the first coordination shell [47]).

profound structural changes, with the two- and threefold coordination dominant from the region near the graphite melting line and a range well into the region near the diamond melting line. This observation is expected to have strong implications on the process of homogeneous nucleation of liquid carbon into diamond as demonstrated in Refs. [30, 74], suggesting that the formation of diamond in Uranus and Neptune is kinetically suppressed.

The availability of a validated and accurate empirical potential such as the LCBOP family provides an optimal starting point for further exploration of properties of carbon materials. Of particular interest would be graphite and diamond nucleation, the (meta)stability of novel carbon phases such as carbyne or M-carbon, the structure, stability and nucleation of condensed nanostructures such as nanotube arrays, layered graphene materials. A detailed atomistic picture of these topics will contribute significantly to fundamental understanding in material and planetary science as well as providing important input for carbon-based technological applications

Acknowledgements We gratefully acknowledge F. Colonna, A. Fasolino, D. Frenkel, J. H. Los, and C. Valeriani for inspiring and useful discussions.

Appendix A. Chemical Potential as a Function of Density

Here, we show how to derive Eq. 1.9. One can always write:

$$\begin{aligned}\beta\mu(\rho) &= \beta\mu(\rho^{\maltese}) + \beta\Delta\mu \\ &= \beta\mu(\rho^{\maltese}) + \frac{\beta}{N}\left[F(\rho) - F(\rho^{\maltese}) + P(\rho)V - P(\rho^{\maltese})V^{\maltese}\right] \\ &= \beta\mu(\rho^{\maltese}) + \frac{\beta}{N}\left[\int_{\rho}^{\rho^{\maltese}} \frac{\partial F}{\partial \rho}\,\mathrm{d}\rho + \frac{P(\rho)}{\rho} - \frac{P(\rho^{\maltese})}{\rho^{\maltese}}\right]\end{aligned} \tag{1.10}$$

Working out the integral:

$$\int_{\rho}^{\rho^{\maltese}} \frac{\partial F}{\partial \rho}\,\mathrm{d}\rho = \int_{\rho}^{\rho^{\maltese}} \frac{PV}{\rho}\,\mathrm{d}\rho = N\int_{\rho}^{\rho^{\maltese}} \frac{P}{\rho^2}\,\mathrm{d}\rho \tag{1.11}$$

From the ansatz $P(\rho) = a + b\rho + c\rho^2$, one finds:

$$\int_{\rho}^{\rho^{\maltese}} \frac{a}{\rho^2} + \frac{b}{\rho} + c\,\mathrm{d}\rho = \left[-\frac{a}{\rho} + \frac{a}{\rho^{\maltese}} + b\ln\rho - b\ln\rho^{\maltese} + c(\rho - \rho^{\maltese})\right] \tag{1.12}$$

Putting this result into Eq. 1.10 and reordering the terms:

$$\begin{aligned}\beta\mu(\rho) &= \beta\mu(\rho^{\maltese}) - \frac{P(\rho^{\maltese})}{\rho^{\maltese}} \\ &\quad + \beta\left[-\frac{a}{\rho} + \frac{a}{\rho^{\maltese}} + b\ln\rho - b\ln\rho^{\maltese} + c(\rho - \rho^{\maltese}) + \frac{a}{\rho} + b + c\rho\right] \\ &= \frac{\beta F^{\maltese}}{N} + \beta\left[2c\rho + b(\ln\rho + 1) - c\rho^{\maltese} - b\ln\rho^{\maltese} + \frac{a}{\rho^{\maltese}}\right] \\ &= \frac{\beta F^{\maltese}}{N} + \beta\left[\frac{a}{\rho^{\maltese}} + b\ln\frac{\rho}{\rho^{\maltese}} + b + c\left(2\rho - \rho^{\maltese}\right)\right] \end{aligned} \quad (1.13)$$

References

1. Whittaker AG (1978) Science 200:4343
2. Smith PPK, Buseck PR (1982) Science 216:984
3. Li Q, Ma Y, Oganov AR, Wang H, Wang H, Xu Y, Cui T, Mao H-K, Zou G (2009) Phys Rev Lett 102:175506
4. Kroto HW, Heath JR, O'Brien SC, Curl RF, Smalley RE (1985) Nature 318:162
5. Iijima S (1991) Nature 354:56
6. Novoselov KS, Geim AK, Jiang SVMD, Zhang Y, Dubonos SV, Grigorieva IV, Firsov AA (2004) Science 306:666
7. Bundy FP (1963) J Chem Phys 38:631
8. Bundy FP (1963) J Chem Phys 38:618
9. Fateeva NS, Vereshchagin LF (1971) Pis'ma Zh Ekps Teor Fiz 13:157
10. Galli G, Martin RM, Car R, Parrinello M (1990) Science 250:1547
11. Shaner JW, Brown JM, Swenson AC, McQueen RG (1984) J Phys (Paris) Colloq 45:C8
12. Grumbach MP, Martin RM (1996) Phys Rev B 54:15730
13. Asinovskii EI, Kirillin AV, Kostanovskii AV (1997) High Temp 35:704
14. Asinovskii EI, Kirillin AV, Kostanovskii AV, Fortov VE (1998) High Temp 36:716
15. Bundy FP, Bassett WA, Weathers MS, Hemley RJ, Mao HK, Goncharov AF (1996) Carbon 34:141
16. Togaya M (1997) Phys Rev Lett 79:2474
17. Korsunskaya IA, Kamenetskaya DS, Aptekar IL (1972) Fiz Metal Metalloved 34:942 English version in: (1972) Phys. Met. Metallogr. (USSR) 34:39.
18. van Thiel M, Ree FH (1993) Phys Rev B 48:3591
19. Fried LE, Howard WM (2000) Phys Rev B 61:8734
20. Glosli JN, Ree FH (1999) Phys Rev Lett 82:4659
21. Glosli JN, Ree FH (1999) J Chem Phys 110:441
22. Wu CJ, Glosli JN, Galli G, Ree FH (2002) Phys Rev Lett 89:135701
23. Bradley DK, Eggert JH, Hicks DG, Celliers PM, Moon SJ, Cauble RC, Collins GW (2004) Phys Rev Lett 93:195506
24. Knudson MD, Desjarlais MP, Dolan DH (2008) Science 322:1822–1825
25. Yin MT, Cohen ML (1983) Phys Rev Lett 50:2006
26. Correa AA, Bonev SA, Galli G (2006) PNAS 103:1204–1208
27. Los JH, Fasolino A (2003) Phys Rev B 68:024107
28. Ghiringhelli LM, Los JH, Meijer EJ, Fasolino A, Frenkel D (2005) Phys Rev Lett 94:145701

29. Los JH, Ghiringhelli LM, Fasolino A, Meijer EJ (2005) Phys Rev B 72:214102. Erratum: Phys Rev B 73:229901 (2006)
30. Ghiringhelli LM, Valeriani C, Los JH, Meijer EJ, Fasolino A, Frenkel D (2008) Mol Phys 106:2011
31. Savvatimsky AI (2005) Carbon 43:1115
32. Bradley DK, Eggert JH, Smith RF, Prisbrey ST, Hicks DG, Braun DG, Biener J, Hamza AV, Rudd RE, Collinds GW (2009) Phys Rev Lett 102:075503
33. Brenner DW, Shenderova OA, Areshkin DA (1998) In: Lipkowitz KB, Boyd DB (eds.) Reviews in computational chemistry, vol 5. Wiley, New York
34. Tersoff J (1988) Phys Rev Lett 61:2879
35. Brenner DW (1990) Phys Rev B 42:9458. Erratum: Phys Rev B 46:1948 (1992)
36. Brenner DW, Harrison JH, White CT, Colton RJ (1991) Thin Solid Films 206:220
37. Brenner DW, Shenderova OA, Harrison JA, Stuart SJ, Ni B, Sinnott SB (2002) J Phys Condens Matter 14:783
38. Stuart SJ, Tutein AB, Harrison JA (2000) J Phys Chem 14:6472
39. Kum O, Ree FH, Stuart SJ, Wu CJ (2003) J Chem Phys 119:6053
40. Ghiringhelli LM (2006) On the nature of the phase transitions in covalent liquids. PhD thesis. It can be downloaded from: http://dare.uva.nl/document/18341/
41. Anwar J, Frenkel D, Noro MG (2003) J Chem Phys 118:728
42. Johnson JK, Zollweg JA, Gubbins KE (1993) Mol Phys 78:591
43. Frenkel D, Ladd JC (1984) J Chem Phys 81:3188
44. Frenkel D, Smit B (2002) Understanding molecular simulation. Academic Press, San Diego, CA
45. Polson JM, Trizac E, Pronk S, Frenkel D (2000) J Chem Phys 112:5339
46. Kofke DA (1993) J Chem Phys 98:4149
47. Ghiringhelli LM, Los JH, Meijer EJ, Fasolino A, Frenkel D (2004) Phys Rev B 69:100101(R)
48. Wang X, Scandolo S, Car R (2005) Phys Rev Lett 95:185701
49. Brygoo S, Henry E, Loubeyre P, Eggert J, Koenig M, Loupias B, Benuzzi-Meounaix A, Rabec le Gloahec M (2007) Nat Mater 6:274–277
50. Strässler S, Kittel C (1965) Phys Rev 139:A758
51. van Thiel M, Ree FH (1992) High Pressure Res 10:607
52. Tanaka H (2000) Phys Rev E 62:6968–6976
53. Ponyatovsky EG (2003) J Phys Condens Matter 15:6123
54. Morris JR, Wang CZ, Ho KM (1995) Phys Rev B 52:4138
55. Eppenga R, Frenkel D (1984) Mol Phys 52:1303
56. Vorholz J, Harismiadis VI, Panagiotopoulos AZ (1996) J Chem Phys 105:8469
57. Car R, Parrinello M (1985) Phys Rev Lett 55:2471
58. CPMD, version 3.3, developed by Hutter J, Alavi A, Deutsch T, Bernasconi M, Goedecker S, Marx D, Tuckerman M, Parrinello M, MPI für Festkörperforschung and IBM Zurich Research Laboratory (1995–1999)
59. Hoover WG (1985) Phys Rev A 31:1695
60. Becke AD (1988) Phys Rev A 38:3098
61. Perdew JP (1986) Phys Rev B 33:8822(R). Erratum Phys Rev B 34:7406 (1986)
62. Blöchl PE, Parrinello M (1992) Phys Rev B 45:9413
63. Harada A, Shimojo F, Hoshino K (2007) J Non-Crystal Solid 353:3519
64. Dacosta PG, Nielsen OH, Kunc K (1986) J Phys C Solid State Phys 19:3163
65. Ghiringhelli LM, Los JH, Fasolino A, Meijer EJ (2005) Phys Rev B 72:214103
66. Mishima O, Stanley HE (1998) Nature 396:329–335
67. Stanley HE, Buldyrev SV, Canpolat M, Mishima O, Sadr-Lahijany MR, Scala A, Starr FW (2000) Phys Chem Chem Phys 2:1551–1558
68. Franzese G, Malescio G, Skibinsky A, Buldyrev SV, Stanley HE (2001) Nature 409:692
69. Marks NA (2001) Phys Rev B 63:035401
70. Marks NA (2002) J Phys Condens Matter 14:2901
71. Galli G, Martin RM, Car R, Parrinello M (1989) Phys Rev Lett 63:988
72. Galli G, Martin RM, Car R, Parrinello M (1990) Phys Rev B 42:7470
73. Marks NA, Cooper NC, McKenzie DR, McCulloch DG, Bath P, Russo SP (2002) Phys Rev B 65:075411
74. Ghiringhelli LM, Valeriani C, Meijer EJ, Frenkel D (2007) Phys Rev Lett 99:055702

Chapter 2
Structure, Stability and Electronic Properties of Nanodiamonds

Giulia Galli[1,2]

Abstract Diamond nanoparticles, or nanodiamonds, have the most disparate origins. They are found in crude oil at concentrations up to thousands of parts per million, in meteorites, interstellar dust, and protoplanetary nebulae, as well as in certain sediment layers on Earth. They can also be produced in the laboratory by chemical vapor deposition or by detonating high explosive materials. Here we summarize what is known about nanodiamond sources; we then describe the atomic and electronic structure, and stability of diamond nanoparticles, highlighting the role of theory and computations in understanding and predicting their properties. Possible technological applications of thin films composed of nanodiamonds, ranging from micro-resonators to substrates for drug delivery, are briefly discussed.

2.1 Introduction

The properties of diamond and its practical and technological applications have been extensively investigated for centuries. Elemental carbon has been known since prehistory, and diamond is likely to have been mined in India already 2,000 years ago. But its utilization could date as far back as 6,000 years ago, as suggested by recent archaeological discoveries of diamond utensils in China [1].

In the last two decades, a 'new' form of diamond has attracted much attention: nanometer sized diamond particles. Such particles have been found in primitive meteorites [2–4], formed before the solar system, in proto-planetary nebulae and in interplanetary dust [5], as well as in several earth sediments dated around 13,000 calendar years before present (B.P.) [6, 7]. The study of the origin and formation of diamond nanoparticles in and from outer space may shed light on several problems, including the understanding of possible impacts of meteorites (carbonaceous chondrites) with our planet [7]. On the other hand, the study of the relative stability

[1]Department of Chemistry, University of California, Davis, CA, USA

[2]Department of Physics, University of California, Davis, CA, USA
e-mail: gagalli@ucdavis.edu

L. Colombo and A. Fasolino (eds.), *Computer-Based Modeling of Novel Carbon Systems and Their Properties*, Carbon Materials: Chemistry and Physics 3,
DOI 10.1007/978-1-4020-9718-8_2,

of diamondoids (hydrocarbon molecules with the diamond structure, terminated by hydrogen atoms) may help assess the ultimate depth of crude-oil preservation on earth [8]. The first diamondoid (adamantane) was identified in crude oil in 1933 [9] but it was only few years ago [10] that higher molecules (with 22 C atoms or more) were recovered from oil.

In addition to geological and astrophysical interest, nanometer-sized diamonds have attracted the interest of the nanoscience and nanotechnology communities [11], for the potential application of nanodiamond films as bright, low voltage (cold) cathodes and light emitters [12], and of diamondoids as electron emitters [13, 14]. The exceptional hardness, fracture strength, and inertness of ultra-nanocrystalline diamond films, together with their smooth surface make them unique materials for miniaturized mechanical systems and devices (MEMS). Such devices include cantilevers and gears, and nanodiamond MEM devices have already been successfully produced [15]. Finally nanodiamonds are finding applications in biology as drug delivery vehicles and as substrates for DNA attachment [16].

The purpose of this brief review is to give a summary of nanodiamond sources and thin film production in the laboratory (Section 2.2), and then to focus on structural, stability and electronic properties of nanodiamonds (Sections 2.3–2.5, respectively), with the intent of highlighting the contributions of theory and computation to the understanding of these fascinating nanostructured materials.

2.2 Nanodiamond Sources

2.2.1 Nanodiamond in and from Outer Space

Nanoscale diamonds were discovered in 1987 by Lewis et al. [2, 3] in meteorites called carbonaceous chondrites. Two major sources of nanodiamonds are the "Allende" chondrite (C3V type), which weighed several tons when it fell on Mexico in 1969, and other types of chondrites, such as the Murchison meteorite (type C2). The nanodiamonds found in these meteorites have a lognormal size distribution, with a median diameter of about 2.5 nm, corresponding to roughly 1,100 atoms per particle. The shape of this distribution has been interpreted as stemming from a growth process in which ensembles of small grains are progressively transformed into larger ones. Meteoritic nanodiamonds were shown to contain impurities, mainly hydrogen, nitrogen, and oxygen in the form of –COOH groups. In some cases, such as in C2-type chondrites, diamond in its nanoscale form is up to five times more abundant than graphite. Additional analysis of meteoritic nanodiamonds by Amarti et al. [4] revealed the nanodiamond content of the C2 meteorite to be about 400 ppm, and the shape of the size distribution was attributed to formation by a condensation process.

An interesting question arising from Lewis' and subsequent discoveries, concerns the origin of nanodiamonds. Daulton et al. [17] performed a detailed

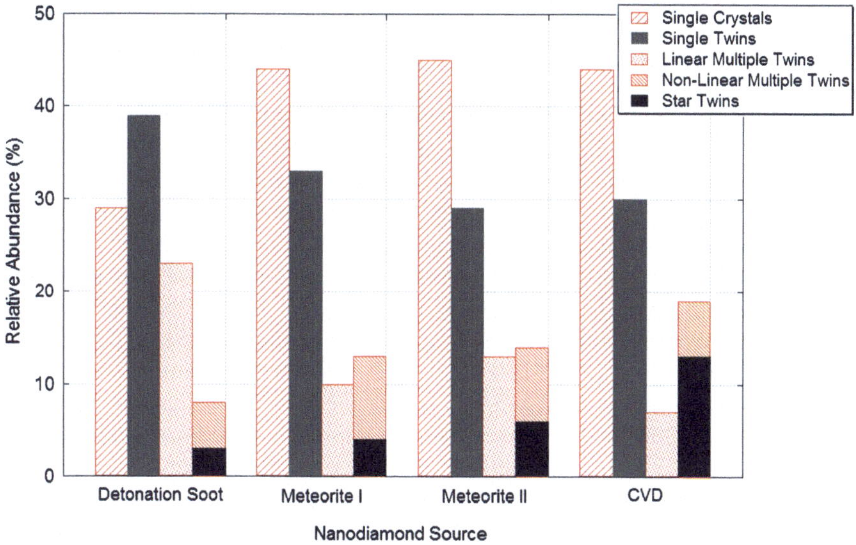

Fig. 2.1 Distribution of the different twinning types found in nanodiamonds with various origins. The meteoritic diamond [2] originated from the Murchison meteorite (I) and the Allende meteorite (II). Note the similarity between the twinning in meteoritic and Chemical Vapor Deposited (CVD) nanodiamonds. From [17]

high-resolution transmission electron microscopy (HRTEM) study of a large quantity of nanodiamonds coming from meteorites, as well as those produced by means of detonation or chemical vapor deposition (CVD) in the laboratory. They concluded that the characteristic of meteoritic nanodiamonds (morphology, type of twinning defects) are much closer to those obtained by low-hydrogen-pressure CVD than by high-temperature–high-pressure detonation, as shown in Fig. 2.1; hence they suggested that meteoritic nanodiamonds are formed by CVD-like processes, and that their growth is possibly seeded by C_{60} or C_{70} molecules. However, recently two novel formation mechanisms have been proposed, one involving diamondoid growth on ices [18] and the other one growth in dust grains [19].

Isotopic anomalies of a fraction of meteoritic diamond indicate an extra-solar source, although recent studies of Dai et al. [20] suggest that nanodiamonds might have been formed within the inner solar system. The discovery, in [20], that only minute quantities of nanodiamond is present in interplanetary dust particles (formed further out in the early Solar System than the parent asteroids from which meteorites may originate) suggests that meteoritic nanodiamond may not be presolar, consistent with data obtained from the accretion disc of young stars [21]. However the presolar origin is not ruled out by the data of Dai et al. [20], and it might still constitute a viable hypothesis; nevertheless, in this case nanodiamond abundance would be expected to decrease with heliocentric distance, a behavior that would open problems in the overall understanding of large scale transport and circulation in the early Solar system.

Nanodiamond signals have also been identified in the emission spectrum of 12 warm supergiants [22], which are carbon-rich protoplanetary nebulae. A broad absorption line from these objects, centered at 21 mm wavelength, was originally attributed to large polycyclic aromatic hydrocarbon (PAH) molecules or to some partially hydrogenated fullerenes; however, more recently Hill et al. [23] re-assigned the line to nanodiamonds. In this study, the observed infrared absorption from the interstellar dust is shown to be very similar to the signal obtained from either nitrogen-rich nanodiamonds, or nanodiamonds containing vacancies and/or interstitial atoms, produced in the laboratory. It was also noted that some absorption features could be explained by assuming that part of the nanodiamond surface attains sp^2 like configurations. By comparing the absorption spectra with those of terrestrial diamonds, Hill et al. [23] estimated the size of interstellar nanodiamonds to be around 2.5–3 nm, exactly the same size range as for meteoritic diamonds. Other indirect, nanodiamond signatures have been found by Van Kerkhoven et al. [21], originating from specific stars; those sources are thought to contain hydrogen-terminated nanoparticles, possibly exhibiting some type of surface reconstruction, and with sizes between 1 and 10 nm.

Recently nanodiamonds were found [6, 7] in the younger Dryas Boundary sediment layer at multiple locations across North America, with sizes ranging from 2 to 300 nm. Interestingly these nanoparticles have the hexagonal lonsdaleite structure and their presence is indicative of multiple air shocks and possibly surface impacts on Earth of the order of 12.9 thousand calendar years B.P. (equivalent to about 10,900 radiocarbon years B.P.).

We now turn to the discussion of nanodiamond produced in the laboratory, either by detonation or by CVD techniques.

2.2.2 *Nanodiamonds in Detonation Soots and Ultra-Dispersed-Diamond (UDD) Films*

Pure TNT (2,4,6-trinitrotoluene) detonation has been shown [24] to produce, among other carbon structures, nanocrystalline (NC) diamonds with diameters of about 10 nm. By mixing TNT with solids such as RDX (cyclotrimethylene trinitramine – $C_3H_6N_6O_6$), or TATB (triaminotrinitrobenzene), and by detonating the mixture in an inert gas atmosphere, spheroidal diamond particles were produced that are $\sim$4 nm in diameter [25]. Following early studies, the synthesis of nanodiamonds by detonation has been optimized and detonation-produced films are now commercially available. They are often called "ultra dispersed diamond" (UDD) films because of the very narrow size distribution of the nanoparticles.

Aleksenski et al. [26, 27] performed a structural study of UDD films using X-ray diffraction and small angle X-ray scattering. They observed a diamond cluster core of about 4.3 nm, the surface of which is covered by a mixture of sp^2- and sp^3-bonded carbons. These authors explained their small-angle X-ray scattering (SAXS)

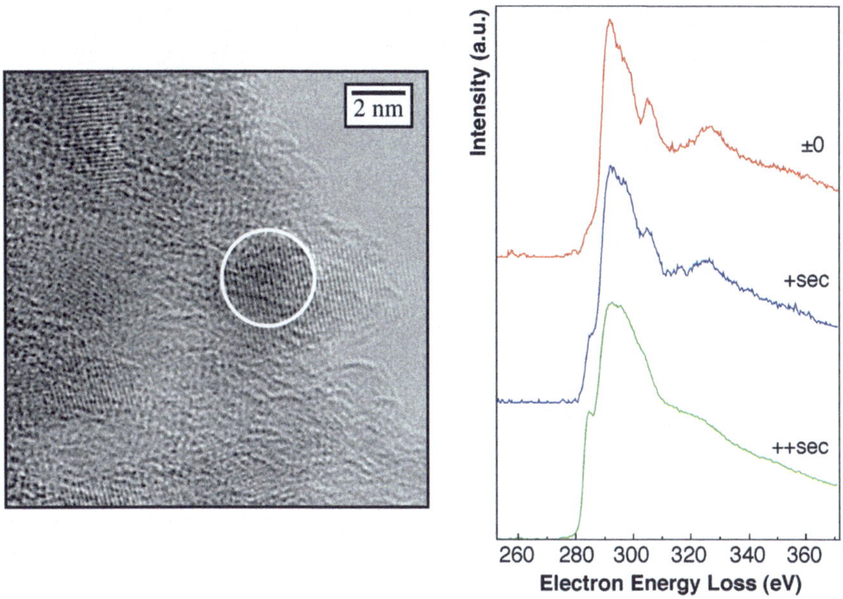

Fig. 2.2 Evolution of the electron energy loss spectrum of a Ultra-Dispersed Diamond (UDD) film (*right hand side*) as a function of the exposure time (Courtesy of T. Van Buuren and J. Plitzko, LLNL). The spectrum rapidly evolves from diamond-like to amorphous-like due to heating produced by the electron beam. A high resolution transmission electron micrograph (HRTEM) of the film is shown on the *left hand side*. A ~3 nm diamond nanoparticle is shown by the *white circle*

measurement with a model in which the diamond core is surrounded by onion-like carbon shells and nanosized graphite platelets. The thickness of these surrounding shells appears to be dependent on the type of detonation synthesis (dry technology–gas cooling, or wet technology–water cooling). Baidakova et al. [28] analyzed the fractal dimension of the external shells of nanodiamonds. Upon annealing around 1,200–1,300 K, they found that the diamond content of the nanoparticles decreases, in favor of the formation of graphite flakes on the surface. Similar transformations occur under electron beam annealing. For example, Fig. 2.2 shows the evolution of the electron energy loss spectroscopy (EELS) of a UDD film as a function of exposure time. The spectrum rapidly evolves from diamond-like to amorphous-like because of heating produced by the electron beam. Other detonation-produced (TNT+RDX) nanodiamonds were analyzed by Chen et al. [29]. In their study, the particles were almost perfectly spherical and 4–6 nm in diameter, but some larger (15 nm) particles were also detected. Their composition analysis showed 87–90% carbon, 0.5–1% hydrogen, 1.6–2.5% nitrogen, and 6–10% oxygen. These authors explained the spherical shape of the particles as a consequence of re-crystallization of liquid-like carbon droplets during the detonation, consistent with an early model proposed by Viecelli et al. [30]. The structure was shown to be diamond, with a small proportion of sp^2-bonded carbon atoms. In UDD, small nanocrystals often

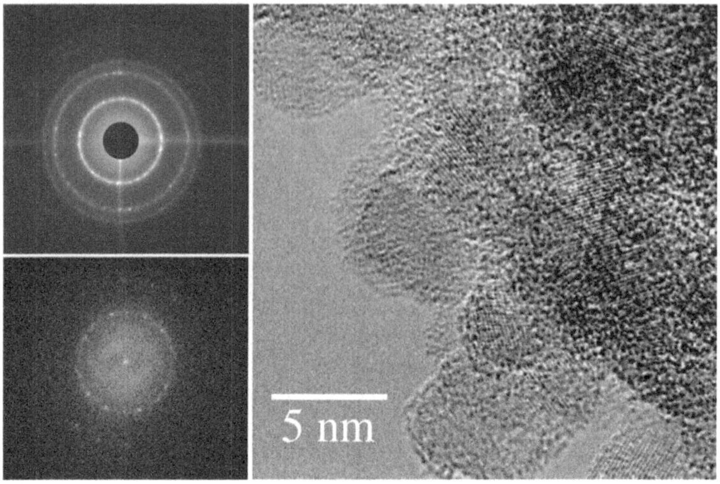

Fig. 2.3 Electron diffraction pattern (*top left*), power spectrum (*bottom left*), and a high resolution transmission electron micrograph (HRTEM) (*right*) of the edge zone of a conglomerate of nanodiamonds in Ultra-Dispersed Diamond (UDD) film (UDD). From [42]

aggregate in larger conglomerates, as shown in the study of Aleksenskii et al. [27], with small distortions of the initial spherical particle shape. A representative electron diffraction pattern is shown in Fig. 2.3.

The structure and defects of UDD have been extensively studied. In [31], infrared spectroscopy showed evidence of O–H, C–H, and C–O–C groups. Other surface groups were detected by nuclear magnetic resonance (NMR). These electron spin resonance measurements indicate that nitrogen is not present as a substitutional (paramagnetic) site inside the particle core. However, some authors [32] have measured a high concentration of paramagnetic centers in UDD films. These are attributed to dangling C–C bonds located at the interface between the diamond core and the graphene sheets forming at the surface. A similar paramagnetism has been observed in [33].

2.2.3 Nanodiamond Produced by Chemical Vapor Deposition and Ultra-Nanocrystalline Diamond (UNCD) Films

While optimizing the CVD technique for the generation of high-quality diamond films, it was found that, under some conditions, the deposited film was no longer a microcrystal assembly, but a smooth film of much smaller diamond particles [12]. A typical CVD experiment would involve the injection of a mixture of methane and hydrogen gas in a plasma reactor. The diamond film, with variable morphologies, is deposited on a silicon substrate. CVD-deposited diamonds films are distinguished into three categories [34]: microcrystalline diamonds (0.5–10 mm), nanocrystalline diamonds (50–100 nm), and ultrananocrystalline (UNCD) diamonds (2–5 nm). Garcia et al. [35] reported diamond particles of 50 nm diameter by applying a negative d.c. bias voltage during the first minutes of the deposition from a mixture of 4% CH_4 in H_2. The nanometric diamonds appear to represent the first stage of diamond growth, as long deposition times show them coalesce to form a micro-crystalline diamond film. Nanocrystalline diamond crystals with a typical size of 50–100 nm are produced by hollow cathode arc plasma CVD, or direct current glow discharge-assisted CVD, among many other experiments.

Gruen et al. [12] devised a procedure to optimize the gas concentration to generate UNC diamonds. Replacing the majority of hydrogen by argon (typical 1% hydrogen) in the plasma yields the deposition of a smooth film of nanoparticles of only a few nanometers in size [36]. The procedure has been perfected over the years and now UCND films for several applications, e.g. for MEMS are available [15]: an example is shown in Fig. 2.4. These UNCD films contain only 2–5% of sp^2-bonded carbons in grain boundaries and less than 1% hydrogen. In other cases [37] a detailed analysis of the proportion of nanodiamonds to graphite has shown that up to 75% of carbon present in the film is sp^3-bonded carbon, in 3- to 5-nm diamond nanocrystals. The remaining sp^2-bonded atoms are likely to be mostly present in the form of a thin graphite layer (150–200 nm) located between the substrate and the UNCD diamond, as well as in grain boundaries. An extensive review of UCND films is given in [12].

In the next session we describe nanometer-sized diamond particles and smaller molecules with a diamond cage, in which the surface is perfectly passivated by hydrogen atoms.

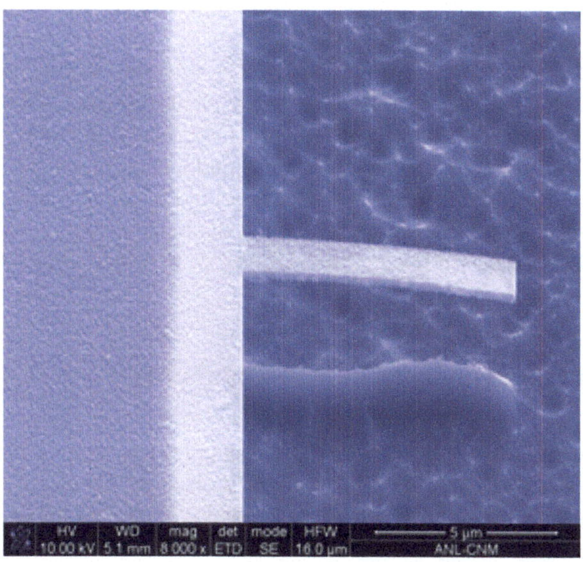

Fig. 2.4 Scanning Electron micrograph (SEM) of a micro-resonator fabricated from ultrananocrystalline diamond films at Argonne National Laboratory. The procedure is described in [15]; the figure is from: http://nano.anl.gov/news/highlights/2009_microresonators.html

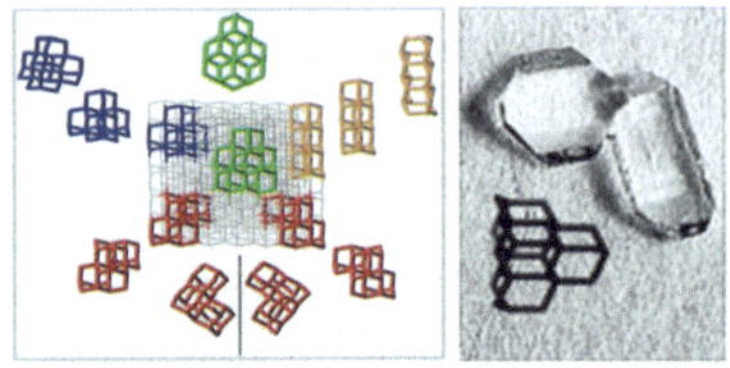

Fig. 2.5 Screw-shaped higher diamondoids (*left hand side*) and an image of the pentamantane crystal (*right hand side*). Adapted from [10]

2.2.4 *Nanodiamond in Oil: Diamondoids*

As mentioned in the introduction, diamondoids are molecules consisting of diamond cages terminated by hydrogen atoms, as shown in Fig. 2.5. The first molecule in the series is adamantane ($C_{10}H_{16}$), discovered in 1933 [9], followed by di-($C_{14}H_{20}$), tri-($C_{18}H_{24}$), tetra-($C_{22}H_{28}$) amantane and so on. These molecules are difficult to synthesize in the laboratory [8, 38] and small traces have been found in petroleum [10, 39, 40] and interstellar gases [5] over decades. In fact both neutral and cationic forms may be present in certain interstellar gases, as recently suggested by theoretical calculations of infrared spectra [5].

The high chemical stability of diamondoids has prompted their use in coal liquefaction studies and as matrices for free-radical reactions. In the Earth, diamondoids outlive other hydrocarbon found in crude oils and it has been suggested that their stability may be used to assess the ultimate depth of crude oil preservations [8].

A major breakthrough in isolating diamondoids in natural gas was made in 2003 [10], and it was also shown that these molecules could be functionalized with several groups, thus increasing their ability to, e.g. stick to surfaces, as achieved in recent experiments [14]. Before experiments probing diamondoids on surfaces, first principles calculations [13, 41] suggested that these hydrogenated carbon clusters are very different from their Si and Ge counterparts and, in particular, that they should exhibit negative electron affinity. Subsequent experiments have confirmed the picture suggested by ab initio calculations. In particular the first experiments of electron emission properties of layers of diamondoids on metal surfaces were recently reported [14]. These diamondoid layers could have interesting applications, e.g. as constituents of low power consumption flat panel displays. Their electronic properties will be discussed in detail in Section 2.5.

2.3 Structural Properties of Nanodiamonds

The crystallinity of a variety of nanodiamond samples has been verified by diffraction experiments, and it has also been directly observed by electron microscopy (see, e.g. enhanced atomic planes in Figs. 2.2 and 2.3). A detailed crystallographic study of nanodiamonds found in meteorites, detonation soots, and CVD grown UNC films has been performed, e.g. by Daulton et al. [17] who have reported the presence of twinning defects in most nanoparticles, in different forms (see Fig. 2.1). Overall, the shape of nanodiamonds has been found to be close to spherical.

While the core of diamond nanoparticles is well characterized and there are little doubts about the presence of a diamond cage, the structure of nanodiamond surfaces is more controversial, as it is much more difficult to characterize experimentally. The presence of hydrogen atoms at the surface appear to be dependent on preparation conditions and will be discussed in more detail in Section 2.4. If hydrogen is not present at the surface, both computer simulations and experiments [26, 27, 42] suggest that graphitic like reconstructions and onion-like structures appear at the surface of nanodiamonds.

In particular, first principles calculations have found fullerene-like reconstructions in the 2–4 nm size range, leading to nanoparticles that have a hybrid structure [42]; these nanoparticles have been called *bucky-diamonds*. Two bucky diamonds as obtained using ab intio simulations and that of a larger one studied with semi-empirical tight-binding methods are reported in Fig. 2.6. We note that the predicted reconstruction induces an effective tensile stress acting on the core of the nanocluster: this effect is at variance with the one found in Si and Ge dots where surface reconstruction induces an effective compression on the nanoparticle core [43]. On the contrary, calculations carried out with empirical potentials (Brenner potential) by Halicioglu et al. [44] found inward relaxation from the top surface layers, the interatomic distance between neighboring atoms decreasing from the center of the cluster to the surface. In *ab-intio* simulations, the barrier between the ideal surface structure and the reconstructed surface found in bucky diamond is size dependent and increases as the size of the nanoparticle is increased, to become of the order of several tens of an eV, as found in bulk diamond [45]. The comparison between ab initio calculations and measured XAS spectra of UDD films [42] indicates that the surface reconstruction of nanodiamond identified by computer simulations is responsible for pre-edge, graphitic-like features observed in absorption spectra. These are shown in Fig. 2.7.

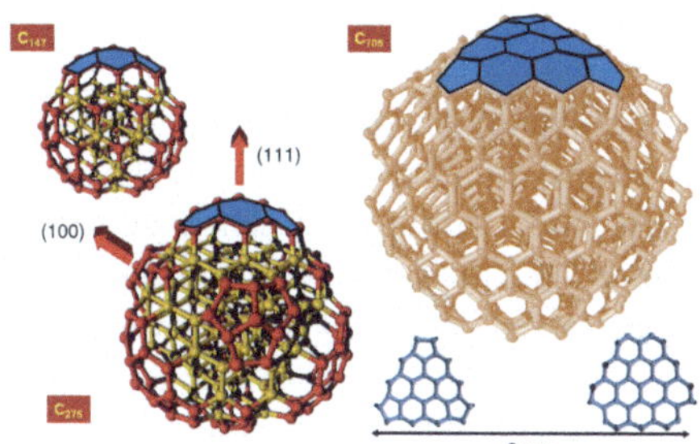

Fig. 2.6 Ball and stick representation of bare nanodiamonds (*bucky-diamonds*) composed of a diamond core (*yellow spheres*) and graphitic-like surfaces (*red spheres*). Some of the reconstructed facets on top of the diamond core are highlighted (*blue*) for clarity. They consist of a mixed pentagon/hexagon network, as represented in the lower *right hand side* of the figure

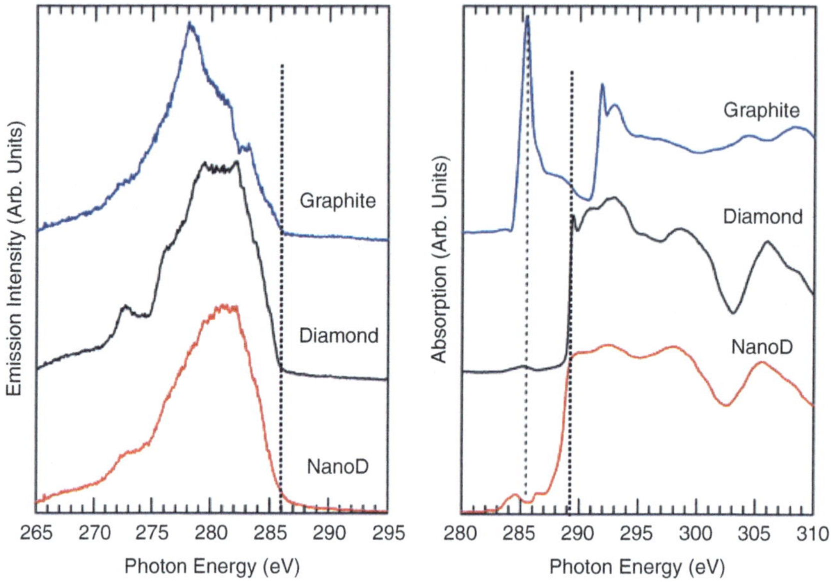

Fig. 2.7 Emission (*left panel*) and absorption (*right panel*) spectra of nanodiamond (*red line*) compared to bulk diamond and highly oriented pyrolitic graphite. From [42]

2.4 Stability of Nanodiamonds

The first attempt to understand the stability of diamonds at the nanoscale was reported by Badziag et al. [46]. These authors computed the binding energy of diamond-like and graphite-like carbon clusters using fixed energy values for carbon–carbon and carbon–hydrogen bonds. They found that below a size of 3–6 nm, diamond clusters become more stable than their graphitic counterparts. After the study of Ref. [46], several groups provided models of the relative stability of diamond and graphite like nanoparticles, with different levels of sophistication. Charged cluster models were presented by Hwang et al. [47] and Jang and Hwang [48], and the crossover between diamond and graphite was predicted to occur for clusters containing 400 atoms. Other charge lattice calculations by Gamarnik [49] predicted instead a reversal of stability between diamond and graphite to occur between 4.3 and 10.2 nm, depending on the temperature. It is worth noting that although these models are based on totally different approximations and neglect the detailed structure of the clusters, such as specific surface reconstructions or degrees of hydrogen passivation, they all reach similar conclusions, namely that nanodiamond becomes more stable than nano-scale graphite below 3–10 nm.

A remarkable property of nanodiamonds produced under different conditions of atmosphere, temperature, and pressure is that most of them have similar size distributions. In particular, extraterrestrial nanodiamonds, some detonation nanodiamonds, and ultrananocrystalline diamonds have sizes that typically range between 2 and 5 nm.

The range of stability of nanodiamond and the reason why diamond is ultra-dispersed at the nanoscale was addressed in Ref. [50] by using first principles calculations [51,52]. In particular, the relative stability of nanodiamonds as a function of size and of the surface hydrogen coverage, in going from a totally passivated surface to a completely bare, fully reconstructed one, was investigated in detail. The ab initio calculations of [50] showed that as the size of diamond is reduced to about 3 nm, it is energetically more favorable for this material to have bare, reconstructed surfaces than hydrogenated surfaces. This inability to retain hydrogen at the surface may then prevent the growth of larger grains. Two ranges of hydrogen chemical potential values were identified, corresponding to two different growth conditions of diamond films: one that favors the formation of bare, reconstructed nanoparticles and thus the formation of UNCD or UDD films, the other favoring the formation of bulk diamond-like films. In the calculations, five core sizes were considered, containing 29, 66, 147, 211 and 275 C atoms; two of these are shown in Fig. 2.8; for all of the different sizes the stability sequence of the particles with different surface structures was found to be the same as a function of the hydrogen chemical potential.

Although the ab initio study of Ref. [50] could not establish the exact size at which the crossover between hydrogenated and bare, reconstructed surfaces occur, it provided a robust, qualitative explanation of why the crossover occurs in the few nanometers range, and why it is the same irrespective of preparation conditions. Based on the simple thermodynamic mechanism presented in Ref. [50], one might

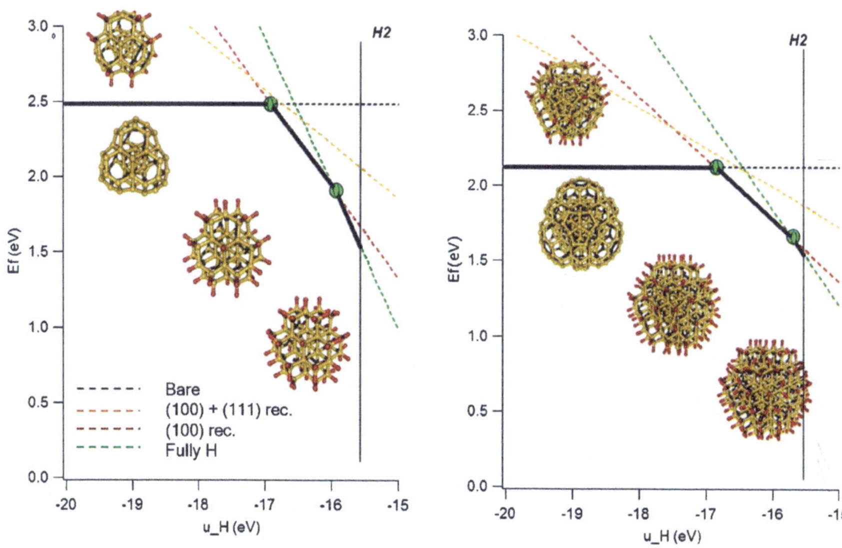

Fig. 2.8 Formation energy per carbon atom (E_f) of nanodiamonds as a function of the hydrogen chemical potential (μ_H), as obtained by first principles calculations. The nanodiamonds contain 66 (*left hand side*) and 275 (*right hand side*) C atoms. The solid black line indicates the most stable structures. From [50]

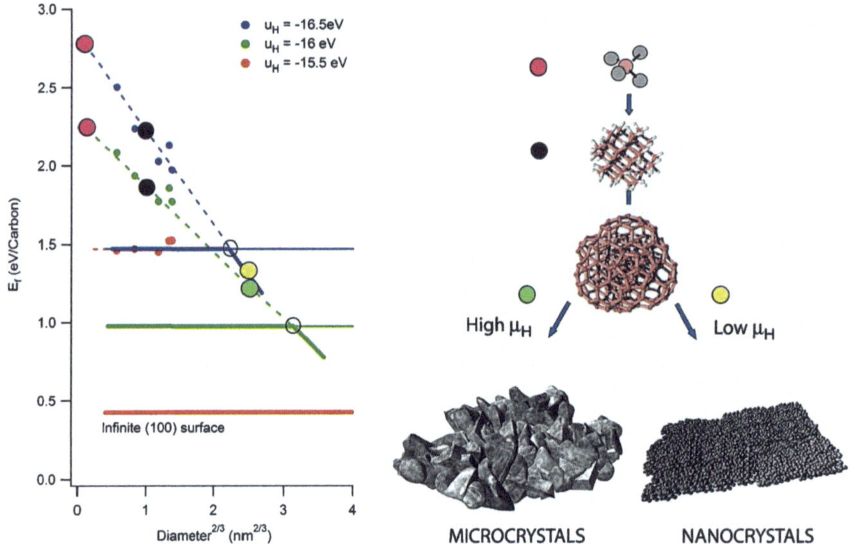

Fig. 2.9 Comparison of the formation of nanodiamonds with that of flat diamond surfaces. The *left hand side* shows the formation energy of nanodiamonds (E_f) as a function of $d^{2/3}$, where d is the particle diameter, for three values of the hydrogen chemical potential (μ_H). The formation energy of a flat (111) diamond surface for the same values of the hydrogen chemical potential, is represented by *dashed lines*. The *thick, solid lines* designate the most stable structure at any stage of a hypothetical diamond growth started from methane gas. The *right hand side* of the figure highlights the formation of different diamond films, depending on the hydrogen chemical potential conditions. From [50]

argue that it would be impossible to grow diamond on a micro- or macroscopic scale. To address this issue, it is necessary to compare the formation energies of nanodiamonds of various sizes with those of flat diamond surfaces. This comparison is shown in Fig. 2.9 where it is seen that at the highest values of the H chemical potential considered in the first principle investigation, the infinite (100) surface is more stable than any nanodiamond, while at lower chemical potential, there exist a critical diameter above which the nanodiamond clusters are the most stable structure. These numerical results helped explain how, by varying the hydrogen pressure (and thus the hydrogen chemical potential) in a CVD reactor, one can deposit either microcrystalline or ultra-nanocrystalline diamond films.

2.5 Electronic Properties of Nanodiamonds

Overall, measuring the opto-electronic properties of nanodiamond and diamondoids has proved to be challenging and has produced several controversial results. In 1999, the results of x-ray absorption near-edge structure studies of diamond films produced by chemical vapor deposition (CVD) [53] were used to infer the evolution of the nanoparticle gap with size. This showed a persistence of quantum confinement effects up to 27 nm, a diameter surprisingly larger than in Si or Ge nanoparticles, where quantum confinement effects disappear above 5–7 nm. In contrast, recent near-edge absorption fine structure studies of diamondoids prepared by hot filament CVD [54] and high-explosive detonation waves [42] showed that quantum confinement effects disappear in particles larger than 4 nm. In particular, in [42] it was shown that there is no shift in valence and conduction band maximum and minimum in detonation nanodiamonds, in comparison with bulk diamonds. This indicates that quantum confinement does not affect the electronic structure of the measured particles, which were in the size range of 4 nm. X-ray emission and absorption spectra of detonation nanodiamonds were found to be very similar to those of bulk diamonds, with an exciton broadening (289.3 eV) and a shallower secondary minimum (302 eV) as pre-edge features. These features were attributed to specific surface reconstructions such as in bucky diamonds, as predicted by theory and described in Section 2.3. These results are consistent with those reported by Aleksenskii et al. [55] on the optical properties of UDD layers studied by optical probes and by XPS. In [55] the band gap was measured to be smaller than in diamond, 3.5 eV, and many energy levels were present in the nanodiamond band gap, contributing to a broad luminescence band (380–520 nm). The optical absorption of the material was attributed to threefold coordinated atoms on the surface.

Raty et al. [42] studied the size dependence of the optical gap of diamondoids using both time-independent and time-dependent density-functional theory [56] calculations, and concluded that quantum confinement effects disappear in nanoparticles larger than 1 nm. They also predicted that the gaps of diamondoids with sizes between 1 and 1.5 nm are *below* the gap of bulk diamond. This is strikingly different from the behavior of H-terminated Si and Ge nanoparticles [43], for which the gaps are always above the bulk band gap. In contrast, Density Functional Theory calculations by McIntosh et al. [57] for the same particles predicted optical gaps 2 eV above the gap of bulk diamond for particles ranging in size from 0.5 to 2 nm.

As discussed in detail below, highly accurate quantum Monte Carlo [58] (QMC) calculations [13] resolved this controversy, showing the disappearance of quantum confinement at about 1 nm. It was also pointed out that the results of [57] may be affected by basis set superposition errors, and these are believed to be responsible for the discrepancy between results obtained with localized and plane wave basis sets. In addition, Drummond et al. [13] predicted that diamondoids exhibit negative electron affinity. Two classes of nanoparticles were studied: (i) diamondoids constructed from adamantine cages: *adamantane*, $C_{10}H_{16}$, *diamantane*, $C_{14}H_{20}$, and *pentamantane*, $C_{26}H_{32}$; (ii) H-terminated, spherical, diamond structure nanoparticles: $C_{29}H_{36}$, $C_{66}H_{64}$, and $C_{87}H_{76}$. Because diamondoids can be extracted in large quantities from petroleum and are highly purified using high-pressure liquid chromatography, one can expect that actual experimental samples consist largely of the high-symmetry structures studied theoretically. This is not the case for Si and Ge nanoparticles, where limitations in current synthesis techniques prevent the routine production of high symmetry nanoparticles.

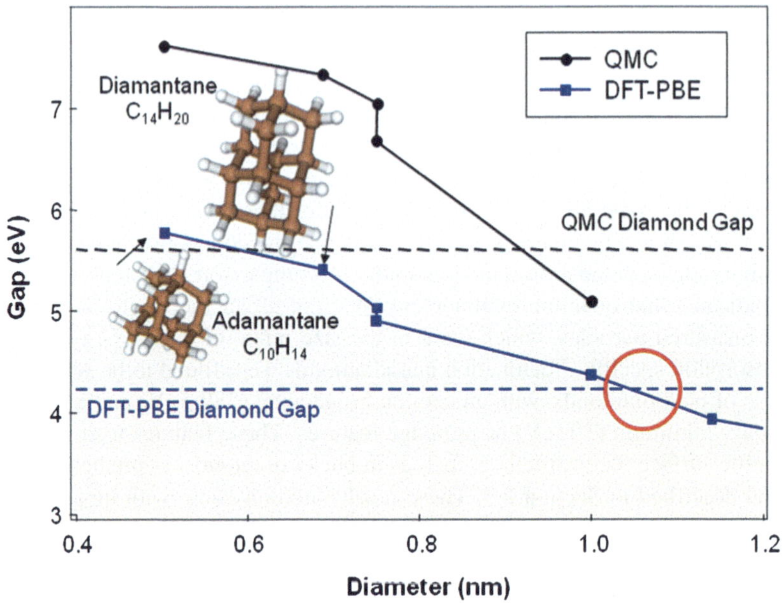

Fig. 2.10 Size dependence of the electronic gap of diamondoids as obtained using Density Functional Theory (DFT) calculations with gradient corrected exchange and correlation functionals (PBE) and Quantum Monte Carlo (QMC) calculations [13]. The *dashed blue and black lines* indicate the values of the gap of bulk diamond computed using DFT-PBE and QMC. The *red circle* indicates that at around 1 nm quantum confinement effects vanish, at both levels of theory

We first discuss the results of Density Functional Theory [51] (DFT) calculations reported in [13]. Figure 2.10 shows the size dependence of the DFT eigenvalue gap between the highest occupied molecular orbital (HOMO) and the lowest unoccupied molecular orbital (LUMO), calculated using plane-wave basis sets. The HOMO-LUMO gap decreases as the size of the nanoparticles increases. For nanoparticles larger than 1 nm, the HOMO-LUMO gap obtained with gradient corrected Density Functional Theory calculations (that we call DFT-PBE [59]) is smaller than the DFT-PBE gap of bulk diamond. It is interesting to analyze the size dependence of the individual HOMO and LUMO eigenvalues obtained from DFT (Fig. 2.11). When the size of the nanoparticles is reduced, the HOMO eigenvalue decreases continuously, as predicted by standard models of quantum confinement. In contrast, the LUMO eigenvalue displays almost no quantum confinement and is nearly independent of size. Once again, this behavior differs from that of H-terminated Si and Ge nanoparticles, in which both the HOMO and LUMO are clearly affected by quantum confinement.

The origin of the anomalous size dependence of the LUMO energy can be understood by analyzing iso-surfaces of these orbitals for specific nanoparticles: Fig. 2.12 shows the case of $C_{29}H_{36}$. The HOMO is localized on the C–C and C–H

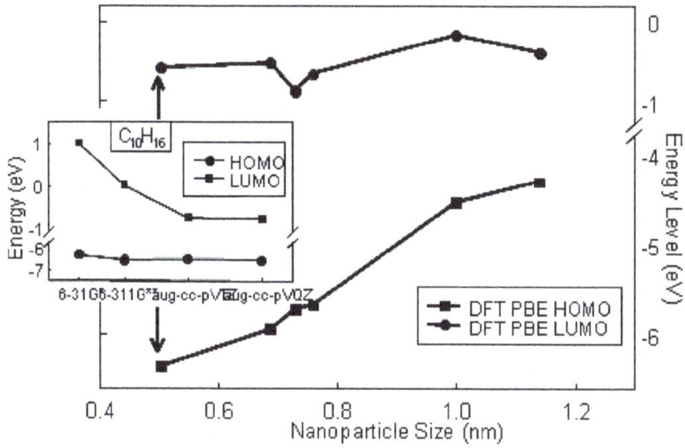

Fig. 2.11 Orbital energies of the Highest Occupied Molecular Orbital (HOMO) and Lowest Unoccupied Molecular Orbital (LUMO) of diamondois as a function of nanoparticle size, as obtained by Density Functional Theory (DFT) calculations within the gradient corrected approximation (PBE), using plane wave basis sets. The inset shows results of DFT calculations using localized basis sets (see text). Adapted from [13]

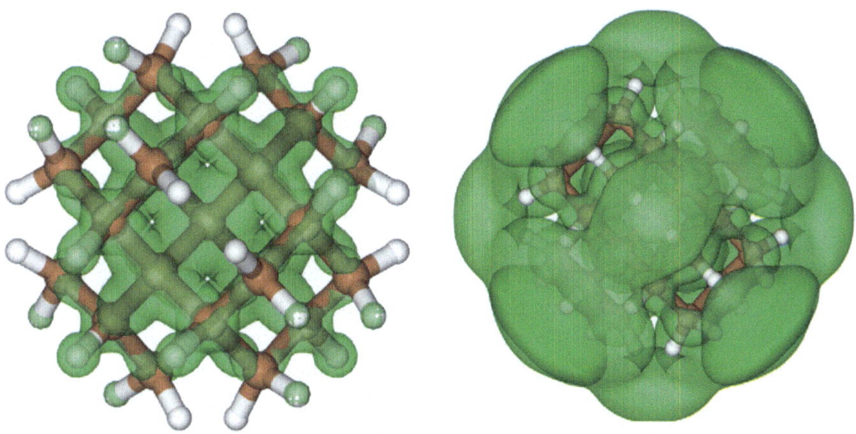

Fig. 2.12 Isosurface plots of the square of the highest occupied (*left hand side*) and lowest unoccupied (*right hand side*) orbitals of $C_{29}H_{36}$ diamondoid. The isosurfaces include 50% of the charge in each orbital. Adapted from [13]

covalent bonds inside the nanoparticle, while the LUMO is a delocalized state with considerable charge outside of the H atoms terminating the surface. As the size of the nanoparticles increases, one expects the HOMO to evolve smoothly into the valence-band maximum of bulk diamond. In contrast, for larger diamondoids, the LUMO is expected to remain in proximity of the surface, closer in nature to a defect or surface level within the gap of the bulk material; therefore the LUMO is not expected to evolve into the conduction band minimum. This surface character of the state causes the optical gaps of the larger diamondoids to lie below the bulk gap. The delocalized nature of the LUMO as predicted by ab initio calculations was confirmed by XAS experiments [60, 61].

The results of quantum monte carlo calculations (QMC) for optical gaps of the diamondoids are plotted against nanoparticle size, together with DFT results discussed above, in Fig. 2.10. QMC calculations do not suffer from the well-known DFT "band-gap problem," as they fully describe the interaction of the valence electrons with the electron excited into the LUMO by the absorption of a photon, so that electron-hole correlation is accounted for. The QMC optical gaps are significantly larger than the DFT HOMO-LUMO gaps, as was found for Si nanoparticles [43]. For example, the QMC optical gap of $C_{10}H_{16}$ is 7.61 [2] eV, while the DFT-PBE gap is 5.77 eV. Nevertheless, the QMC results confirm the qualitative trend for the size dependence of the gap predicted by mean field theory. The QMC calculations predict that quantum confinement will be observed only in diamondoids smaller than 1 nm. This rapid decay of quantum confinement is consistent with the small exciton Bohr radius in diamond (1.6 nm), compared, e.g. to Si (4.9 nm). We note that the optical gap of $C_{66}H_{64}$ is below the bulk-diamond gap, consistent with the results of [42]. These QMC optical-gap calculations support the interpretation of the X-ray absorption measurements [42], which found no quantum confinement effects in 4 nm nanodiamonds.

First principle calculations were also applied to study the electron affinity of diamondoids. Both electron affinity (EA) and ionization potential (IP) of representative nanoparticles were evaluated at the DFT and QMC level of theory, by computing the total energy differences EA $= E_N - \mathrm{E}_{N+1}$ and IP $= E_{N-1} - E_N$, where E_N is the ground-state energy of a neutral molecule with N electrons. In contrast to the optical gaps, where the QMC values are significantly larger than the DFT values, the QMC electron affinities agree well with the DFT values.

First principles calculations pointed at the interesting possibility of coating surfaces with diamondoids in order to produce, in a simple and economic manner, electron emitters. These predictions were recently verified experimentally [14], by depositing self-assembled monolayers (SAMs) of tetramantane-6-thiol on to a gold surface. These SAM were found to display a uniquely sharp and strong electron emission peak. Interestingly, the magnitude of this peak is much higher than that found for hydrogen-terminated diamond surfaces, that are known to exhibit a negative electron affinity (NEA). The extremely intense peak was interpreted as a signature of NEA in diamondoid SAMs. This finding indicates that nanometer-sized diamondoids may be superior to bulk diamond in electronic applications.

Not only diamondoid SAMs but also single-molecules were recently deposited on metal surfaces [62, 63]. In particular Wang et al. deposited tetramantane diamondoids on Au(111) and reported a detailed scanning tunnelling microscopy and spectroscopy study. It was shown that the diamondoid electronic structure and electron–vibrational coupling exhibit unique and unexpected spatial correlations characterized by pronounced nodal structures across the molecular surfaces.

Ab initio calculations revealed that much of the observed electronic and vibronic properties of diamondoids are determined by surface hydrogen terminations, a feature having important implications for designing future diamondoid-based molecular devices.

A very promising way to incorporate nanodiamond films in technological applications is by doping them with, e.g. nitrogen [64]. The goal is to introduce carrier levels into the diamond gap (this level is located 1.7 eV below the conduction band minimum in bulk diamonds) to increase conductivity and to lower the electron emission voltage threshold. Nitrogen is a major impurity of natural diamond. As mentioned earlier, nanodiamonds produced by detonation contain a high percentage (1–2%) of nitrogen. This nitrogen is brought into the diamonds from the trinitrotoluene reactant and cannot be removed from the nanoparticles [65]. The precise location of the nitrogen atoms in the nanoparticles is uncertain, but several magnetic studies have been unable to see any trace of substitutional nitrogen (known as P1 center) [31]. This is surprising as nitrogen is present in N-doped CVD-produced microcrystals as a substitute to core carbon atoms [66]. A theoretical study of the grain boundaries of UNC diamonds in the presence of nitrogen impurities has been performed by Zapol et al. [67]. This tight-binding, DFT study shows that nitrogen substitution to carbon in the grain boundary is energetically more favorable than in the crystalline core. The conductivity increase of the N-doped film was attributed to an increase in threefold coordinated carbon atoms caused by the nitrogen impurity. Preliminary results from ab initio molecular dynamics simulations indicate that for small diamond clusters, nitrogen is preferentially present as a substitute to surface carbon atoms. The tensile stress at the surface found in these nanoparticles facilitates nitrogen inclusion in comparison to bulk diamonds, where intentional nitrogen doping has proven to be difficult. However, nitrogen incorporation is found to require more energy with increasing particle size.

Experimentally, nitrogen doping of UNC films has been studied by Bhattacharyya et al. [68] who introduced nitrogen gas (1–20%) into the mixture fed to the CVD reactor, and achieved the highest carrier concentration and electrical conductivity measured so far for a phase-pure diamond thin film. The increase of N_2 gas concentration causes increasingly larger grain boundaries to be deposited, and it is believed to be responsible for the increase of the conductivity. However much remains to be done in order to achieve controllable doping of nanocrystalline diamond films, as well as to understand the details of the conduction mechanisms.

2.6 Conclusions and Outlook

In this brief review I have summarized structural, stability and electronic properties of nanodiamonds, and I have emphasized the role of ab initio calculations in understanding and predicting the properties of these nanoparticles. In particular I have highlighted the understanding of stability and formation in terms of

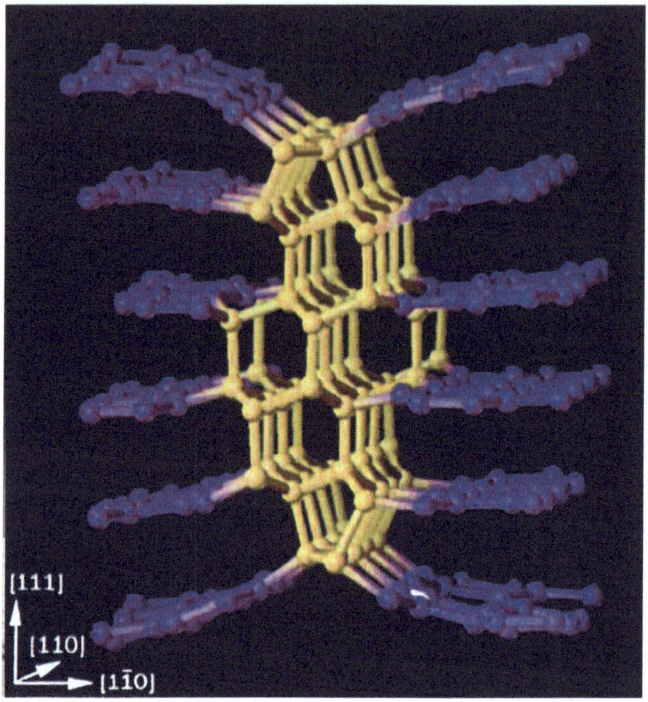

Fig. 2.13 Atomistic model of a diamond-graphite interface, as formed during graphitization of a diamond surface. Adapted from [45]

hydrogen content at the surface, and predictions about graphitic-like surface reconstructions and electronic properties. These predictions have recently been confirmed by experiments.

Of course Carbon will not stop to surprise us any time soon and new carbon based (nano)-materials are expected to be discovered and/or engineered in the future. Although the discovery of nanodiamonds is more than 20 years old and the utilization of nanodiamond films is about a decade old, certainly not all properties of these nanostructured materials have been unraveled and understood, and the general expectation is that a number of new technological applications of nanodiamonds and diamondoids lie ahead of us. Much effort is currently dedicated to another form of nanoscale Carbon: graphene and its derivatives. It will be interesting to see if nanodiamond and graphene or graphane may be combined (perhaps in a fashion similar to configurations found in diamond–graphite interfaces [45], shown in Fig. 2.13), and interfaces formed at the nanoscale. This might originate new nanoscale objects, with novel properties.

Acknowledgments This work was supported by grant No DOE/BES. DE-FG02-06ER46262

References

1. Lu PJ et al (2005) Archaeometry 47:1
2. Lewis RS, Ming T, Wacker JF, Anders E, Steel E (1987) Nature 326:160
3. Lewis RS, Anders E, Draine BT (1989) Nature 339:117
4. Amari S, Lewis RS, Anders E (1994) Geochim Cosmochim Acta 58:459
5. Bauschlicher CW, Liu YF, Ricca A, Mattioda AL, Allamandola LJ (2007) Astrophys J 671:458
6. Kennett DJ et al (2009) Science 323:94
7. Kennett DJ et al (2009) Proc Natl Acad Sci USA 106:12623
8. Schoell M, Carlson RMK (1999) Nature 399:15
9. Landa S, Machacek V (1933) Collect Czechoslovkian Chem Commun 5:1
10. Dahl JE, Liu SG, Carlson RMK (2003) Science 299:96
11. Drummond ND, Nat Nano (2007) 2:462
12. Shedorova OA, Gruen DM (ed) (2006) Ultrananocrystalline diamond: synthesis, properties, and applications. William Andrew Publishing, Norwich, NY
13. Drummond ND, Williamson AJ, Needs RJ, Galli G (2005) Phys Rev Lett 95:096801
14. Yang WL et al (2007) Science 316:1460
15. Adiga VP et al (2009) Phys Rev B 79:245403
16. Shimkunas RA et al (2009) Biomaterials 30:5720
17. Daulton TL, Eisenhour DD, Bernatowicz TJ, Lewis RS, Buseck PR (1996) Geochim Cosmochim Acta 60:4853
18. Kouchi A, Nakano H, Kimura Y, Kaito C (2005) Astrophys J Lett 626:L129
19. Freund MM, Freund FT (2006) Astrophys J 639:210
20. Dai ZR et al (2002) Nature 418:157
21. Van Kerckhoven C, Tielens A, Waelkens C (2002) Astronom Astrophys 384:568
22. Buss RH et al (1990) Astrophys J 365:L23
23. Hill HGM, Jones AP, d'Hendecourt LB (1998) Astronom Astrophys 336:L41
24. Vanthiel M, Ree FH (1987) J Appl Phys 62:1761
25. Greiner NR, Phillips DS, Johnson JD, Volk F (1988) Nature 333:440
26. Aleksenskii AE, Baidakova MV, Vul AY, Siklitskii VI (1999) Phys Solid State 41:668
27. Aleksenskii AE et al (2000) Technic Phys Lett 26:819
28. Baidakova MV, Siklitsky VI, Vul AY (1999) Chaos Solitons Fractals 10:2153
29. Chen PW, Ding YS, Chen Q, Huang FL, Yun SR (2000) Diamond Related Mater 9:1722
30. Viecelli JA, Bastea S, Glosli JN, Ree FH (2001) J Chem Phys 115:2730
31. Iakoubovskii K et al (2000) Diamond Related Mater 9:861
32. Shames AI et al (2002) J Phys Chem Solid 63:1993
33. Beloborv P, Gordeev A, Petrakovskaya E, Falaleev O (2000) Doklady Phys 46:459
34. Fayette L et al (1998) Phys Rev B 57:14123
35. Garcia MM et al (1998) Appl Phys Lett 72:2105
36. Jiao S et al (2001) J Appl Phys 90:118
37. Heiman A et al (2001) J Appl Phys 89:2622
38. McKervey MA (1980) Tetrahedron 36:971
39. Dahl JE et al (1999) Nature 399:54
40. Lin R, Wilk ZA (1995) Fuel 74:1512
41. Drummond ND, Williamson AJ, Needs RJ, Galli G (2005) Phys Rev Lett 95:096801
42. Raty JY, Galli G, Bostedt C, van Buuren TW, Terminello LJ (2003) Phys Rev Lett 90:037401/1
43. Williamson AJ, Grossman JC, Hood RQ, Puzder A, Galli G (2002) Phys Rev Lett 89:196803
44. Halicioglu T (1997) Phys Status Solidi B Basic Res 199:345
45. De Vita A, Galli G, Canning A, Car R (1996) Nature 379:523
46. Badziag P, Verwoerd WS, Ellis WP, Greiner NR (1990) Nature 343:244
47. Hwang NM, Hahn JH, Yoon DY (1996) J Crystal Growth 162:55
48. Jang HM, Hwang NM (1998) J Mater Res 13:3536
49. Gamarnik MY (1996) Phys Rev B 54:2150
50. Raty JY, Galli G (2003) Nat Mater 2:792

51. Kohn W, Sham LJ (1965) Phys Rev A 140:1133
52. Car R, Parrinello M (1985) Phys Rev Lett 55:2471
53. Chang YK et al (1999) Phys Rev Lett 82:5377
54. Tang YH et al (2003) Chem Phys Lett 372:320
55. Aleksenskii AE et al (2001) Phys Solid State 43:145
56. Runge E, Gross EKU (1984) Phys Rev Lett 52:997
57. McIntosh GC, Yoon M, Berber S, Tománek D (2004) Phys Rev B 70:045401
58. Towler MD, Hood RQ, Needs RJ (2000) Phys Rev B 62:2330
59. Perdew JP, Burke K, Ernzerhof M (1996) Phys Rev Lett 77:3865
60. Willey TM et al (2005) Phys Rev Lett 95:113401
61. Willey TM et al (2006) Phys Rev B (Condensed Matter and Materials Physics) 74:205432
62. Wang Y et al (2008) Nat Mater 7:38
63. de Lozanne A (2008) Nat Mater 7:10
64. Achatz P et al (2006) Phys Rev B (Condensed Matter and Materials Physics) 74:155429
65. Maillard-Schaller E et al (1999) Diamond Related Mater 8:805
66. Nokhrin S, Rosa J, Vanecek M, Badalyan AG, Nesladek M (2001) Diamond Related Mater 10:480
67. Zapol P, Sternberg M, Curtiss LA, Frauenheim T, Gruen DM (2001) Phys Rev B 65:045403
68. Bhattacharyya S et al (2001) Appl Phys Lett 79:1441

Chapter 3
Hexagon Preserving Carbon Nanofoams

Gotthard Seifert[1], Agnieszka Kuc[2], and Thomas Heine[2]

Abstract Carbon foams are hypothetical carbon allotropes that contain graphene (sp^2 carbon) segments, connected by sp^3 carbon atoms, resulting in porous structures. These structures may represent novel stable carbon modifications with $sp^2 + sp^3$ hybridization. Carbon foams show high structural stability at very low mass density. The construction principles, the structures, the energetics as well as the electronic and mechanical properties of carbon foams is discussed. The study is restricted to foam structures with sixfold rings and hexagonal cross-sections (hexagon preserving foams). The relation to defected graphite and the so-called Glitter structures is briefly discussed. The similarity of electronic band structure and the electronic DOS (density of states) of foams to carbon nanotubes is explained.

3.1 Introduction

Carbon science has been revolutionized by the discovery and synthesis of fullerenes [26] and nanotubes [18]. This revealed that solid carbon, formally believed to exist only as graphite and diamond, was capable to form novel structures. Carbon atoms can form various types of chemical bonds due to hybridization of s and p orbitals. Four valence electrons in carbon atoms can form the sp^1, sp^2 and sp^3 bond configurations. Thus, carbon has the ability to form various allotropes, i.e. linear, planar and three-dimensional structures. Therefore, carbon nanostructures can be classified into seven groups: sp^1, sp^2, sp^3, $sp^1 + sp^2$, $sp^1 + sp^3$, $sp^2 + sp^3$, and $sp^1 + sp^2 + sp^3$. At ambient conditions of pressure and temperature, the most stable crystalline carbon allotrope is hexagonal graphite (sp^2). Diamond (sp^3), the second stable allotrope of carbon, is at the same conditions nearly as stable as graphite.

Graphite is well known as a moderator in nuclear reactors. Under irradiation the defined layered structure of graphite changes significantly and many defects are

[1]Technische Universitaet Dresden, Physikalische Chemie, Bergstr. 66b, 01062 Dresden, Germany
e-mail: gotthard.seifert@chemie.tu-dresden.de

[2]Jacobs University Bremen, School of Engineering and Science, Campus Ring 1, 28759 Bremen, Germany
e-mail: a.kuc@jacobs-university.de; t.heine@jacobs-university.de

L. Colombo and A. Fasolino (eds.), *Computer-Based Modeling of Novel Carbon Systems and Their Properties*, Carbon Materials: Chemistry and Physics 3,
DOI 10.1007/978-1-4020-9718-8_3,

created. Often occurring defects, vacancies or interstitials can merge into extended defects called dislocation loops or lines. Aggregation of interstitials cause the interlayer distance expansion due to the formation of new layers. Vacancy lines, on the other hand, produce the basal contraction [45, 46].

Recently, considerable advances have been made in the synthesis and theoretical predictions of various carbon nanostructures (for reviews see e.g. [1, 3, 12–15, 19, 20, 22, 38, 42]). These nanosystems are often referred to as new carbon allotropes, in addition to the classical allotropes: diamond, graphite and carbyne (sp^1). Among the carbon structures graphenes, [36] onions, [17, 25] diamondoids, [10, 33] peapods, [43] scrolls, [29], etc. have been synthesized or isolated, and many others were proposed on the basis of theory [16, 30–33, 49].

In addition to pure sp^2 or sp^3 crystals, several experimental and theoretical works focus on the properties of new possible carbon forms, namely those with co-existing sp^2 and sp^3 hybridization [2, 21, 23, 24, 48, 50]. For instance, the $sp^2 + sp^3$ group includes a number of nanoforms with various structures and properties, e.g. fullerene polymers, nanotube assemblies, diamond-like crystallites and their combinations. In addition, diamond–graphite hybrids, vacancies in graphite or carbon foams have been studied. These investigations include for example pressure–temperature phase transitions of graphite into a cubic diamond structure [39]. It was shown that some of the new carbon forms, such as a super-hard carbon phase of C_{60} fullerenes with co-existing sp^2 and sp^3 hybridization, appears to have hardnesses higher than that of the (100) and (111) diamond faces [4, 41, 47].

A very interesting family of carbon nanostructures are so-called Carbon Nanofoams.

Carbon nanofoams can be thought of as assembly of graphene layers of various width, interconnected by sp^2 and sp^3 carbon junction atoms. In carbon nanofoams some of the sp^2 carbon atoms are transformed into sp^3 carbon atoms, which allows diamond-like fragments to co-exist with graphene fragments. Dislocations in graphite [45, 46] lead to other type of foam-based carbon nanostructures, where the graphene planarity is lost, but two types of systems can be found: those with pure sp^2 hybridization or those with co-existing $sp^2 + sp^3$ hybridization of atoms.

Single-walled carbon foams might be formed by hierarchical self-assembly processes from layered graphite [44] or by a cold compression of carbon nanotubes [53]. The experiment of Wang et al. [53], where a sample of carbon nanotubes has been coldly compressed in a diamond anvil cell, shows the transformation into what is believed to be a novel carbon allotrope. It was discussed on the basis of theoretical studies [6] that the new form of carbon obtained by Wang et al. [53] can be described as a carbon foam structure. The results indicate that kinetically stabilized products, such as low-density carbon foam materials, may have possibly been formed in this experiment. It was also shown that the new crystalline material is a hard carbon phase with a high (calculated) bulk modulus [6].

In 1987, Vanvechten et al. [51] showed that dense packing of C_{11} clusters consisting of three fivefold rings in a condensed phase can result in a foam-like system. The structure of carbon foams was, however, proposed only later by Karfunkel et al. [21] and Balaban et al. [2] Recently, experimental [23, 24, 53] and theoretical [6, 37, 50] investigations have shown that carbon-foam-like materials can be formed by rather simple syntheses. For example, the mesophase pitch precursor is molten at high temperatures resulting in so-called graphitic foams. Although these systems are no single-wall carbon foams, as they are discussed in this work, it is possible to achieve similar structures experimentally.

Recently, Braga et al. [5] have shown, using molecular dynamics simulations, that after applying compression small diameter carbon nanotubes (CNTs) undergo polymerization. The resulting structures belong to the $sp^2 + sp^3$ group and are similar to carbon foams. Nanotube-derived foams are also discussed by Ding et al. [11]. The new carbon forms are designed theoretically based on the welding technique using cross-linkers. Such linkers create covalently bound nanotube arrays with high porosity and well-separated CNTs. Cross-linkage of nanotubes was also successfully performed in experiments [28].

Another interesting carbon allotrope of the sp^2+sp^3 group is the so-called glitter [7]. It can be thought of 1,4-cyclohexadienoid motifs connected into a 3D structure, forming a 3D network of channels, in contrast to carbon nanofoams, which can have either 1D or 2D pore networks. Glitter is a hypothetical carbon allotrope but can be viewed as a plausible model of n-diamond. This system is predicted to be as hard as diamond. The cross section of glitter reminds one of the possible carbon foam structures, therefore it is discussed in this work as well.

In this chapter, single-walled carbon foams [27] (see Fig. 3.1), defected graphite and glitter are discussed. Due to the pattern of open edges, two types of such

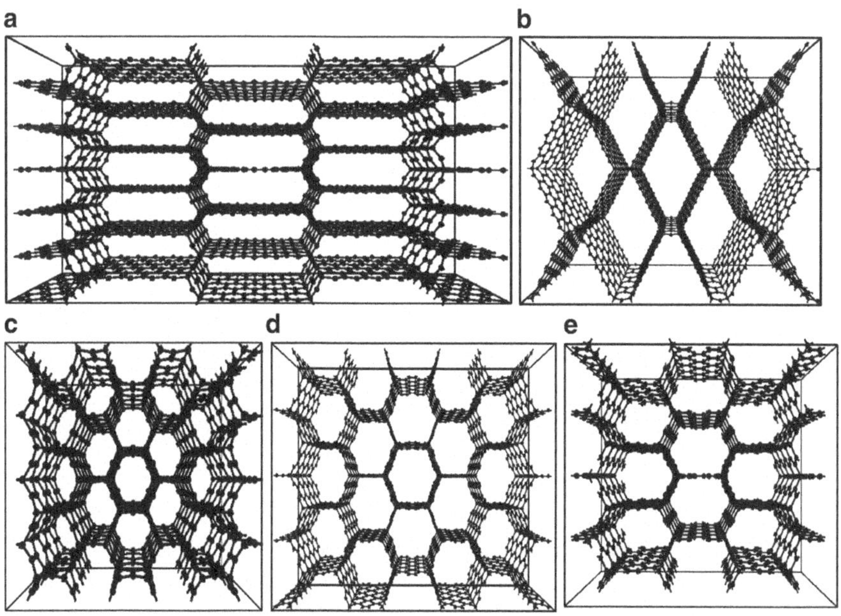

Fig. 3.1 Exemplary carbon foam structures: (5,1) (**a**), (1,5) (**b**), (1,1) (**c**) zig-zag carbon foams: (3,3) (**d**), (3,2) (**e**) armchair carbon foams (see text for the nomenclature)

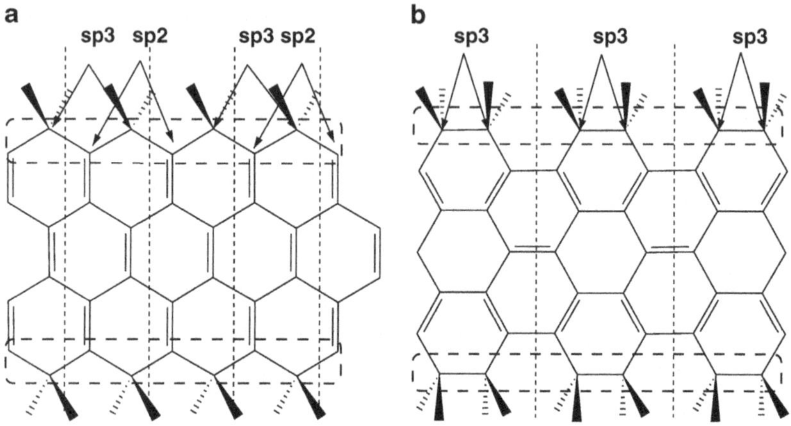

Fig. 3.2 The boundary atoms of zig-zag (**a**) and armchair (**b**) carbon foams. The unit cells are indicated by *vertical lines* [27]

structures can be built, so-called armchair and zig-zag foams, in analogy to the nomenclature of carbon nanotubes. The pore size is defined by a pair of integer numbers (N,M), which indicate the number of hexagonal units between the junctions. The junctions consist of the sp^3 boundary-atom chains (cf. Fig. 3.2). The two numbers are necessary to distinguish between the possible symmetric carbon foams of size $N = M$ and asymmetric ones with $N \neq M$. This nomenclature will be use in the following.

3.2 General Considerations and Computational Methods

The binding energies of the carbon foams can be considered as follows: The elementary cell consists of n_x carbon atoms at the boundaries (junctions) and n_i atoms inside the graphene fragments (stripes). Thus, the energy of a carbon foam can be written as:

$$E_{\text{bind}} = n_i \varepsilon_\infty + n_x \varepsilon_x, \tag{3.1}$$

where ε_∞ is the binding energy of the infinite graphene layer, whereas ε_x describes the binding energy of the boundary atoms in the unit cell. Since the total number of atoms $n = n_i + n_x$, the energy can be expressed as follows:

$$E_{\text{bind}} = n\varepsilon_\infty - n_x \varepsilon_\infty + n_x \varepsilon_x. \tag{3.2}$$

Defining $\Delta\varepsilon = \varepsilon_x - \varepsilon_\infty$, the energy per atom becomes:

$$\frac{E_{\text{bind}}}{n} = \varepsilon_\infty + \frac{n_x \Delta\varepsilon}{n}, \tag{3.3}$$

where the number of boundary atoms n_x per unit cell is hold constant at $n_x = 8$.

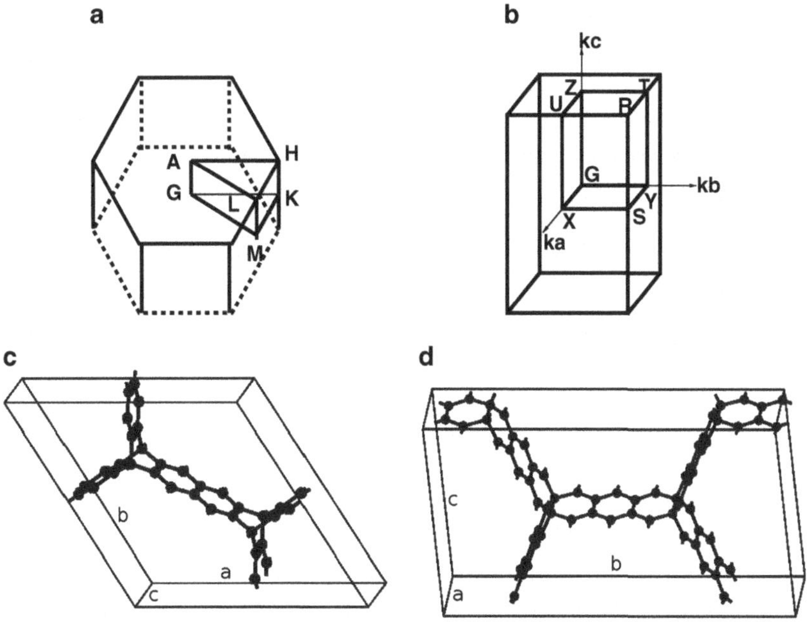

Fig. 3.3 The hexagonal (**a**) and orthorhombic (**b**) Brillouin zones for the respective elementary unit cells (**c** and **d**) of the (3,3) zig-zag carbon foam [27]

The detailed and quantitative studies of the stability, electronic and mechanical properties of different carbon foams were performed using the Density Functional based Tight-Binding [40] (DFTB) method. Periodic boundary conditions were used to calculate the infinite crystalline solid state. The conjugate-gradient scheme was chosen for the geometry optimization. The number of **k**-points was determined by reaching convergence for the total energy as a function of **k**-points according to the scheme proposed by Monkhorst and Pack [35]. Band structures were computed along lines between high symmetry points of the Brillouin zone. The first Brillouin zones with the highly symmetric points for hexagonal and orthorhombic unit cells (Fig. 3.3c and d) are shown in Fig. 3.3a and b.

The mechanical properties were studied in detail by estimation of the bulk and shear moduli. The elastic constants (stiffness) c_{ij} were calculated using the finite-difference scheme (the derivatives of the total energy with strain $\varepsilon_{i,j}$) as:

$$\frac{\partial}{\partial \varepsilon_i}\left(\frac{\partial E}{\partial \varepsilon_j}\right) = c_{ij}. \tag{3.4}$$

The matrix of constants was further used to obtain the bulk modulus:

$$\mathbf{B} = \frac{1}{9}\left[c_{11} + c_{22} + c_{33} + 2\left(c_{12} + c_{13} + c_{23}\right)\right] \tag{3.5}$$

for orthorhombic lattices or:

$$\mathbf{B} = \frac{\Delta \times c_{33} + 2c_{13}}{\Delta + 2} \tag{3.6}$$

for hexagonal unit cells, where

$$\Delta = \frac{c_{11} + c_{12} - 2c_{13}}{c_{33} - c_{13}}. \tag{3.7}$$

The shear modulus **G** can be calculated according to:

$$\mathbf{G} = \frac{1}{15}\left[(c_{11} + c_{22} + c_{33} - c_{12} - c_{13} - c_{23}) + 3\,(c_{44} + c_{55} + c_{66})\right]. \tag{3.8}$$

3.3 Structures

As described above, the Carbon Nanofoams are three-dimensional porous structures that contain both sp^2 and sp^3 hybridized atoms (see Fig. 3.1). They can be thought of as constructed from graphene planes interconnected rigidly with one another at 120°, forming a linear chain of sp^3 bonded atoms along the junctions (see Fig. 3.4). At these junctions, always three graphene layers meet (cf. Fig. 3.4c and f)

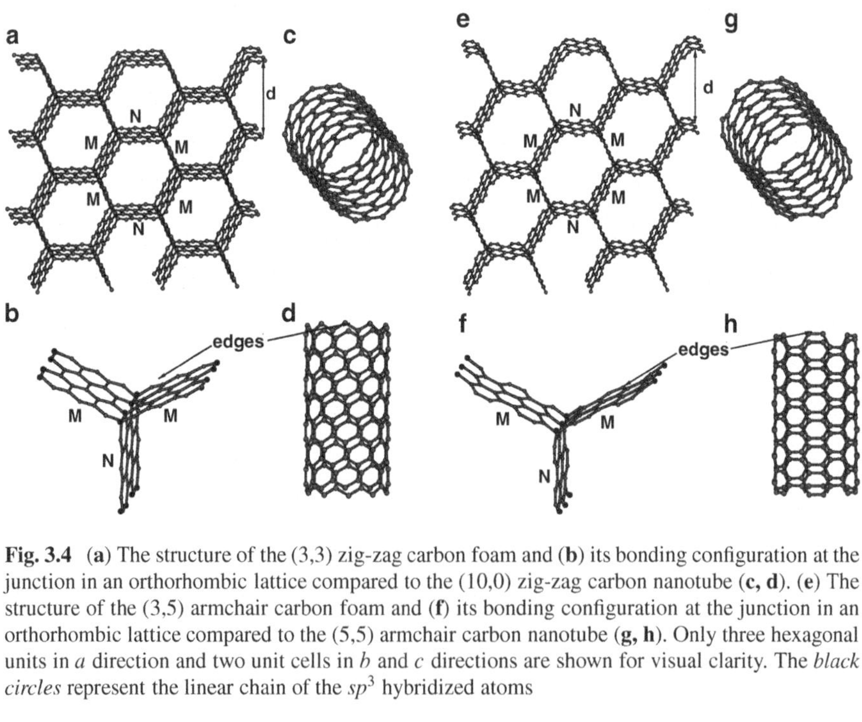

Fig. 3.4 (**a**) The structure of the (3,3) zig-zag carbon foam and (**b**) its bonding configuration at the junction in an orthorhombic lattice compared to the (10,0) zig-zag carbon nanotube (**c, d**). (**e**) The structure of the (3,5) armchair carbon foam and (**f**) its bonding configuration at the junction in an orthorhombic lattice compared to the (5,5) armchair carbon nanotube (**g, h**). Only three hexagonal units in *a* direction and two unit cells in *b* and *c* directions are shown for visual clarity. The *black circles* represent the linear chain of the sp^3 hybridized atoms

resulting in the honey-comb-shaped cross-section (CS) of the foam. If the graphitic segments are connected to each other at different angles than 120°, the systems with non-hexagonal CS are obtained. The present discussion was restricted to foams that preserve sixfold rings and hexagonal cross-sections [27].

Carbon foams discussed here can also be considered as AA-stacked graphite structures with a significantly increased interlayer distance. Thus, N describes the length of graphene fragments linked together by M-long graphene stripes. In other words, the N-stripes are functionalized by M-fragments. In this way, M determines the interlayer spacing as $d = \sqrt{3}M$. The ABAB-stacked forms are possible as well, as e.g. in the case of defected graphite (see Section 3.6). The sp^2 carbon atoms, which are transformed into sp^3 atoms, make rigid interconnections between the graphene layers.

In Fig. 3.4a and e perspective views of the (3,3) zig-zag and the (3,5) armchair carbon foams are given, in comparison to the (10,0) zig-zag (Fig. 3.4c and d) and (5,5) armchair (Fig. 3.4g and h) carbon nanotubes.

For a given pore size, the elementary unit cells of zig-zag and armchair carbon foams differ in the number of atoms (n). As an example, the (2,2) carbon foam consists of 44 and 20 carbon atoms for the zig-zag and the armchair foam, respectively. All structures can be represented in orthorhombic 3D carbon networks. Moreover, the zig-zag foams, with $N = M$, can be described within hexagonal lattices as well.

The interlayer distances d (Fig. 3.4) vary in the range of 4.7 Å [(1,1) zig-zag foam] and 32.3 Å [(9,9) armchair foam]. The orthorhombic unit cell length in a direction can be as short as 2.46 and 4.27 Å for armchair and zig-zag carbon foams, respectively. The pore size is determined by the unit cell parameters b and c (see Fig. 3.3d for definition). Keeping a at its minimum and reducing the width in the b and c directions to zero ($N, M = 0$), the structure reduces to a network of fourfold coordinated carbon atoms, namely that of cubic diamond (from armchair carbon foams) or hexagonal diamond (isodiamond; from zig-zag carbon foams) [2, 52]. On the other hand, increasing the system in a and b directions gives, in the $a, b \longrightarrow \infty$ limit, the structure of an isolated graphene layer.

The structures of zig-zag and armchair foams also differ in the types of connections (bonds): three types of covalent bonds, sp^2–sp^2, sp^2–sp^3, and sp^3–sp^3, can be found in the zig-zag arrangement (cf. Table 3.1), whereas armchair foams have two kinds of sp^2–sp^2 bonds (single and double; cf. Table 3.2) and sp^2–sp^3 bonds (there are no direct sp^3–sp^3 connections along the a axis). Comparing the geometries of carbon foams with the corresponding data for graphite and diamond (Table 3.1), one can find that the bond lengths are in between the values for both carbon allotropes. The sp^3–sp^3 bonds at the junctions are only slightly distorted from the ideal tetrahedral bonds in diamond. Moreover, the angles at the threefold and fourfold coordinated atoms are the same as those in graphite and diamond, respectively. The sp^2–sp^2 bond lengths in the zig-zag foams are very close to the bond lengths in graphite, whereas in armchair foams 'single' and 'double' bonds between sp^2 carbon atoms exist (cf. Table 3.2). The difference between 'single' and 'double' bonds decreases with increasing size, approaching the value of graphite.

Table 3.1 The geometry parameters of the symmetric zig-zag carbon foams compared to the calculated and the experimental data for graphite and diamond (values in parenthesis) [27]

Structure	Bond length (Å)		Bond angle (°)	
Graphite	$sp^2–sp^2$	1.420 (1.421)	$sp^2–sp^2–sp^2$	120.11 (120.00)
Diamond	$sp^3–sp^3$	1.541 (1.545)	$sp^3–sp^3–sp^3$	109.47 (109.47)
(1,1)	$sp^2–sp^2$	1.359	$sp^2–sp^2–sp^2$	120.00
	$sp^2–sp^3$	1.536	$sp^3–sp^2–sp^3$	109.11
	$sp^3–sp^3$	1.557		
(2,2)	$sp^2–sp^2$	1.418	$sp^2–sp^2–sp^2$	120.00
	$sp^2–sp^3$	1.534	$sp^3–sp^2–sp^3$	109.16
	$sp^3–sp^3$	1.585		
(3,3)	$sp^2–sp^2$	1.420	$sp^2–sp^2–sp^2$	120.00
	$sp^2–sp^3$	1.530	$sp^3–sp^2–sp^3$	109.17
	$sp^3–sp^3$	1.567		
(4,4)	$sp^2–sp^2$	1.424	$sp^2–sp^2–sp^2$	120.00
	$sp^2–sp^3$	1.531	$sp^3–sp^2–sp^3$	109.16
	$sp^3–sp^3$	1.576		
(5,5)	$sp^2–sp^2$	1.425	$sp^2–sp^2–sp^2$	120.00
	$sp^2–sp^3$	1.532	$sp^3–sp^2–sp^3$	109.17
	$sp^3–sp^3$	1.575		

Table 3.2 The geometry parameters of the symmetric armchair carbon foams [27]

Structure	Bond length (Å)		Bond angle (°)	
(2,2)	$sp^2–sp^2$	1.352, 1.448	$sp^2–sp^2–sp^2$	120.87
	$sp^2–sp^3$	1.523	$sp^3–sp^2–sp^3$	110.46
(3,3)	$sp^2–sp^2$	1.372, 1.442	$sp^2–sp^2–sp^2$	120.63
	$sp^2–sp^3$	1.518	$sp^3–sp^2–sp^3$	110.07
(4,4)	$sp^2–sp^2$	1.384, 1.436	$sp^2–sp^2–sp^2$	120.41
	$sp^2–sp^3$	1.515	$sp^3–sp^2–sp^3$	109.23
(5,5)	$sp^2–sp^2$	1.394,1.443	$sp^2–sp^2–sp^2$	120.27
	$sp^2–sp^3$	1.515	$sp^3–sp^2–sp^3$	108.81

Furthermore, the results indicate that the optimized unit cell sizes of carbon foams correspond to mass densities smaller than that of graphite ($\rho = 2.27$ g cm^{-3}) and diamond ($\rho = 3.55$ g cm^{-3}) (cf. Fig. 3.5). The only exception was found for the (1,1) zig-zag foam with a mass density of $\rho = 2.42$ g cm^{-3}.

The smallest armchair foams $(N,1)$ with an initial distance between the graphitic segments smaller than the van der Waals interlayer distance of graphite become very interesting systems after a full geometry optimization (Fig. 3.6). During the optimization, the sp^3 hybridized atoms of these structures open one of the four bonds and bind strongly to three neighbors only. This results in a porous system, built of sp^2 carbon atoms (in the following these forms are called 'sp^2 carbon foams'). To obtain such structures a larger unit cell in a direction is required (at least three times the original lattice) to allow the breaking of $sp^3–sp^3$ connections.

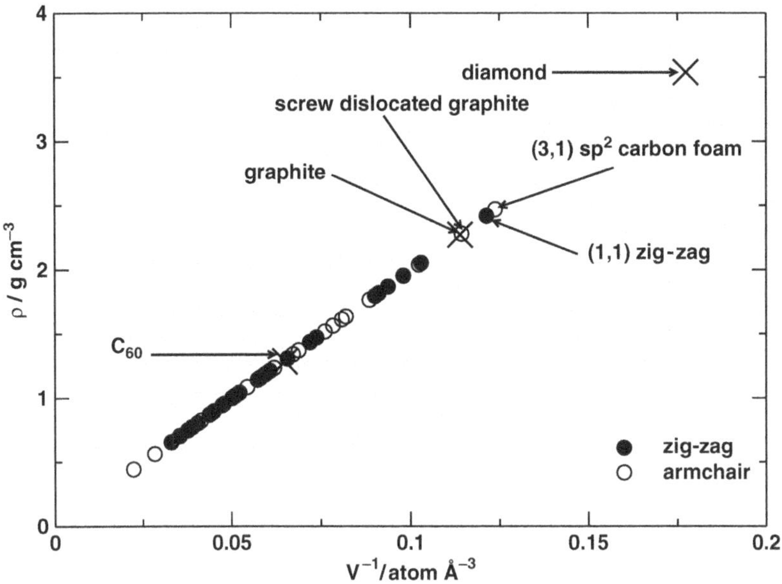

Fig. 3.5 The mass densities versus V^{-1} (V – the atomic volume) of carbon foams compared to C_{60}, graphite and diamond [27]

Otherwise the optimization leads to a typical diamond system. As an example the (3,1) sp^2 armchair foam is shown in Fig. 3.6a. Its distance $d = 3.24$ Å between the graphitic fragments as well as the bond lengths are similar to those of layered graphite, although the arrangement of the atoms is different. The mass density ($\rho = 2.47$ g cm^{-3}) of this foam is about the same as for the (1,1) zig-zag carbon foam. Analogous sp^2 foams were also discussed by Umemoto et al. [50] with similar conclusions.

Similar to the sp^2 foams, carbon foams can be constructed by screw twisting of graphite layers. They can be related to the structure of 'Screw Dislocated Graphite (SDG) [45]' that is shown in Fig. 3.6b. SDG has the same interlayer distance and bond lengths as layered graphite. However, there are covalent bonds present in c direction that connect the neighboring graphene fragments parallel to the ab planes. By construction, these bonds are formed by providing atoms within the graphene layers with additional neighbors in c direction, locally removing planarity without changing the hybridization of the carbon atoms. The system – shown in Fig. 3.6b – corresponds to the SDG (7,1) armchair carbon foam.

Both types of sp^2 foams have different kinds of nanopores, but both form two-dimensional interconnected channels between the pores: the sp^2 carbon foams have direct connections between pores along the b direction (Fig. 3.6a bottom), while SDG connections are rather twisted (Fig. 3.6b bottom). On the other hand, the sp^2+sp^3 carbon foams have closed nanopores (one-dimensional channels), similar to nanotubes.

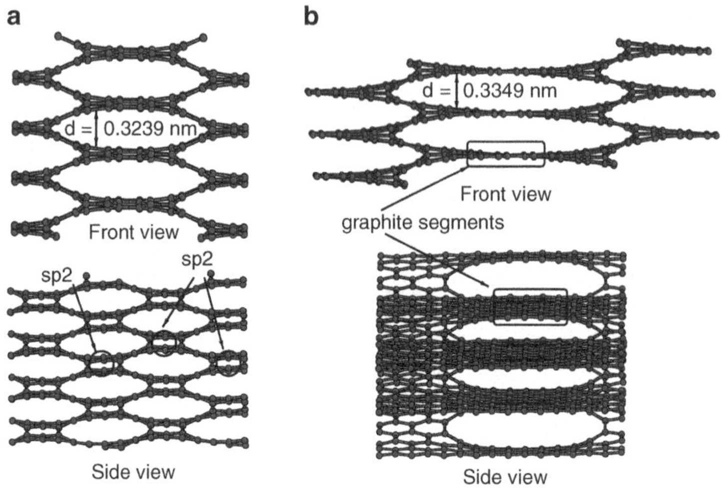

Fig. 3.6 The structure of sp^2 carbon foams: (**a**) the (3,1) armchair carbon foam and (**b**) SDG – screw dislocated graphite

3.4 Energetics and Mechanical Stability

The energetic and mechanical stability of different carbon foams was studied following the discussion in Section 3.2. The binding energy (per atom) as a function of n (n is the number of atoms per unit cell) is shown in Fig. 3.7. The energy of carbon foams asymptotically approaches the binding energy of a graphene layer. According to the proposed model consideration (Section 3.2) the cohesive energy should follow a linear trend with respect to n^{-1}. Figure 3.8 shows that the energy of the investigated carbon foams indeed increases nearly linearly with n^{-1}, as expected from Eq. 3.3. The deviations from the linearity (cf. Fig. 3.8) may be explained by the differences in the types of boundary atoms (n_x) for a given type of structures. In zig-zag systems there are eight sp^3 junction atoms per elementary unit cell, while four sp^2 and four sp^3 carbon atoms (per unit cell) occur at the junctions in armchair structures (cf. Fig. 3.2). The sp^2 carbon atoms in zig-zag foams form graphene stripes with a fully delocalized π-electron system. In armchair foams, the π-electron delocalization is distorted by the π bonds at the sp^2 atoms at the boundary of graphene-like stripes. This has obviously a stronger influence in smaller structures. For larger systems the bonding behavior of all boundary atoms becomes very similar, i.e. all structures have nearly the same n^{-1} size dependence. This is also confirmed by the observation, that the sp^2–sp^2 bond lengths in armchair systems depend on the size: The single and double bond lengths become similar with increasing the system size (Table 3.2). They slowly approach the values of bond lengths in graphite.

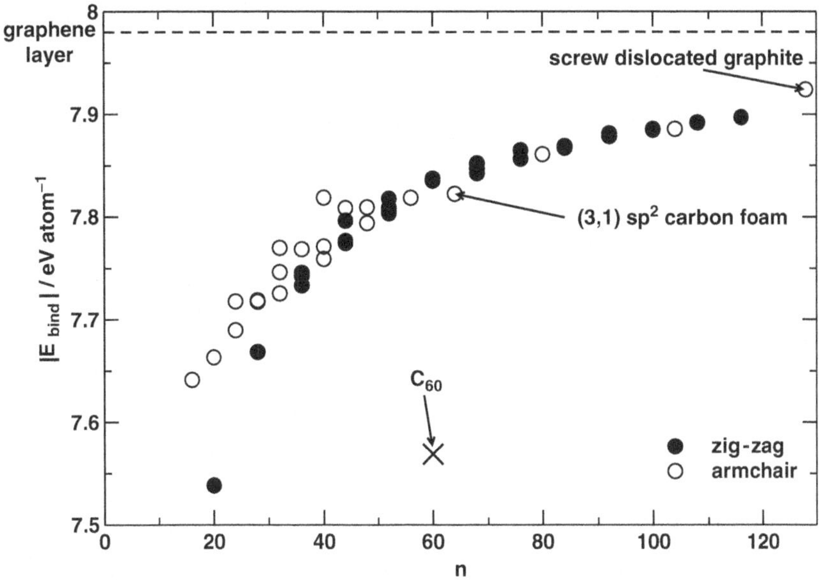

Fig. 3.7 The binding energy of the investigated carbon foams as a function of n (n – the number of atoms per unit cell). The corresponding energy of a single graphene layer is given as a reference (*dashed line*) [27]

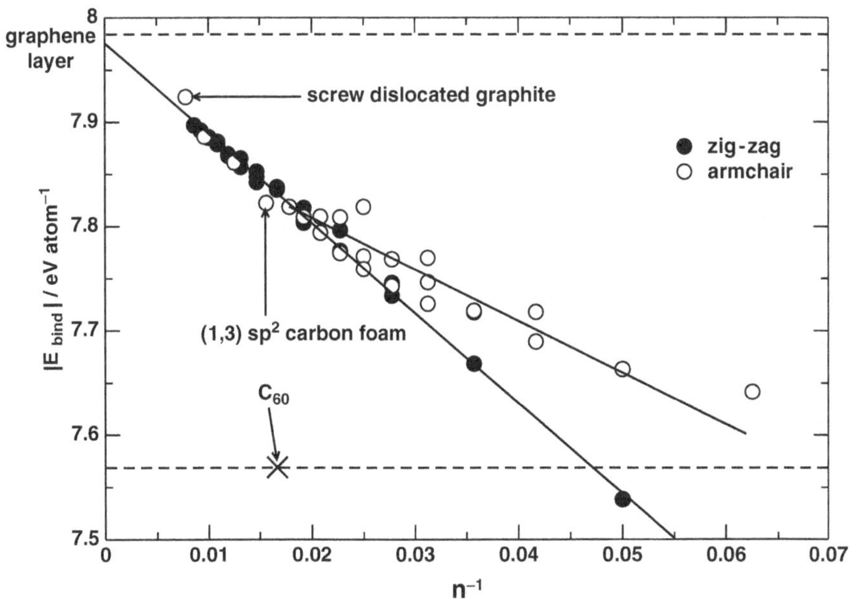

Fig. 3.8 The binding energy of carbon foams as a function of n^{-1}. The corresponding energy of a single graphene layer is given as a reference (*dashed line*) [27]

The calculations indicate that carbon foams are quite stable systems compared to the other well-known carbon allotropes. The origin of their favorable stability is the fact that the carbon foams discussed here do not contain bents, but only straight graphitic planes, in contrast to fullerenes and nanotubes that are purely sp^2 bonded but exhibit curved graphitic fragments. The largest studied carbon foams [(5,5) zig-zag ($n = 116$) and (7,7) armchair ($n = 104$)] are almost as stable as graphite and diamond. Their cohesive energies were found to be smaller by only $\sim$0.09 eV atom^{-1} than that of a graphene layer ($E_{\mathrm{bind}} = 7.986$ eV atom^{-1}). The least stable (1,1) zig-zag carbon foam ($n = 20$) with an energy of 7.538 eV atom^{-1} is as stable as the (5,5) armchair carbon nanotube ($E_{\mathrm{bind}} = 7.539$ eV atom^{-1}) and similarly stable to the (10,0) zig-zag carbon nanotube ($E_{\mathrm{bind}} = 7.501$ eV atom^{-1}). Furthermore, it was found that, except for the (1,1) zig-zag carbon foam, all structures are more stable than the isolated C_{60} cage ($E_{\mathrm{bind}} = 7.569$ eV atom^{-1}) by at least $\sim$0.08 eV atom^{-1}. The sp^2 carbon foams are very stable, as well. As an example, the (3,1) foam is less stable by $\sim$0.16 eV atom^{-1} than the most stable carbon allotropes. The stability of SDG (7.924 eV atom^{-1}) and graphite/diamond is about the same. These results suggest that carbon foams can be stable once they have been formed.

The mechanical properties were studied by estimation of the bulk (**B**) and shear (**G**) moduli according to Eqs. 3.5 and 3.8. The results (Tables 3.3 and 3.4) show that with increasing size of the pores (N and/or M) the bulk modulus decreases. The smallest **B** belongs to the (1,5) zig-zag foam (4.25 GPa). The shear moduli have a similar tendency and the smallest value was found for the (7,7) armchair system

Table 3.3 The bulk and the shear moduli (**B** and **G**) of zig-zag carbon foams given in GPa [27]

M	B					G				
N	1	2	3	4	5	1	2	3	4	5
1	285.1	89.0	21.5	8.25	4.25	176.95	53.1	15.3	4.3	3.6
2	265.5	157.7	54.85	18.3	6.95	105.8	32.3	13.5	6.6	3.2
3	225.2	172.9	97.0	69.5	21.0	122.7	17.9	8.7	8.1	4.6
4	148.6	137.3	124.8	73.9	69.1	142.8	11.6	6.3	4.6	4.6
5	183.5	107.9	109.4	98.9	75.5	76.5	7.95	5.0	3.8	2.9

Table 3.4 The bulk and the shear moduli (**B** and **G**) of armchair carbon foams given in GPa [27]

M	B				G			
N	2	3	4	5	2	3	4	5
1	266.7	78.2	28.8	12.6	64.35	24.65	12.45	7.9
2	213.3	156.3	57.4	23.4	26.2	16.8	9.2	6.4
3	140.6	162.1	95.7	47.2	12.2	8.7	4.7	4.3
4	89.6	114.8			9.0	4.3		
5	65.0	88.2	100.6	92.3	3.6	3.7	2.4	0.9

(0.1 GPa). The most stiff carbon foam is the (1,1) zig-zag with **B** = 285.13 GPa and **G** = 176.95 GPa, because this structure is closest to the diamond structure. On the other hand, it is interesting to note that for a given pore size the armchair carbon foams seem to be mechanically more stable than the zig-zag structures. Another tendency is that the bulk as well as the shear moduli become smaller going from systems with $M = 1$ to $M = 5$.

As expected for structures built from graphite and diamond segments containing both sp^2 and sp^3 hybridized atoms, the calculated bulk moduli of the carbon foams vary over a wide range: ~5 up to ~300 GPa, i.e. ranging between that of graphite (5.5 GPa) and nearly approaching that of diamond (442 GPa [34]). Moreover, the (3,1) armchair and SDG carbon foams also possess rather large bulk moduli of 48.5 GPa and 20.2 GPa, respectively, compared to graphite.

The **G** values of carbon foams are, however, clearly smaller than those of diamond (621 GPa) and for larger structures they become close to that of graphite (3 GPa). Evidently carbon foams are mechanically rather stable concerning the bulk moduli. However, it is important to notice that larger foams (Tables 3.3 and 3.4) could become unstable against shear forces, because of their small shear moduli. This behavior comes from the fact that carbon foams are highly anisotropic systems.

Furthermore, the properties of the investigated carbon foams with $M = 2$ and increasing N were studied to search for size dependent trends. The results are shown in Table 3.5. Indeed, these carbon foams slowly approach the properties of layered graphite. The bulk modulus reaches the maximum value at the (3,2) structure and decreases continuously with increasing sizes. As the foam with increasing size starts to mimic the structure of layered graphite, its binding energy increases as well. However, the mass densities are much smaller, because the distance between graphitic segments is over two times larger than in layered graphite.

Table 3.5 Calculated mass densities (ρ), bulk moduli (**B**), band gaps (Δ) and binding energies per atom (E_{bind}) of zig-zag carbon foams with $M = 2$

(N,M)	ρ (g cm^{-3})	**B** (GPa)	Δ (eV)	E_{bind} (eV atom^{-1})
(1,2)	1.77	89.0	1.48	7.745
(2,2)	1.46	157.7	1.56	7.796
(3,2)	1.30	172.9	0.0	7.817
(4,2)	1.20	137.3	1.11	7.836
(5,2)	1.14	107.9	0.78	7.852
(6,2)	1.09	84.1	0.0	7.863
(7,2)	1.06	74.2	0.63	7.873
(8,2)	1.03	66.6	0.50	7.881
(9,2)	1.01	58.8	0.0	7.887
(10,2)	0.99	51.4	0.44	7.893
Graphite	2.27	5.5	0.0	7.986

3.5 Electronic Properties

In this section the densities of states (DOS) and band structures of the investigated carbon foams are discussed. It is interesting to point out, that the calculated band gaps of zig-zag foams indicate a similar size dependence as for zig-zag carbon nanotubes. They are metallic, if the distance between two junctions is a multiple of three hexagonal units:

$$(N, M) = [3 \times m, M] \tag{3.9}$$

and/or

$$(N, M) = [N, 3 \times m] \tag{3.10}$$

with $m = 1, 2, 3, \ldots$; otherwise the foams are semiconducting with a gap size in similar range as for semiconducting carbon nanotubes (0.6–1.0 eV). Similar to armchair carbon nanotubes, the armchair carbon foams are all metallic independent of their size.

Figure 3.9 shows the band structures and densities of states for the symmetric zig-zag carbon foams calculated in hexagonal lattices. These results are compared with the band structure and DOS of AA-stacked hexagonal graphite. Although the dispersion along the lowest conduction band as well as the highest valence band is very small, the systems were recognized as indirect-gap semiconductors. A distinct dispersion appears along the K–H lines (Fig. 3.9b and c). The electronic structure reveals that mostly the atoms in the direct vicinity of the sp^3 carbon chains contribute to the bands near the Fermi level. The (3,3) zig-zag carbon foam is metallic with bands crossing the Fermi level at the K point of the Brillouin zone.

Some examples of band structures and DOS of the orthorhombic ($N \neq M$) zig-zag carbon foams are shown in Fig. 3.10a and b. These foams are metallic as it was stated in Eqs. 3.9 and 3.10. There is a visible large band dispersion along the $k_a - k_b$ plane (see Fig. 3.3 for definition of k_a and k_b), similar to that of a graphene monolayer.

Band structures and DOS of some armchair carbon foams are shown in Fig. 3.10c and d. This group of metallic structures has bands crossing the Fermi level along the X–S and Y–Γ lines. Large dispersions of valence and conduction bands are visible as well.

PDOS of the junction atoms (sp^3) is highlighted in Figs. 3.9 and 3.10 for the metallic systems. It can be seen that the line of junction atoms has an insulating character and the metallic properties of the carbon foams are restricted to the graphene-like stripes with sp^2 hybridized carbon atoms. Thus, for larger semiconducting carbon foams the band gap will decrease with increasing size (cf. Table 3.5).

The family of sp^2 armchair foams is also metallic. The densities of states of the (3,1) structure and the SDG are shown in Fig. 3.11. The band structures are very complicated and therefore not shown here. The DOS of both systems are similar to that of layered graphite.

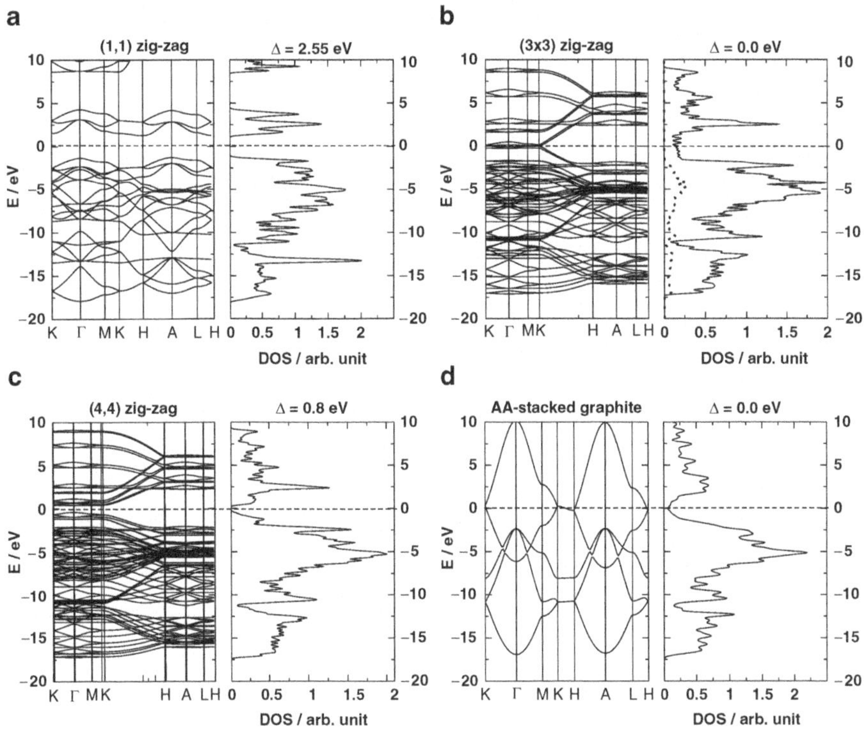

Fig. 3.9 The band structures and the densities of states of the symmetric zig-zag carbon foams (**a–c**) and graphite (**d**) in a hexagonal lattice representation. The *dotted line* in (**b**) denotes the PDOS of the sp^3 carbon atoms along the junction. Δ denotes band gaps. The Fermi level is shifted to 0.0 eV (*horizontal dashed lines*) [27]

3.6 Defected Graphite

In this chapter, two of the recently reported [45] dislocations are investigated: the zig-zag and the armchair prismatic edge dislocations. Under strong irradiation the original layered structure of graphite ($d = 3.35$ Å) is lost. Two types of highly defected graphite structures are shown in Fig. 3.12. Both are formed from ABAB-stacked graphite.

In the zig-zag prismatic edge dislocated system the neighboring layers are interconnected locally by sp^3 carbon atoms (Fig. 3.12a). This structure is similar to the carbon foams discussed above. Unlike in the carbon foams, the sp^3 carbons connect only two graphitic stripes with sp^2–sp^3 C bonds of 1.491 Å. The occurrence of sp^3 hybridized atoms causes a reorientation of the single layers leading to the formation of connected double layers (containing loops or cavities). The cavities have a hight of maximum 6.7 Å. Each double layer is repeated with about 10 Å distance, thus the minimum distance between two graphitic fragments is as large as in layered graphite (3.3 Å).

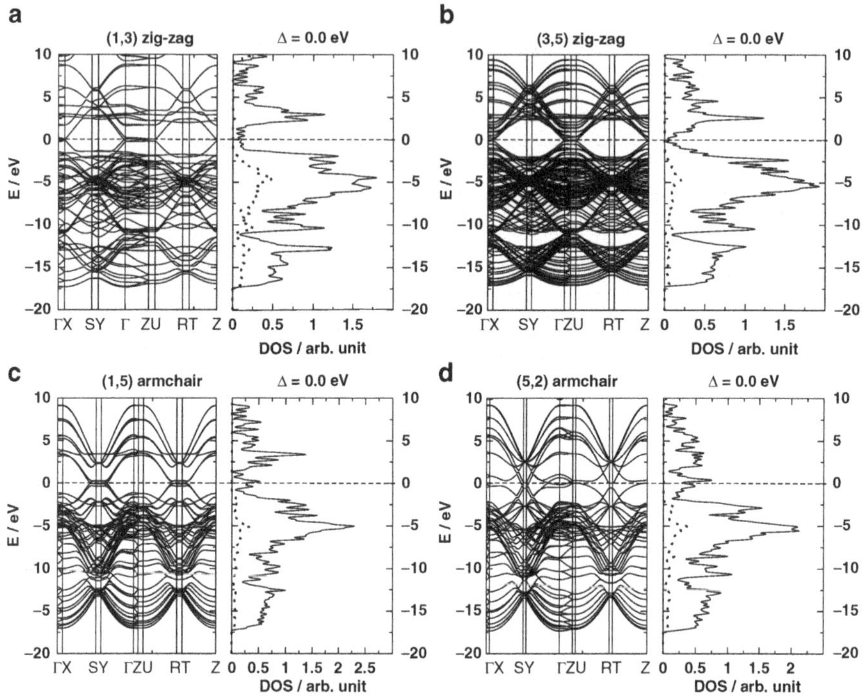

Fig. 3.10 The band structures and the densities of states of zig-zag (**a, b**) and armchair (**c, d**) carbon foams in an orthorhombic lattice representation. The *dotted lines* denote the PDOS of the sp^3 carbon atoms along the junction. Δ denotes band gaps. The Fermi level is shifted to 0.0 eV (*horizontal dashed lines*) [27]

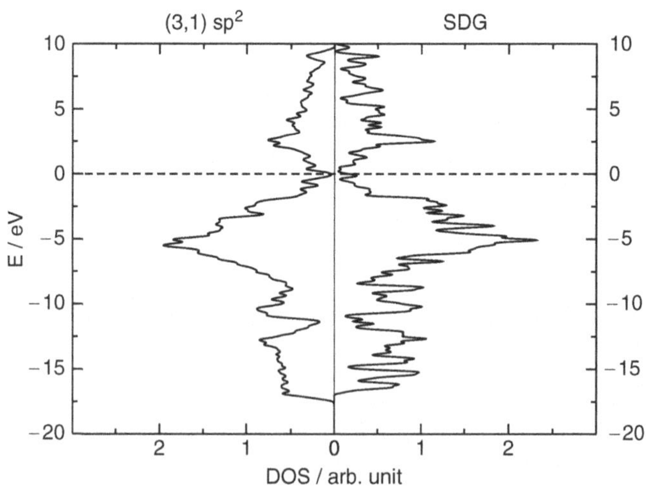

Fig. 3.11 The densities of states of the (3,1) armchair carbon foam (*left*) and the screw dislocated graphite (*right*). The Fermi level is shifted to 0.0 eV (*horizontal dashed line*)

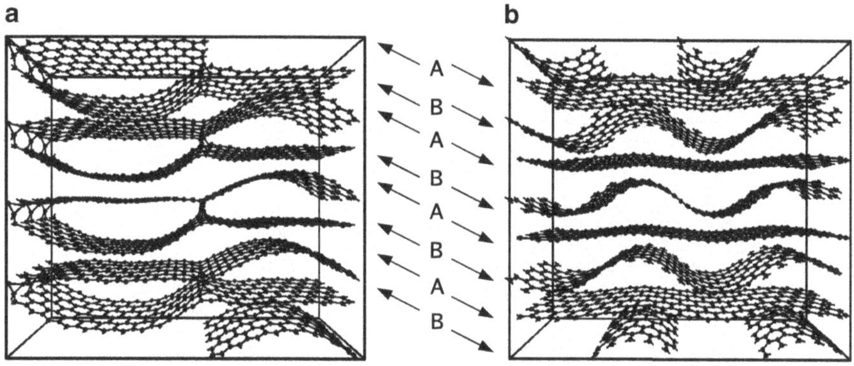

Fig. 3.12 (**a**) The zig-zag and (**b**) the armchair prismatic edge dislocations in layered graphite. A and B indicate the stacking of layers

On the other hand, the armchair prismatic edge dislocations are characterized by formation of wrinkled not bonded graphene layers (Fig. 3.12b). In this case, well-defined cavities and a d-expansion appear due to the bends of the graphene layers. No direct connections between the neighboring layers exist and thus, the system is built only from sp^2 hybridized carbon atoms. This structure can be considered as an intercalated graphite. Here the intercalant is simply a curled graphene layer. The bond lengths are typical for graphite. The maximal separation between the flat and the curled layers is about 6.8 Å, whereas the minimum is around 3.2 Å. The bent layers can be considered as spacers in the AA-stacked graphite with an interlayer distance of 10.0 Å.

As both structures have pores with diameters larger than the van der Waals distance in graphite, the mass densities decrease from $\rho = 2.27$ g cm^{-3} (graphite) to $\rho = 1.62$ g cm^{-3} for both dislocated graphite modifications. They are very stable having binding energies of 7.88 and 7.94 eV atom^{-1} for zig-zag and armchair dislocations, respectively.

3.7 Glitter

Glitter was first proposed in 1994 by Bucknum and Hoffmann [7] as a potential allotrope of carbon. The original intent was to combine the archetypal trigonal planar, 3-connected bonding of carbon in graphite with archetypal tetrahedral, 4-connected bonding of carbon in diamond. Such a material is known as a 3-, 4-connected carbon network and belongs to the $sp^2 + sp^3$ group. Glitter is a hypothetical structure constructed from a structural basis constituted by a 1,4-cyclohexadienoid motif. The extended structure of glitter together with its unit pattern is shown in Fig. 3.13. This system resembles the carbon nanofoam structure but with 3D interconnected channels. In some sense, glitter represents intermediate carbon form between graphite and diamond, similar to small size carbon foams.

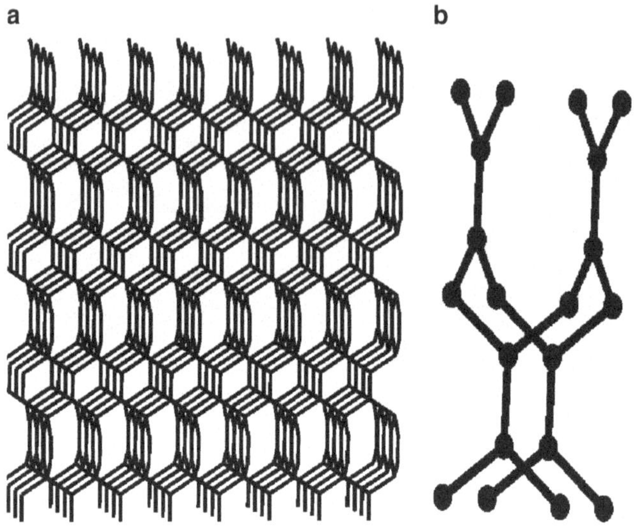

Fig. 3.13 An extended view (**a**) and the unit pattern (**b**) of glitter

Table 3.6 Calculated unit cell parameters (a, b, c), mass densities (ρ), C–C bond lengths, and bulk moduli (**B**) of glitter

Property	DFTB	DFT [7–9]
a = b	2.564 Å	2.560 Å
c	6.064 Å	5.925 Å
ρ	3.00 g cm^{-3}	3.12 g cm^{-3}
sp^2–sp^2	1.348 Å	1.350 Å
sp^2–sp^3	1.534 Å	1.510 Å
B	397 GPa	440 GPa

In Table 3.6 the results from DFTB calculations are presented. The results agree very well with the previously published DFT calculations of glitter [7–9]. Mass density of this tetragonal structure is 3.00 g cm^{-3}, intermediate between that of graphite (2.27 g cm^{-3}) and diamond (3.55 g cm^{-3}). The C–C bonds correspond to the single (1.534 Å) and double (1.348 Å) carbon bond lengths. The bulk modulus calculated at the DFTB level here is compared to the number obtained from Cohen's semiempirical formula [7]. This number suggests that glitter is almost as stable as diamond (experimental 442 GPa [34]). Glitter, with its binding energy of 7.52 eV atom^{-1}, is by 0.47 eV atom^{-1} less stable than graphite (7.99 eV atom^{-1}). At the DFT calculations [7–9] the authors show that this difference is also around 0.5 eV atom^{-1}.

Electronic structure calculations at the DFTB and DFT levels [7–9] suggest metallic character of the system. The band structure and DOS, with highlighted contributions of sp^3 and sp^2 carbon atoms, are shown in Fig. 3.14. It can be seen that the π* band dips down into the occupied bands of glitter at symmetry point M in the reciprocal space.

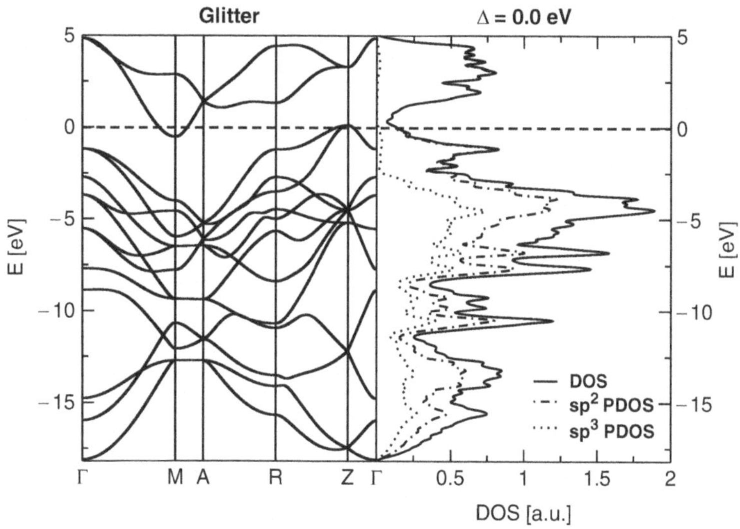

Fig. 3.14 The calculated band structure (*left*) and density of states (*right*) of glitter. The *black, green and red lines* denotes the total DOS, partial DOS of the sp^3 and sp^2 carbon atoms, respectively. Δ denotes band gaps. The Fermi level is shifted to 0.0 eV (*horizontal dashed line*)

3.8 Conclusions

In this work hypothetical carbon allotropes, called carbon nanofoams, have been discussed concerning the stability and the electronic properties. The construction shows that the chain of the sp^3-hybridized atoms along the junction are connected covalently with layers of graphite stripes having either zig-zag or armchair types of edges.

The results of DFT-based computations confirm high stability of carbon foams as compared to the most stable carbon allotropes (graphite and diamond). These systems may represent novel porous carbon modifications with sp^2–sp^3 hybridization and high structural stability at low mass density. The foams have large bulk moduli, although they might become less resistant against shear forces, when the size of the pores is increased. The stiffness of carbon foams can be improved by intercalation with e.g. carbon nanotubes. The electronic properties of the investigated foams are very similar to those of carbon nanotubes. Zig-zag carbon foams are metallic only, if one of the walls has a size, which is a multiple of three hexagonal units. Otherwise, the zig-zag foams are semiconducting. Armchair foams have metallic character independent of their pore size.

The investigations on carbon foams are still in progress but the results should encourage experimental investigations for the synthesis of such new carbon systems. High porosity, crystal-like structure and low mass density are very attractive features, in particular for guest–host interactions.

Acknowledgements Support of this research is acknowledged to Stiftung Energieforschung BW and Deutsche Forschungsgemeinschaft DFG. The authors thank also M. Heggie and I. Suarez-Martinez for data about screw dislocated graphite. Figures were made using GTK Display Interface for Structures 0.89.

References

1. Dresselhaus MS, Dresselhaus G, Eklund PC (1996) Science of Fullerenes and Carbon Nanotubes. Academic Press, London
2. Balaban AT, Klein DJ, Folden CA (1994) Chem Phys Lett 217:266–270
3. Benedek G, Milani P, Ralchenko VG (2001) Nanostructured carbon for advanced applications. Springer, Berlin
4. Blank VD, Buga SG, Serebryanaya NR, Denisov VN, Dubitsky GA, Ivlev AN, Mavrin BN, Popov MY (1995) Phys Lett A 205:208–216
5. Braga SF, Galvao DS (2007) Molecular dynamics simulation of single wall carbon nanotubes polymerization under compression. J Comput Chem 28(10):1724–1734
6. Bucknum MJ, Castro EA (2006) J Chem Theory Comput 2:775–781
7. Bucknum MJ, Hoffmann R (1994) A hypothetical dense 3,4-connected carbon net and related b2c and cn2 nets built from 1,4-cyclohexadienoid units. J Am Chem Soc 116(25): 11, 456–11,464
8. Bucknum MJ, Stamatin I, Castro EA (2005) A chemically intuitive proposal for the structure of n-diamond. Mol Phys 103(20):2707–2715
9. Bucknum MJ, Pickard CJ, Stamatin I, Castro EA (2006) On the structure of i-carbon. J Theoret Comput Chem 5(2):175–185
10. Dahl JE, Liu SG, Carlson RMK (2003) Isolation and structure of higher diamondoids, nanometer-sized diamond molecules. Science 299:96–99
11. Ding F, Lin Y, Krasnov PO, Yakobson BI (2007) Nanotube-derived carbon foam for hydrogen sorption. J Chem Phys 127(16):164703
12. Gogotsi Y (2006) Nanomaterials handbook. Taylor and Francis Group, Boca Raton
13. Gonzalez-Aguilar J, Moreno M, Fulcheri L (2007) Carbon nanostructures production by gas-phase plasma processes at atmospheric pressure. J Phys D-Appl Phys 40:2361–2374
14. Harris RJF (1999) Carbon nanotube and related structures: new materials for the twenty-first century. University Press, Cambridge
15. Hirsch A (1999) Fullerenes and related structures. Springer, Berlin
16. Hoffmann R, Hughbanks T, Kertesz M, Bird PH (1983) J Am Chem Soc 105:4831–4832
17. Iijima S (1980) Direct observation of the tetrahedral bonding in graphitized ccarbon-black by high-resolution electron-microscopy. J Cryst Grow 50:675–683
18. Iijima S (1991) Helical microtubules of graphitic ccarbon. Nature 354:56–58
19. Ivanovskii A (2008) Assemblies of carbon and boron-nitrogen nanotubes and fullerenes: structure and properties. Rus J Inorgan Chem 53(14):2083–2102
20. Jorio A, Dresselhaus G, Dresselhaus MS, Eklund P (2008) Carbon nanotubes. Advanced topics in synthesis, structure, properties and applications. Springer, Berlin
21. Karfunkel HR, Dressler T (1992) J Am Chem Soc 114:2285–2288
22. Katsnelson MI (2007) Mater Today 10:20–27
23. Klett J, Hardy R, Romine E, Walls C, Burchell T (2000) Carbon 38:953–973
24. Klett JW, McMillan AD, Gallego NC, Walls CA (2004) J Mater Sci 39:3659–3676
25. Kroto HW (1992) Carbon allotropes: cabron onions introduce new flavor to fullerene studies. Nature 359:670–671
26. Kroto HW, Heath JR, Obrien SC, Curl RF, Smalley RE (1985) Nature 318:162–163
27. Kuc A, Seifert G (2006) Phys Rev B 74:214104

28. Leonard AD, Hudson JL, Fan H, Booker R, Simpson LJ, O'Neill KJ, Parilla PA, Heben MJ, Pasquali M, Kittrell C, Tour JM (2009) Nanoengineered carbon scaffolds for hydrogen storage. J Am Chem Soc 131(2):723–728
29. Li JL, Peng QS, Bai GZ, Jiang W (2005) Carbon scrolls produced by high energy ball milling of graphite. Carbon 43:2830–2833
30. Liu AY, Cohen ML (1992) Phys Rev B 45:4579–4581
31. Liu AY, Cohen ML, Hass KC, Tamor MA (1991) Phys Rev B 43:6742–6745
32. Mailhiot C, Mcmahan AK (1991) Phys Rev B 44:11578
33. McIntosh GC, Yoon M, Berber S, Tomanek D (2004) Diamond fragments as building blocks of functional nanostructures. Phys Rev B 70:045401
34. McSkimin HJ, Andreatc P Jr (1972) Elastic-moduli of diamond as a function of pressure and temperature. J Appl Phys 43(7):2944
35. Monkhorst HJ, Pack JD (1976) Phys Rev B 13:5188–5192
36. Novoselov KS, Geim AK, Morozov SV, Jiang D, Katsnelson MI, Grigorieva IV, Dubonos SV, Firsov AA (2005) Two-dimensional gas of massless dirac fermions in graphene. Nature 438:197–200
37. Park N, Ihm J (2000) Phys Rev B 62:7614–7618
38. Pokropivny VV, Ivanovskii AL (2008) New nanoforms of carbon and boron nitride. Uspekhi Khimii 77(10):899–937
39. Ribeiro FJ, Tangney P, Louie SG, Cohen ML (2005) Phys Rev B 72:214109
40. Seifert G, Porezag D, Frauenheim T (1996) Int J Quantum Chem 58:185–192
41. Serebryanaya NR, Blank VD, Ivdenko VA, Chernozatonskii LA (2001) Solid State Comm 118:183–187
42. Shenderova OA, Zhirnov VV, Brenner DW (2002) Carbon nanostructures. Crit Rev Solid State Mater Sci 27:227–356
43. Smith BW, Monthioux M, Luzzi DE (1998) Encapsulated c-60 in carbon nanotubes. Nature 396:323–324
44. Ströbel R, Garche J, Moseley PT, Jorissen L, Wolf G (2006) Hydrogen storage by carbon materials. J Pow Sour 159:781–801
45. Suarez-Martinez I (2007) Theory of diffiusion and plasticity in layered carbon materials, PhD thesis, University of Sussex
46. Suarez-Martinez I, Savini G, Heggie MI (2006) First principles modelling of dislocations in AA graphite, Conference-Carbon 2006, Aberdeen
47. Talyzin AV, Dubrovinsky LS, Oden M, Le Bihan T, Jansson U (2002) Phys Rev B 66:165409
48. Telling RH, Ewels CP, El-Barbary AA, Heggie MI (2003) Nat Mater 2:333–337
49. Tomanek D (2002) Physica B 323:86–89
50. Umemoto K, Saito S, Berber S, Tomanek D (2001) Phys Rev B 64:193409
51. Vanvechten JA, Keszler DA (1987) Phys Rev B 36:4570–4573
52. Wang CX, Yang YH, Liu QX, Yang GW, Mao YL, Yan XH (2004) Appl Phys Lett 84: 1471–1473
53. Wang ZW, Zhao YS, Tait K, Liao XZ, Schiferl D, Zha CS, Downs RT, Qian J, Zhu YT, Shen TD (2004) Proc Nat Acad Sci USA 101:13,699–13,702

Chapter 4
Simulations of the Structural and Chemical Properties of Nanoporous Carbon

Johan M. Carlsson[1,2]

Abstract Nanoporous carbon (NPC) has porosity on the nanometer scale that leads to a large surface area and significantly enhanced chemical activity. These properties set NPC apart from the crystalline forms of carbon, such as graphite or diamond. However, the porous structure of NPC is very complex on the atomic scale and the actual geometry depends on the synthesis process. Large efforts have been spend on characterizing the structure of NPC synthesized under different conditions to better understand the relationship between the structure and the chemical properties of NPC. This chapter reviews these experimental and theoretical investigations and derives the common factors that can be summarized as the characteristics of NPC. A large section is devoted to the structural modeling that together with characterization experiments have provided the picture that NPC consists of crumpled sp^2-bonded graphene-like sheets. Further characterization has revealed that there is a considerable amount of non-hexagonal rings in the lattice. These non-hexagonal rings have a deciding role in the structure and properties of NPC as they induce curvature and combinations of non-hexagonal rings are centers of increased chemical activity. Three chemical reactions are finally presented in detail to exemplify the interaction between molecules and the active centers in the NPC.

4.1 Introduction

Carbon has a unique ability to form a plethora of structures with a rich variety of properties as is described in this volume. Graphite and Diamond are the thermodynamic stable forms of pure carbon. More recent members of the family such as fullerenes and nanotubes are long term metastable as the barriers for graphitization

[1]Fritz-Haber-Institut der Max-Planck-Gesellschaft, Faradayweg 4-6, D-14195 Berlin, Germany

[2]Accelrys Ltd, 334 Cambridge Science Park, Cambridge CB4 OWN, United Kingdom
e-mail: jcarlsson@accelrys.com

L. Colombo and A. Fasolino (eds.), *Computer-Based Modeling of Novel Carbon Systems and Their Properties*, Carbon Materials: Chemistry and Physics 3,
DOI 10.1007/978-1-4020-9718-8_4,

are large, since the structural transformations require breaking of the strong bonds. All of these crystalline forms of pure carbon are very stiff along the bonding directions and relatively inert towards chemical reactions. Less ordered forms, such as porous or amorphous carbon on the contrary are softer and chemically more reactive. Particularly the porosity and large surface area sets porous carbons apart from defect-free forms of sp^2-bonded carbon, manifesting that the structure and porosity play a major role for the chemical reactivity of carbon materials.

Microporous and nanoporous carbon have a large importance in a number of applications. Nanoporous carbon (NPC) for instance has a long tradition of being used as molecular sieves for gas purification. A typical example is the separation of oxygen and nitrogen from air [49]. It appears that NPC has an optimal pore size that allows the absorption of a larger amount of O_2 molecules than N_2 molecules. NPC can also be seen as an atomic scale warehouse, where a variety of species can be stored. NPC is ideal as bulk material for charge storage to develop electrochemical double layer capacitors (EDLC) [17, 43, 73]. NPC has similarly been suggested as a candidate for a high capacity hydrogen storage material [112]. The porosity in addition makes NPC suitable for gas microsensors based on the surface acoustic wave (SAW) method. A thin NPC film can selectively absorb large amounts of a gas, which gives a clearly detectable change in the vibration frequency [96]. The huge surface area furthermore makes NPC useful as catalyst support [47] and NPC has even been used as catalysts in its own right [75].

It is intriguing that a material made from a single element can have such a variety of applications. Surprisingly, considering its importance, the atomistic structure of NPC is not well characterized yet. This is due to the structural complexity of these materials. The structure of NPC can roughly be described as intermediate between amorphous carbon and perfectly crystalline graphite. An additional complication is that there is not a unique structure for NPC. The exact geometry of an individual piece of NPC is dependent on the history of the material, so one should preferably consider NPC as a class of carbon materials rather than a well defined, crystalline form of carbon.

Building structural models for NPC is truly challenging and this chapter intends to the review the present understanding of NPC and give answers to the questions: What is the structure of NPC? How stable is actually NPC? What is the origin of the chemical activity? The chapter contains four theoretical sections, where the first section contains a large review of the structural models that have been suggested to describe the microscopic geometry of NPC. The following theoretical sections treat the mechanical, vibrational and electronic properties of NPC. This leads finally to the chemical properties of NPC, where three reactions are considered: graphitization, gas adsorption in the pores related to gas separation and oxidation of NPC. Each section covers a description of the theoretical methods that have been used and summarizes the important results. However, the chapter starts with a review of the experimental literature to summarize the experimental knowledge available about NPC.

4.2 Experimental Investigations of the Structure of NPC

Nanoporous carbon can be synthesized in various ways and this section summarizes a few common methods, such as pyrolysis of carbonaceous materials and the arc discharge method. The microstructure of the resulting NPC material is highly dependent on the details of the process and shows a large variation, but there are also a few common features that can be summarized as characteristic of NPC.

4.2.1 Pyrolysis of Hydrocarbons

The most common method to synthesize NPC is carbonization of carbonaceous materials, such as hydrocarbons [34,35,49,84] or carbides [44,54,62,98,111]. Pyrolysis of polymers, sugar and organic waste for instance, is widely used to produce porous carbons that are used as molecular sieves.

During pyrolysis, the carbonaceous feedstock material is heated in an inert atmosphere in multiple steps. The material is first pre-heated to intermediate temperatures around a few hundred degrees Celsius and the temperature is then raised in a second heating step to a final level in the range 400–1,200°C. The heating leads to two processes, carbonization and graphitization, that are competing in the generation of porosity in the synthesized NPC material. The elevated temperature used in the pyrolysis process leads to evaporation of non-carbon atoms out of the carbonaceous material. The concentration of carbon atoms increases as the relative amount of non-carbon atoms decreases and the removal of the non-carbon atoms leave voids in the lattice. The carbonization process therefore leads to an increasingly porous NPC material. However, the carbon atoms are also more mobile during the pyrolysis process and loosely bound C-atoms may diffuse around until they are able to find a vacancy where they can attach. The voids are thereby successively healed out leading to an increasing degree of graphitization that decreases the porosity of the NPC material [49, 84].

Monitoring the structural evolution and chemical composition of NPC materials synthesized at different temperatures can give insight into the competition between the carbonization and graphitization processes. It was early noted that the carbon materials obtained by pyrolysis could be divided into graphitizing and non-graphitizing materials [24]. However, it should be noted that various authors use the word graphitization differently. The graphitizing materials in the early classification originated from pure hydrocarbons and had high density, no porosity and obtained a graphite structure when heated to temperatures above 1,700°C. The graphitization was here meant the formation of more or less perfectly ordered parallel, flat graphene sheets as in graphite [24]. The non-graphitizing carbon materials originated from carbon materials with oxygen groups and are characterized by lower density and larger porosity. The non-graphitizing carbon are kinetically trapped into a porous structure and restrains transforming into graphite even when heated to very high temperatures above 3,000°C [24]. The initial model of the graphitization process considered that the building blocs were small graphite crystallites and the distinction was how the crystallites were ordered. The graphitizing carbon obtained the crystallinity due to alignment of the crystallites, while the crystallites remained disordered in non-graphitizing carbon.

The word graphitization could in more recent years actually be replaced by "graphenization" as the focus has moved to how the individual graphene sheets are formed during the pyrolysis process. A detailed analysis of the pyrolysis of poly(furfuryl alcohol) (PFA) at different temperatures was able to follow the formation of the graphene sheets during the pyrolysis process [84]. The elemental analysis of the synthesized NPC confirmed that increasing the pyrolysis temperature led to an increased carbon content. The NPC materials synthesized at 400°C had a carbon content of around 80% (H/C ratio around 0.2) and the carbon content increased to 98% for materials synthesized at 1,200°C which is an almost complete carbonization [84]. The structural order in the synthesized NPC material was further monitored via the radial distribution function (RDF) obtained from Neutron scattering experiments and the structural evolution was observed by transmission electron microscopy (TEM) [49, 84]. The graphitization was here defined by the occurrence of more pronounced peaks in the RDF at the distances characteristic of the nearest neighboring distances in a graphene sheet. Comparing the RDF for NPC derived from PFA at different temperatures in Fig. 4.1 revealed that increasing the pyrolysis temperature led to an increasing number of peaks in the RDF [49, 84]. The RDF for the sample synthesized at 400°C only showed one significant nearest neighbor peak, while the material synthesized at 800°C showed peaks corresponding to the nearest neighboring distances in the hexagonal ring indicating that a hexagonal ring

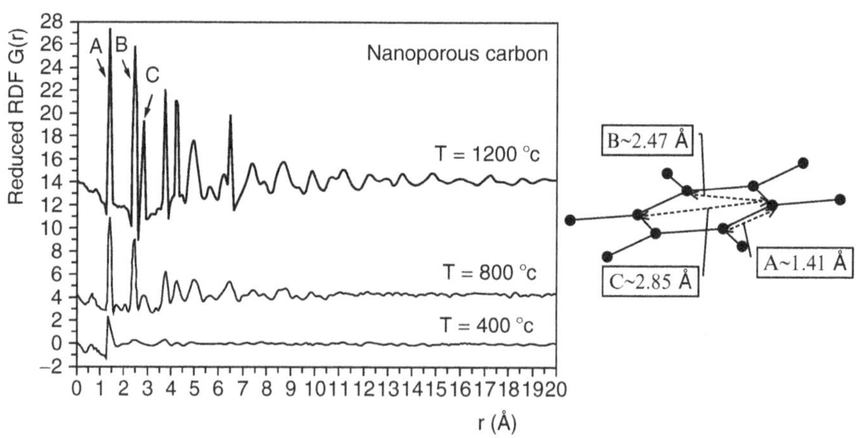

Fig. 4.1 The radial distribution function (RDF), G(r) for NPC produced by pyrolysis of PFA at three different temperatures, T = 400°C, T = 800°C and T = 1,200°C. The three nearest neighboring distances, A, B, and C are shown in the right illustration. Reprinted with permission from Ref. [84]. Copyright 1999 Taylor & Francis Ltd. The original article can be found at http://www.informaworld.com

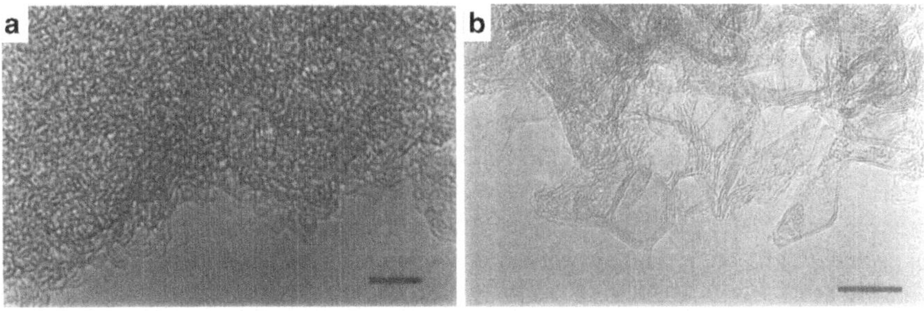

Fig. 4.2 Transmission electron micrograph of NPC synthesized by (**a**) pyrolysis of sucrose at 700°C and (**b**) further heated to 2,600°C. Reprinted with permission from Ref. [34]. Copyright 1997 Taylor & Francis Ltd. The original article can be found at http://www.informaworld.com

structure had been established. The RDF for NPC synthesized at 1,200°C finally showed peaks beyond the nearest neighbors in the hexagonal ring, which indicates a long range ordering characteristic of a graphene sheet.

The structure of the NPC generated by pyrolysis of PFA can be summarized as that the material generated at 400°C still carry a significant resemblance of the polymer structure of the feedstock material and does not show the characteristic features of the hexagonal rings in a graphene sheet. The carbonization increases in materials synthesized at 800°C leading to purer carbon materials, even though impurities are still present. The graphitization leads to the establishment of the hexagonal structure, but the sheets still contains a lot of defects such that they are crumpled. Increasing the pyrolysis temperature further leads to an almost complete carbonization and graphitization. NPC materials synthesized at 1,200°C can rather be characterized as disordered graphite, where defective graphene layers are randomly stacked [84]. This characterization is also supported by the fact that the oxygen uptake is completely blocked in NPC materials synthesized at 1,200°C indicating that the porosity is to large extent removed [49]. Heating of NPC, generated by pyrolysis of sucrose at 700°C, to 2,100–2,600°C showed that the initially disordered graphene sheets successively align, but they do not align perfectly like in graphite. The graphene sheets rather formed curved shapes similar to multilayered fullerenes as can be seen in Fig. 4.2 [34, 35]. This suggests that the defects are not completely healed out and lead to remaining non-hexagonal rings in the structure. It was then proposed that the non-hexagonal rings are the obstacle that prevent the alignment of the graphene sheets into the perfect graphite structure [34]. This also suggests that NPC belong to the non-graphitizing carbon materials as they would not form perfect graphite irrespectively of the annealing temperature.

4.2.2 The Template Method

The pyrolysis method has the disadvantage of producing quite disordered NPC even though the pore size distribution is quite narrow. The structural coherency can be improved by performing the pyrolysis in the presence of a template that has an ordered porosity. Similar templating methods have been widely used to obtain a variety of well-defined micro- and nanostructured materials [71] and this approach has lately also been applied to the synthesis of NPC to generate a more regular porosity than is spontaneously obtained [33, 47, 48, 61, 88].

The first attempts to synthesize templated NPC proceeded through a multi step process [61, 88]. The first step was to identify and synthesize an appropriate template material. Zeolites have an ordered structure with a narrow pore size distribution and MCM-48 [88] and SBA-15 [48] have successfully been used as templates for nanoporous carbon. The carbon precursor, for instance sucrose dissolved in water [88] or furfuryl alcohol [47], was then entered into the prefabricated template zeolite. Recently a simplified nanocasting method has been proposed, where the carbon precursor (sucrose solution) and the zeolite precursor were mixed in solution and dried for several months to obtain a composite gel [33]. The composite system of the zeolite filled with the carbon precursor is in either case heated in an inert atmosphere to temperatures in the range 1,073–1,373 K, which leads carbonization of the carbon precursor. The resulting carbon material obtains a porous structure that has the inverse form of the template zeolite as it fills the voids in the zeolite structure. A freestanding NPC is finally obtained by removing the template zeolite by etching with a strong acid such as hydrofluoric acid.

The resulting nanoporous carbon has an order porosity with a well-defined pore diameter of 3 nm related to the wall thickness in MCM-48 [88] or 4.5 nm corresponding to the wall thickness in SBA-15 [48]. A Brunauer–Emmett–Teller (BET) analysis [10] shows that the surface area is very large ranging from 1,380 m^2/g for NPC grown in MCM-48 [88] to 1,970 m^2/g for NPC grown by the nanocasting technique [33]. Unfortunately, the atomistic structure has not been resolved for these templated NPCs, but x-ray diffraction (XRD) measurements suggest that the walls are quite disordered. This is supported by Raman spectroscopy that shows both D-mode and G-modes. The G-mode is the hallmark of sp^2-bonded graphene sheets as is further discussed in Section 4.4.2, while the D-mode is connected to defect scattering. The presence of both the G- and D-peak suggests that the walls can be described as defective sp^2-bonded sheets [48].

4.2.3 Arc Discharge

The analysis of the NPC generated by pyrolysis at different temperatures indicates that the carbonization and graphitization processes are competing processes. It is therefore quite difficult to synthesize a high purity nanoporous carbon material by removing the non-carbon atoms through heating without obtaining a highly graphitized material at the same time. The pyrolysis method consequently has the drawback that there is either a smaller or larger amount of impurities remaining in the material or the material is loosing its porosity due to graphitization. A different approach to the synthesis of NPC, which minimizes the impurity content, is to use pure carbon as feedstock. This may be achieved via arc discharge between two graphite electrodes. This process has been highly successful in producing fullerenes [57] and nanotubes [41], but fullerene soot and amorphous carbon are also generated as bi-products.

In the arc discharge method, a voltage is applied between two graphite electrodes in an inert atmosphere of helium or argon. A discharge occurs when the voltage is around 18 V and the distance between the two electrodes is decreased below 1 mm. Carbon atoms are ejected from the smaller electrode during the discharge and accumulate on the larger electrode. The gas pressure has a deciding influence on the structure of the deposited material as amorphous carbon is generated at gas pressures below 100 torr, while high quality multiwalled carbon nanotubes are grown at gas pressures above 500 torr [20]. In the intermediate pressure regime a mixture of nanotubes, fullerenes and fullerene soot is grown. The soot fraction was initially dismissed as uninteresting, but it was later discovered that the fullerene fragments agglomerate into a nanoporous structure as can be seen in Fig. 4.3 [9]. An analysis of TEM pictures of the nanoporous carbon material synthesized at a gas pressure of 200 torr, suggested that it consisted of a mixture of large fullerenes with diameters in the range 3–10 nm and denser amorphous carbon. The amorphous carbon contained a random distribution of non-hexagonal rings giving both positively and negatively curved regions. A comparison of fullerene soot found in different regions of the chamber revealed that the material deposited outside the funnel could be characterized as amorphous carbon, while the material in the plasma region had polygonal shapes like fullerenes [8]. This indicates that the temperature is an important factor for the final structure. This hypothesis was supported by subsequent annealing of fullerene soot generated by arc discharge [36]. Heating the amorphous carbon to 1,700°C generated long, interlocking microfibrils enclosing large voids similar to

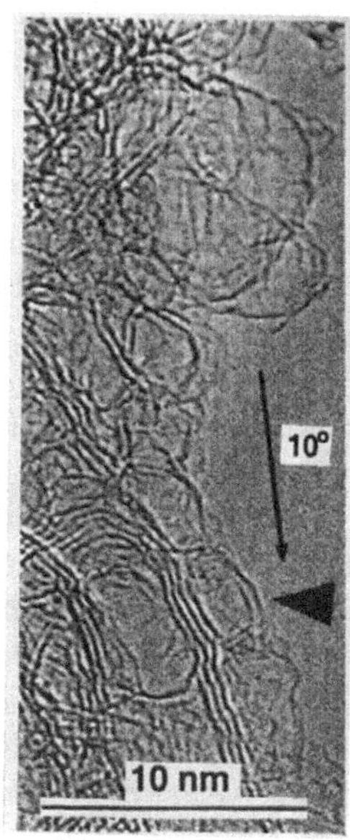

Fig. 4.3 Transmission electron micrograph of NPC synthesized by arc discharge. Reprinted with permission from Ref. [9]. Copyright 1997 Taylor & Francis Ltd. The original article can be found at http://www.informaworld.com

glassy carbon. Increasing the annealing temperature to 2,100°C converted some of the fibrils into spherical fullerenes with two to three layers similar to nano-onions. Increasing the temperature further increased the nano-onion content such that materials annealed to 2,400°C contained almost exclusively nano-onions. Summarizing these observations suggested that the arc discharge process could be described by a fast nucleation of small amorphous carbon fragments that occurs at low temperatures. An additional slower agglomeration of the fragments into larger sheets occurs in the hot plasma region and this process eventually leads to the closure of the sheets into large fullerenes [9]. It is interesting to note that the NPC generated by pyrolysis of sucrose in Fig. 4.2 and the NPC generated by arc discharge in Fig. 4.3 show significant similarities. In particular after heating to temperatures above 2,000°C.

4.2.4 Supersonic Cluster Beam Deposition Method

A related method that is able to produce nanoporous carbon films with a good structure control is the supersonic cluster beam deposition method (SCBD). This method utilizes a well-focused molecular beam of carbon molecules that is aimed at a target substrate [7, 77]. The carbon molecules are generated by an arc discharge between two graphite electrodes like in the arc discharge method mentioned above, but the carbon molecules are not left on the electrodes, but instead carried away from the plasma region by a stream of inert carrier gas. The gas stream passes through a nozzle and is then expanded in a chamber, such that only the carbon molecules with a selected molecular weight are continuing straight and hit the substrate. The acceleration energy can furthermore be varied to choose if the molecules collapse or are kept intact at the impact at the target substrate. The molecules are in either case successively agglomerating into a nanostructured carbon film as the number of molecules increase on the substrate. The structure of the deposited carbon film is consequently dependent on the molecular size and impact energy. Subsequent annealing can be used to fine-tune the porosity of the film further.

The nanoporous carbon films grown by the SCBD method were characterized by scanning electron microscopy (SEM), time of flight (TOF) and Raman spectroscopy [77]. The TOF spectroscopy showed that the molecular beam contained carbon clusters with masses up to 2,500 atoms per cluster, but the size distribution had its maximum around 500 atoms. A typical impact energy for a medium sized molecule was around 300 eV and the beam intensity was 10^{15} molecules/s giving an average growth rate for the carbon films of around 25 Å/min. The SEM pictures showed a sponge-like morphology where the carbon molecules were randomly stacked into a nanoporous structure. The Raman spectrum showed three distinct peaks located at 1,180, 1,360 and 1,585 cm^{-1} together with a broader feature around 2,700 cm^{-1}. These peaks can be assigned to a distorted sp^2-bond, the D-mode, the G-mode and the 2D-mode the respectively. The large D-peak indicates that a large amount of defects is initially present in the films grown by the SCBD. However, the ratio between the D-mode and G-mode, I_D/I_G, decreases after annealing, which suggests that the film is graphitizing upon heating. Summarizing the analysis of the nanoporous carbon films grown by the SCBD method indicates that the films may be characterized as an agglomerate of fullerene fragments, i.e. assemblies of defective, curved sp^2-bonded graphene sheets that contain non-hexagonal rings in the structure. These non-hexagonal rings have an important influence on the electronic properties of the NPC films as optical spectroscopy measurements show a bandgap of 0.6 eV, that is distinctly different compared to the semi-metallic band structure of graphene.

4.3 Structural Modeling of NPC

The structural diversity and complexity of nanoporous carbon that has been presented in the previous section provides a real challenge for modeling and simulations. Semi-disordered crystalline systems such as nanoporous carbon have a short range crystallinity, but loose their long range periodicity, such that NPC can not be described by a periodic supercell model in an easy way. The modeling efforts have consequently been divided into several distinct branches that are focusing on individual aspects of NPC. These may divided into semi-empirical methods, thermodynamic approaches and synthesis simulations. The semi-empirical methods rely on experimental data to determine structural models for NPC. The thermodynamic approaches, in contrast, takes its inspiration from the TEM pictures of nanoporous carbon discussed in Section 4.2. The aim is to generate curved sp^2-bonded carbon lattices that are local energy minima and resemble the combinations of hexagons and non-hexagonal rings observed in the TEM pictures of NPC. However, there are two different perceptions of the origin of the pores in the material. The "defect approach" views the pores as voids in the lattices that are derived from multiple-atom vacancies giving pore sizes in the range 0.5–1 Å. The other perception is that the pores are enclosed by graphene sheets like a slit pore or in a nanotube. In either case do the non-hexagonal rings play a vital role either directly as voids or to induce negative curvature in the lattice to form the pore channels. The synthesis simulations finally focus on the fact that the structure of NPC is dependent of its history. These methods aim to simulate the actual synthesis and thereby obtain an appropriate model for NPC generated under certain conditions.

Common for all methods is the importance of the non-hexagonal rings in the otherwise hexagonal structure of sp^2-bonded carbon. This section therefore starts by a review of the tiling theory for two-dimensional surfaces made out of polygons to lay down the foundations for the modeling of the NPC structure.

4.3.1 Tiling Theory for sp^2-Carbons

There is only a limited number of combinations of regular polygons that are able to cover a two-dimensional surface as we know from the patterns of cobbled stones

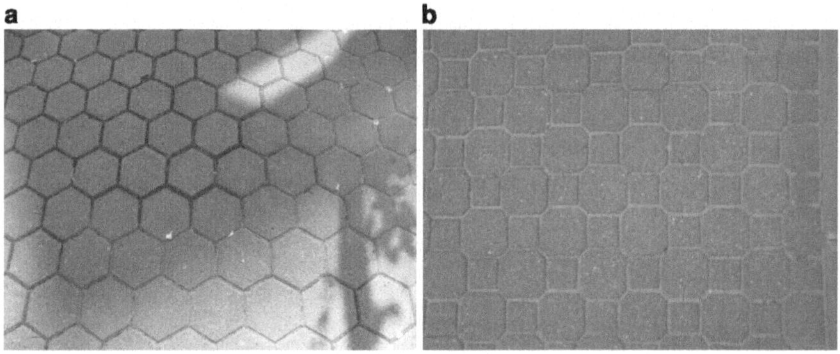

Fig. 4.4 Example of tilings in our every day life, the cobbled stones on the pavement. (**a**) an example of a tiling with a single polygon. (**b**) An example of a tiling with two different polygons, squares and octagons

on the pavement as shown in Fig. 4.4. The triangular, square and hexagonal lattices are examples where only one type of polygons is required to form a lattice that cover a two-dimensional surface [27]. A sp^2-bonded graphene sheet for instance, where every carbon atom has 120° bond angles, naturally form a flat hexagonal lattice. It should be noted that graphene may become buckled at finite temperatures due to thermal fluctuations [22, 76], but for the sake of simplicity we consider a perfect hexagonal graphene sheet as flat. Certain combinations of polygons are also able to form strictly two-dimensional lattices. One example is a lattice made out of squares and octagons as in Fig. 4.4, but incorporation of non-hexagonal rings in a hexagonal lattice in general induces curvature. The curvature may be described by the Gaussian curvature classification. The curvature of a surface can be divided into two orthogonal principal directions, κ_1 and κ_2, as illustrated in Fig. 4.5b. The Gaussian curvature K, is defined as the product of the curvature of the two principal axes $K = \kappa_1 \cdot \kappa_2$. It is positive $K > 0$, if the principal curvatures has the same sign in both direction, as the case is at an extreme point. If on the other hand the principal curvatures have opposite signs in the different directions $\kappa_1 > 0$ and $\kappa_2 < 0$, like at a saddle point, the Gaussian curvature is defined as negative $K < 0$.

Incorporation of pentagons in the hexagonal lattice leads to a positive Gaussian curvature. The simplest example is the nanocone shown in Fig. 4.5a, which is a graphene sheet with a conical shape, due to a single pentagon positioned at the apex of the cone [15, 56]. Larger rings such as heptagons instead induce a negative Gaussian curvature in a graphene sheet as these rings tend to take a boat shape when incorporated in the hexagonal lattice as can be seen in Fig. 4.5b. The non-hexagonal rings determine the shape of the lattice and the topology of the resulting structure can be predicted by Euler's rule. The general Euler's rule counts the number of faces F, edges E, and vertices V, of the polygons that form the lattice and relates these numbers to the genus, g, of a surface as follows.

$$V - E + F = 2(1 - g) \qquad (4.1)$$

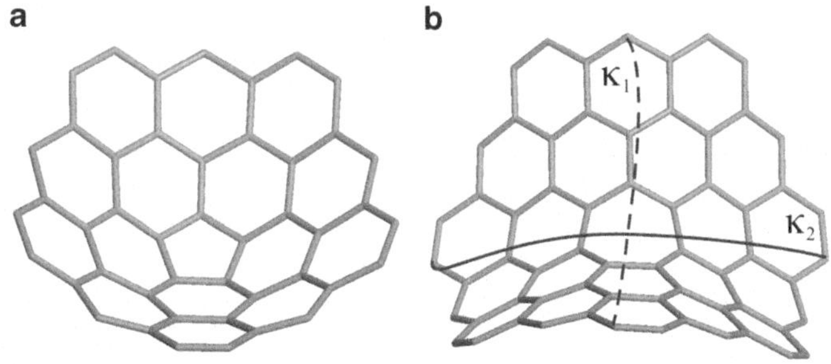

Fig. 4.5 Graphene sheets with non-hexagonal rings relaxed with the COMPASS force field. (**a**) Graphene with a pentagon, also known as the nanocone. (**b**) Graphene with a heptagon. The two principle curvature axes are marked as κ_1 and κ_2

The genus is the number of channels that penetrate the surface per unit cell. A sphere has genus zero as it is completely closed and a cylinder has genus one as it has one channel going from unit cell to unit cell.

The vertices in the graphene lattice are the carbon atoms and the polygons or atomic rings are defined by the bonds between nearest neighbors. This means that three polygons meet at each vertex and two polygons share an edge in sp^2-bonded carbon. A lattice with a total of F faces or atomic rings made up of n_5 pentagons, n_7 heptagons, n_8 octagons, n_9 nanogons, n_{10} decagons and an arbitrary number of hexagons is

$$F = n_5 + n_7 + n_8 + n_9 + n_{10} \tag{4.2}$$

$$2E = 5n_5 + 7n_7 + 8n_8 + 9n_9 + 10n_{10} \tag{4.3}$$

$$3V = 5n_5 + 7n_7 + 8n_8 + 9n_9 + 10n_{10} \tag{4.4}$$

Entering the values for F, E and V into Eq. 4.1 gives Euler's rule for graphene based materials:

$$\sum_{n_{m_{ring}}} (6 - m_{ring}) n_{m_{ring}} = n_5 - n_7 - 2n_8 - 3n_9 - 4n_{10} \ldots = 12(1 - g) \tag{4.5}$$

where m_{ring} is the number of atoms in the non-hexagonal ring and $n_{m_{ring}}$ is the number of non-hexagonal rings in the lattice. Euler's rule states that the genus of a carbon lattice is determined by the sum of non-hexagonal rings per unit cell of the structure. The pentagons have a positive prefactor as they give positive curvature, while the larger rings with more than six atoms have a negative prefactor as they contribute a negative curvature. The best-known example of a sp^2-bonded carbon surface that contain non-hexagonal rings is of course the fullerenes. Euler's rule states that 12 pentagons are needed to obtain a sphere of genus zero. The fullerenes contains exactly 12 pentagons that close the graphene sheet into a spherical carbon

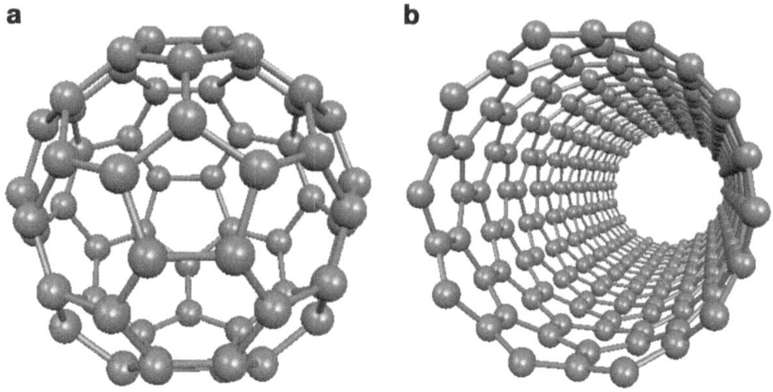

Fig. 4.6 The atomic structure of carbon structures with (**a**) genus = 0, the C_{60}-fullerene, (**b**) genus = 1, a (10,0) nanotube

cage as can be seen in Fig. 4.6. However, the fullerenes can contain an arbitrary number of hexagons changing the radius, as the hexagons do not change the curvature.

A conservation rule that reveals which combinations of non-hexagonal rings that can be incorporated into a hexagonal lattice without changing the genus, can be derived from Euler's rule [11]:

$$n_5 = \sum_{n(m_{ring}>6)} (m_{ring} - 6) n_{m_{ring}} \tag{4.6}$$

This relation states that the negative curvature induced by incorporating large rings with more than six carbon atoms in the hexagonal lattice can be compensated by the positive curvature of a few adjacent pentagons. The exact number of pentagons required to compensate the negative curvature equals the excess number of carbon atoms in the large rings. This can be exemplified by heptagons, where it is possible to motivate the conservation rule by Hückel's aromatic rule. A consequence of the aromatic rule, which will be discussed in more detail in Section 4.4.3.1, is that pentagons have an acceptor character, while larger rings such as the heptagons have a donor character. The conservation rule corresponds to a summation of non-hexagonal rings of donors and acceptors character that adds up to zero.

The simplest example of a genus conserving combination of non-hexagonal rings in the hexagonal lattice is the edge dislocation core in graphene that consists of a heptagon connected to a pentagon as can be seen in Fig. 4.7a. The heptagon adds an additional row of atoms into the lattice and a pentagon is formed next to the heptagon to saturate the bonds between the C-atoms that are bend away from each other to insert the additional row of atoms. This confirms that edge dislocations do not affect the global shape of a graphene sheet even though they might lead to a local buckling around the dislocation. Another well-known example of a genus preserving

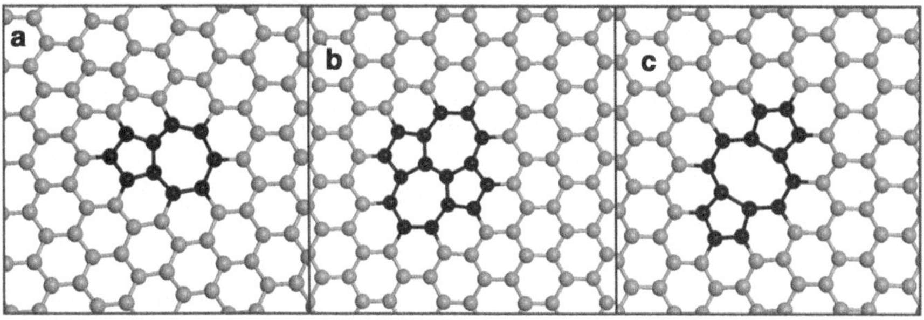

Fig. 4.7 The atomic structure of genus conserving defect motifs in graphene. The C-atoms that belong to non-hexagonal rings are highlighted as black. (**a**) The edge dislocation core in graphene is a 5-7-ring combination. (**b**) The Stone–Wales defect contains two pentagons and two heptagons. (**c**) The divacancy forms an eight-atom ring that is surrounded by two pentagons

combination is the Stone–Wales defect, shown in Fig. 4.7b. The SW-defect may be characterized as two adjacent, oppositely directed dislocation cores, as it contains two pentagons and two heptagons [100].

Larger rings with more than seven atoms are introduced by vacancies in the graphene sheet. Ab initio calculations have shown that the undercoordinated C-atoms at the periphery of the vacancies rebind in pairs to remove dangling bonds [11]. The rebonding is complete at even vacancies (a vacancy where an even number of carbon atoms are missing), such that C-atoms surrounding the vacancy are forming a large ring of three-coordinated C-atoms. In contrast, the rebonding is incomplete at vacancies with an odd number of removed atoms. The remaining, undercoordinated, C-atom at the odd vacancies has a configuration similar to the C-atoms at a zigzag edge, such that it has a dangling bond. The incomplete hexagons surrounding the vacancy are finally rebonded into pentagons. This rebonding of the undercoordinated atoms at the periphery of the vacancies leads to a transformation where motifs are formed that consists of a combination of non-hexagonal rings. The vacancy itself form a large void or pore in the shape of a large ring incorporated in the lattice and the pore is surrounded by a number of pentagons as exemplified by the divacancy in Fig. 4.7c) that has a 5-8-5 ring structure. A four atom vacancy form a nine atom ring with a diameter of $\approx$4 Å and the 12 atom vacancy form 12 atom ring with diameter of $\approx$5.5 Å.

The relative location of the pentagons is also important and Kroto provided a simple rule to sort different combinations of pentagons with respect to the energy cost they induce in the fullerenes [58]. He observed that the most stable fullerene structures have the lowest number of adjacent pentagons in the lattice and deduced that the energy penalty increases as more pentagons are located as neighbors. This led to a classification of the combinations of pentagons as follows: An isolated pentagon surrounded by hexagons constitute type I and has the lowest energy. Two pentagons next to each other, as in a Stone–Wales defect is type II. Type III and IV are clusters of three and four pentagons that increase the energy further. This rule, although formulated for fullerenes, appears to have a general validity for sp^2-bonded carbon materials, such that may as well be applied to nanotubes or graphite.

The genus-conservation rule in Eq. 4.6 imply that the Stone–Wales transformation is a genus preserving transformation. This can easily be understood by examining the effect on the four rings that participate in the SW-transformation. The general SW-transformation decreases the number of C-atoms by one in the two rings that share the bond that is rotated, while the two rings that are opposite to the rotated bond are gaining one C-atom each. The genus is therefore conserved as the four rings involved in the SW-transformation pairwise exchange one C-atom, but no C-atoms are added or removed from the four rings in total. It should however be noted that the pentagons and heptagons have a different aspect ratio as hexagons, so that they induce a local curvature even though the graphene sheet remain flat on average.

This implies furthermore that an arbitrary number of SW-transformations can be applied to a sp^2-lattice without changing the character or shape of the surface. Ab initio calculations have for instance shown that a divacancy that initially has a 5-8-5 ring structure shown in Fig. 4.7c), can be transformed into a 555-777 ring structure by a SW-transformation. Which one of these combinations that has the lower energy depend on the curvature. The 555-777 ring structure has the lower energy in a flat graphene sheet, while the 5-8-5 has the lower energy in a curved sheet like in a nanotube [3]. If the SW-transformations are taken to their extreme, a whole graphene-like sheets made out of pentagons and heptagons could be formed. However, it should be noted that it is strictly speaking not possible to cover a surface by perfect pentagons and heptagons, so the bond angles are slightly bent compared to ideal pentagons and heptagons to fit into a two-dimensional lattice. These graphene-like sheets made from pentagons and heptagons are flat and they have been named Haeckelites [103] after the drawings of Radiolaria made by Ernst Haeckel, a German physician and zoologist [28]. Radiolaria are one-cell plankton with a silicon-based skeleton that is made of pentagons, hexagons and heptagons. The Haeckelites can theoretically be generated from a graphene sheet by applying periodic Stone–Wales transformations to every twelfth C–C bond, but the barrier for SW-transitions is very high such that is unlikely to happen in a defect-free graphene sheet. This is supported by Monte-Carlo simulations of graphene at finite temperature, which did not show any occurrence of SW-defects [22]. Other synthesis methods that start from pentagonal and heptagonal building blocks are necessary to generate Haeckelites. This concludes the section about tiling theory that has established the potential building blocks of nanoporous carbons.

4.3.2 Semi-empirical Methods

Semi-empirical methods provide a route to determine the structure of NPC from experimental data. In particular the reverse Monte-Carlo method has been used by Gubbins and co-workers to obtain structure models for micro and nanoporous carbon [85, 105]. The reverse Monte-Carlo method aims to reconstruct the atomic structure from an experimental fingerprint, such as the pair-correlation function $g(r)$ or the structure factor $S(q)$ [74]. The method uses an iterative approach to refine the atomic structure until the theoretical spectrum calculated from the tentative structure agrees with the experimental spectrum. The calculation begins by generating a starting atomic configuration. The theoretical spectrum is obtained from the atomic arrangement and the deviation between the theoretical and experimental spectrum is calculated and used as target function for the minimization. The deviation target function is then successively minimized by stochastically moving the atoms around to improve the theoretical structure. The Metropolis algorithm is used to determine if a move should be accepted or rejected and the procedure continues until a satisfactory agreement between the theoretical and experimental spectrum is obtained.

The method was first applied to determine the structure of an activated mesocarbon microbead (a-MCMB). The starting structure was a random distribution of graphene plates with a mean size of 13 hexagonal rings. The interior of these building blocks was kept fixed and only the size, in terms of number of rings, and their relative orientation was optimized during the reverse Monte-Carlo run. The resulting structure for the a-MCMB shown in Fig. 4.8a had a fine, quite homogeneous, highly connected pore morphology, where parallel oriented plates constituted the pore walls. The smallest pore size was around 4.5 Å and this is more than 1 Å larger than the layer spacing in graphite. The radial distribution function in Fig. 4.8b showed perfect agreement beyond 5 Å, where the spectrum is quite featureless, but significant deviations for shorter distances that correspond to internal, nearest neighbor distances. This indicates that this structure model is able to capture the disorder between neighboring sheets, but that there is also internal disorder in terms on non-hexagonal rings that was not taken into account by keeping the atoms in the graphene plates in a perfect hexagonal order.

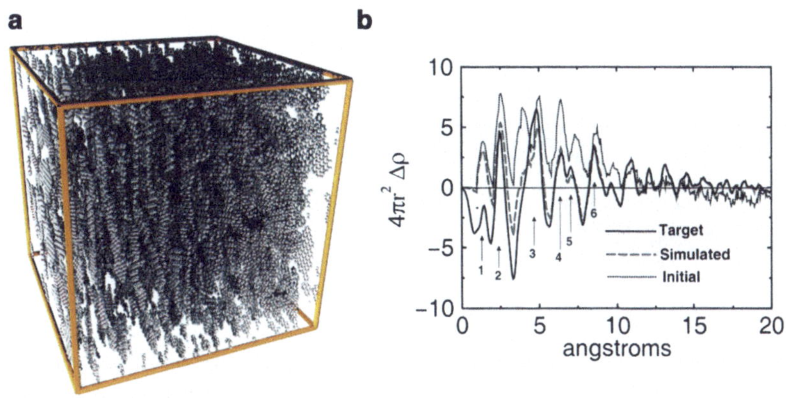

Fig. 4.8 (**a**) Structure of activated mesocarbon microbead (a-MCMB) obtained by the reverse Monte Carlo(RMC) method. (**b**) C–C radial distribution function (RDF) for the RMC model of a-MCMB. The *solid line* is the experimental RDF, the *dotted line* show the initial and the *dashed line* show the converged RDF for the RMC calculation. Reprinted with permission from Ref. [105]. Copyright 2000 American Chemical Society

Smith et al. took the opposite approach to obtain models for NPC by focusing on the role of the non-hexagonal rings in the structure [97]. These authors used a perturbative approach to generate NPC models by constructing graphene sheets with non-hexagonal rings incorporated by hand using the Cerius molecular modeling tools [14]. Pentagons or heptagons were taken as nucleation sites and hexagons were added around the non-hexagonal ring to construct defective graphene sheets. The ratio of non-hexagonal to hexagonal rings was varied to simulate different defect concentrations, but an equal number of pentagons and heptagons were used to preserve the genus and to assure an overall sheet-like structure. Three dimensional structures models were furthermore built by randomly stacking different defective graphene sheets on top of each other mimicking a turbostratic ordering to achieve an appropriate density corresponding to NPC. These NPC model structures were finally optimized by molecular mechanics using the Dreiding force field.

The resulting NPC models shown in Fig. 4.9a have the shape of buckled graphene sheets as each pentagon induces a local conical region similar to Fig. 4.5a and the heptagons similarly form local saddle regions as in Fig. 4.5b. The pair distribution function (PDF) was calculated for NPC models with a varying concentration of non-hexagonal rings to compare to experimental results from NPC synthesized by pyrolysis of PFA [49,84]. Analyzing the PDFs for different NPC models in Fig. 4.9b revealed that a higher concentration of non-hexagonal rings leads to more disorder in the lattice as expected. This is reflected in a more rapidly decaying PDF with fewer peaks at longer atom separations. The comparison to the experimental PDF from NPC synthesized from PFA by pyrolysis at 800°C showed a reasonable agreement with NPC models with 4% non-hexagonal rings. Comparison with the

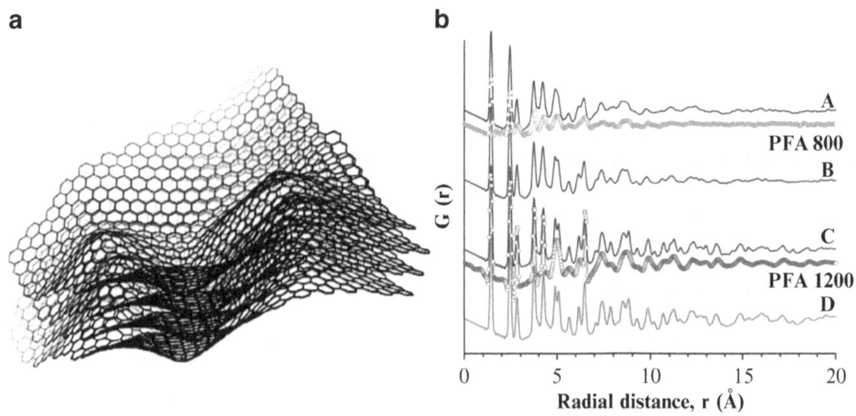

Fig. 4.9 Structure of NPC generated by randomly incorporating non-hexagonal rings into a hexagonal lattice. (**a**) Turbostratically ordered NPC sheet forming a three dimension NPC structure. (**b**) Pair distribution functions (PDF) calculated with A = 3.9%, B = 2.8%, C = 1% and D = 0.5% non-hexagonal rings in the lattice. Experimental PDFs obtained from NPC synthesized by pyrolysis of PFA at 800°C and 1,200°C are marked by PFA 800 and PFA1200. Reprinted from Ref. [97] with permission from Elsevier

PDF for NPC synthesized at 1,200°C showed better agreement with models with 1% non-hexagonal rings. These results supports the interpretation that the NPC synthesized at 1,200°C can really be considered as graphene-like with some non-hexagonal rings incorporated into the structure, while the NPC synthesized at 800°C is still not enough graphitized to be successfully described by this type of perturbation model.

4.3.3 Synthesis Simulations

The fact that NPC is a metastable material and the structure depends on the details of synthesis is a motivation to simulate the actual process in detail. Acharya et al., have developed a method to simulate the growth of NPC by pyrolysis of poly(furfuryl alcohol) (PFA) [1]. This process has been studied in great detail by experimental methods [49, 84] as described in Section 4.2.1. One interpretation of the experimental studies is that the process can be described as assembling small graphene fragments into larger sheets. The intention was to mimic this procedure and thereby generate appropriate models for NPC derived by pyrolysis of PFA at different temperatures. The structure generation algorithm had six steps. The initial step was to identify the appropriate building blocks, which have the correct H/C ratio that corresponds to a NPC material generated at a particular temperature. The building blocks resembled irregular polyaromatic hydrocarbons (PAH) as the graphene fragments were saturated by hydrogen on the edges. The size of the building blocks therefore determined the H/C ratio, since the fraction of the perimeter to the area decreases with increasing size and thereby the number of hydrogen to carbon atoms. The size was also used to simulate variation in temperature, since the materials synthesized at higher temperature have a higher degree of carbonization. Using larger building blocks that have a smaller H/C ratio simulated a higher pyrolysis temperature.

The actual simulation started by selecting a building block at random. The potential binding sites on the building block and the as-grown sheet were determined by identifying the undercoordinated edge sites. The attachment site was then selected as the place where there were sufficient undercoordinated carbon atoms that could connect the building block and the sheet to form a pentagon, hexagon or a heptagon. This procedure continued until a nanoporous sheet was formed. The NPC models were subsequently optimized by a force field calculation to minimize the energy. The optimization process also involved decreasing the H/C ratio to simulate an increasing rate of carbonization. The calculated cohesive energy increased as the H/C ratio decreased confirming that the more carbonized structures are more stable.

The resulting NPC structures can be characterized as curved graphene sheets that contain pentagons and heptagons as shown in Fig. 4.10. Interestingly, these non-hexagonal rings do appear in the generated NPC structure even though they are not present in the building blocks. The pentagons and heptagons are instead the result of "bad connections" between the graphene fragments, like grain boundaries occur between misoriented crystal domains during crystal growth. The density of

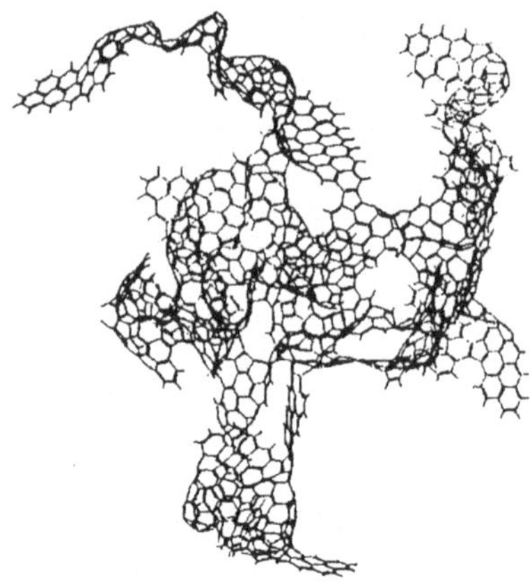

Fig. 4.10 Structure of NPC generated by randomly assembling random graphene flakes according to the method of Acharya et al. Reprinted with permission from Ref. [1]. Copyright 1999 Taylor & Francis Ltd. The original article can be found at http://www.informaworld.com

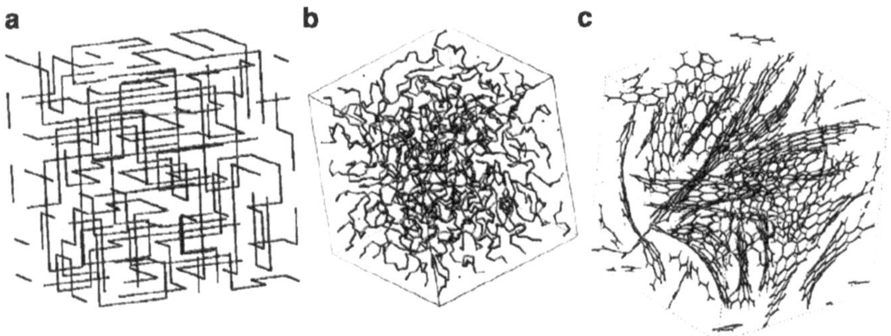

Fig. 4.11 Illustration of three states of the simulation for the Monte Carlo simulations of the pyrolysis of PFA. Reprint from Ref. [59] with permission from Elsevier

the models generated at 400–600°C agrees quite well with measurements of NPC derived from PFA at the corresponding temperature. The radial distribution function for a model generated at 500°C shows two dominant peaks for separations that correspond to the nearest and next nearest neighbor distances. This is similar to the experimental PDFs obtained by pyrolysis of PFA at 800°C [84]. It appears that this quite simple approach is able to capture the essential features of NPC synthesized by pyrolysis.

A more realistic starting point for modeling of the pyrolysis process would be to start from the actual polymer and follow the carbonization process towards a nanoporous carbon material step by step as indicated in Fig. 4.11. Kumar et al. have attempted to use this approach to model the pyrolysis of PFA by canonical Monte-Carlo (MC) simulations [59]. The starting configuration was an atomistic model of the backbone of PFA shown in Fig. 4.11a. All carbon atoms that were not three-fold coordinated with other carbon atoms were assumed to be potential connection sites to grow NPC-sheets. To facilitate the rebonding of the carbon atoms, all non-carbon atoms, i.e. hydrogen and oxygen, were removed from the polymer so the actual backbone model of PFA contained only carbon atoms. A pseudo amorphous model of the PFA polymer melt was then constructed by inserting these polymer backbone molecules randomly into a simulation box. At the end of the simulation all remaining non-three fold coordinated carbon atoms were assumed to be saturated by hydrogen atoms, such that the H/C ratio was calculated as the ratio between the three-coordinated and the undercoordinated carbon atoms. This model system was equilibrated at a particular temperature by a two-step constant volume and temperature (NVT ensemble) Monte-Carlo method to simulate the transformation into NPC. The first step was a "vibration step" of max 0.2 Å to simulate the thermal movement. The acceptance ratio was regulated by a Metropolis criterion using the Dreiding force field for the total energy evaluation. A second longer "bond breaking" step was applied to a subset of randomly selected atoms. The bonds to the neighboring atoms were broken and the atom was randomly moved to a new position within a distance of maximum 3.82 Å. Formation of bonds at the new position was determined by a Metropolis criterion based on a Boltzmann distribution for the bond energy. More details about these type of Monte-Carlo methods can be found in the excellent book about molecular simulations by Frenkel and Smit [25].

The resulting structure model for NPC shown in Fig. 4.11c has the shape of curved graphene sheets with pentagons and heptagons, but larger voids are also visible. Simulated TEM pictures show a porosity that is similar to pictures taken from NPC synthesized by pyrolysis of PFA at 800°C [84]. The pair correlation function (PDF) for NPC also shows good agreement with the PDF from the same sample. The H/C ratio on the other hand is higher than the experimentally observed values indicating that this MC method underestimates the connectivity by a too strict bonding criterion. It is also interesting to note that variation of the initial conditions do not change the average porosity, even though the actual atomic structures differ. This is an important test of the methodology as it is tricky to find local minima by the MC method unless care is taken to avoid going to the ground state. However, this NVT-MC method appeared to be able to find metastable structures of NPC as long as the step length is kept short and the acceptance criterion is strict to prevent complete graphitization.

Another approach to directly follow the experimental procedure is to use molecular dynamics(MD). Donadio et al. have used this method to simulate the growth of nanoporous carbon films grown by supersonic cluster beam experiments as discussed in Section 4.2.4. In these simulations carbon clusters ranging from small linear clusters of 1–10 atoms up to fullerenes were accelerated towards a diamond (001)-surface. The motion of the clusters and the substrate was described by atomistic molecular dynamics and the atomic interactions were described by the Tersoff potential [104]. The acceleration energy was chosen in the range 0.1–1 eV in order to match the experimental conditions [77]. More details about molecular dynamics methods can be found in Ref. [25].

These MD-simulations revealed that the structure of the grown carbon film was strongly dependent on the type of cluster and the acceleration energy. Small clusters tended to form porous structures where the porosity decreased with increasing energy as the clusters have more energy available to rearrange in the film at higher

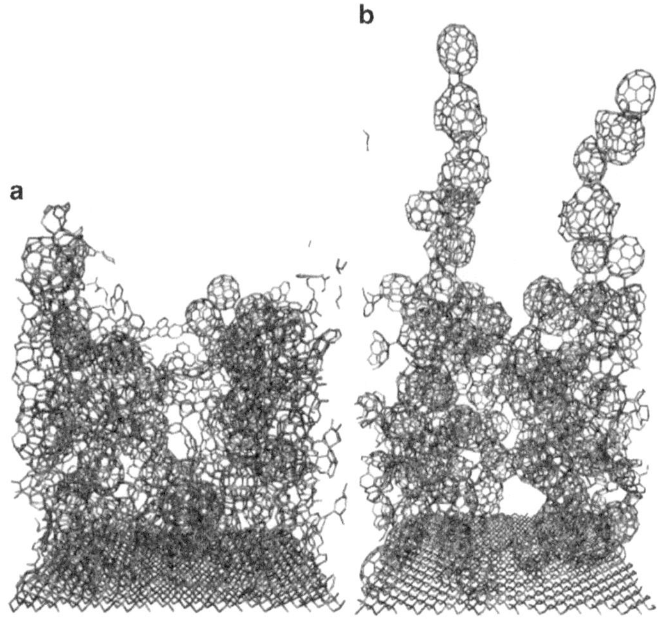

Fig. 4.12 The atomic structure of nanoporous carbon films generated by atomistic MD simulations of the SCBD process. The clusters used in the MD-simulation have a bimodal distribution, small clusters with 1-23 C-atoms and larger fullerenes with 46-120 C-atoms. The ratio between the two cluster types is in (**a**) 5:1, while in (**b**) the ration is 1:10. Reprinted with permission from Ref. [18]. Copyright (1999) by the American Physical Society

acceleration energies. Fullerenes are more stable than the smaller clusters, such that the fullerenes remain intact to a larger degree after impact at corresponding acceleration energies. The NPC films based on fullerene beams had the character of stacked and interconnected fullerenes as can be seen in Fig. 4.12. This indicates that the carbon films prepared by the SCBD method can be tailored by carefully choosing the cluster size distribution and the acceleration energy.

4.3.4 Schwarzites

The fact that nanoporous carbon is dominantly sp^2-bonded, such that it form sheet-like structures have inspired a number of researcher to use periodic minimal surfaces (PMS) to derive models for NPC. A PMS is a mathematical concept that defines a periodic, negatively curved, surface that divides space into two congruent regions, the "inside" and the "outside" region [91]. A simple example is an infinite cylinder that can be seen as a periodic repetition of a cylindrical segment. The cylinder encloses an internal volume and separates it from the exterior region, such that it can be defined as a PMS with genus one. More complex PMS, such as the cubic P-surface or the "Plumber's nightmare" in Fig. 4.13, can contain a number of channels that

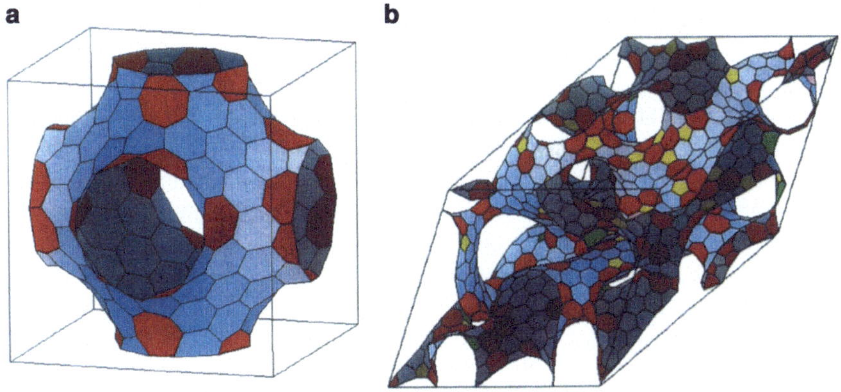

Fig. 4.13 The atomic structure of two Schwarzites. (**a**) The P-surface with genus g = 3, containing 216 C-atoms. The hexagons are blue and the heptagons are orange. (**b**) The atomic structure of a random Schwarzite having genus g = 12. The structure contains 1,248 atoms arranged into 38 pentagons (*yellow*), 394 hexagons (*blue*), 155 heptagons (*orange*), 12 octagons (*green*) and one nanogon (*pink*). Reprinted with permission from Ref. [106]. Copyright (1992) by the American Physical Society

connect neighboring volumes into a porous network similar to a sponge or a Swiss cheese. Each PMS is characterized by the genus or the number of channels per unit cell and how the channels are connected. The PMS are therefore divided into primitive (P-surfaces), gyroid (G-surfaces) and diamond-like (D-surfaces) depending on the symmetry of the connections. The P-surfaces have one joint per unit cell that is an intersection at right angles between three channels along the (100), (010) and (001)-directions as can be seen in Fig. 4.13a. The D-surfaces have tetrahedral connections like the bonds in the diamond lattice. The G-surfaces finally have channels along the (100) and (111) direction meeting at an angle 70.5° [92].

Considering the porosity and periodic character of the PMS makes them an attractive starting point to generate models for nanoporous carbon [66, 67, 82, 101, 102, 106–108]. These models for NPC have been called Schwarzites [66] in honor of the pioneering work on PMS by the German mathematician H. A. Schwarz at the end of the nineteenth century [94]. The procedure to generate a Schwarzite model for nanoporous carbon proceeds by selecting a PMS surface that has an appropriate porosity. A graphene lattice is projected into the PMS surface, but a perfect hexagonal lattice would not fit onto a PMS. The exception is the cylindrical PMS with genus equal one, which generates a normal nanotube in Fig. 4.6 that only contain hexagons. In general it is necessary to include non-hexagonal rings into the lattice to obtain the negative curvature of the underlying PMS. Euler's rule in Eq. 4.5 provides a convenient method to calculate how many non-hexagonal rings need to be included in the lattice per unit cell to fit onto a particular PMS. A P-Schwarzite has genus = 3 and can be generated using $n_7 = 24$ heptagons or $n_8 = 12$ octagons per joint, such that the smallest P-surface constructed with only heptagons contains 56 atoms. This is the parallel to the smallest fullerene, the dodecahedron C_{20}, which is constructed by 12 pentagons. However, the Schwarzites are usually inflated with hexagons to diminish the strain in the structure, similarly to the fullerenes where the C_{60} have 20 hexagons inserted in addition to the 12 pentagons. The P-surface in Fig. 4.13a have 80 hexagons in addition to the 24 heptagons, such that the unit cell contains 216 atoms. The D-surfaces have tetrahedral connections like the bonds in the diamond lattice, which means that each joint have two channels entering and leaving the joint. The diamond lattice is constructed by two joints per primitive cell and eight joints in the cubic unit cell, such that the primitive cell have genus $g = 3$ while the unit cell has genus $g = 9$. Euler's rule states that the primitive cell for the D-surface contains $n_7 = 24$ heptagons or $n_8 = 12$ octagons, such that the minimal D-surface constructed from 24 heptagons contains 56 atoms. However, the D-surface has been inflated by 56 hexagons to contain 168 atoms [108] or 80 hexagons to contain 216 atoms [66, 106]. The final structure of the Schwarzite model is obtained by optimizing the geometry by various methods such as force fields [108], tight-binding and DFT calculations [107].

The resulting Schwarzite models for NPC are highly symmetric and have a channel diameter around 5 Å. This value agrees well with pore sizes measured for many types of NPC, but the high symmetry of the Schwarzites is somewhat contradictory to the experimental observations of disordered curved, graphene sheets discussed in Section 4.2. Table 4.1 shows that, apart from some scattering due to methodological differences, the cohesive energy of the Schwarzites containing octagons (P8,D8 and G8) and heptagons (P7,D7 and G7) is comparable to or even smaller than some existing carbon polymorphs, such as the C_{60} crystal. This indicates that the Schwarzites are thermodynamically possible as metastable carbon structures. However, it has been very difficulty to synthesize such symmetric nanoporous carbon structures so the Schwarzites should be considered as idealized models for NPC until methods have been established to synthesize them. The template method in Section 4.2.2 could be a way to generate such periodic NPC if the appropriate template material is used. A more realistic model for NPC synthesized by pyrolysis or the arc discharge method discussed in Section 4.2 is probably the random Schwarzite proposed by Townsend et al. [106]. The authors used a three dimensional fcc Ising model to define a complex landscape with two distinct states. The spin up and spin down regions were used to define the "inside" and "outside" regions in the cell, such that the domain boundaries corresponded to the dividing surface. This dividing surface was then covered by a graphene lattice containing 38 pentagons, 394 hexagons, 155 heptagons, 12 octagons and one nanogon giving genus $g = 12$. The resulting surface in Fig. 4.13b shows a complex structure with an irregular pore structure that resembles TEM pictures of NPC or even amorphous carbon [9,34,49].

4.3.5 Defect Motifs in NPC

Some genus preserving pores and voids are also observed in the interior of the crumpled graphene sheets that make up the nanoporous carbon. These pores can be

Table 4.1 Summary of the structural properties of Schwarzites. The models are divided into D-, P- and G-surfaces and characterized by the genus, the number of non-hexagonal rings per cell, $n_{x \neq 6}$. N_{atoms} is the number of atoms in the unit cell and the ρ is the density. The cohesive energy is presented with respect to perfect graphite E_{coh}, and bulk modulus is denoted B. The method used in the calculations is indicated as FF for force field methods, TB for tight binding and DFT for density functional theory

Name	Genus	$n_{x \neq 6}$	N_{atoms}	$\rho(\text{g/cm}^3)$	E_{coh} (eV/atom)	B (GPa)	Method	Refs.
P8	3	$n_8 = 12$	192					[67]
P8	3	$n_8 = 12$	192	1.16	0.19	133	FF	[66]
P192	3	$n_8 = 12$	192	1.16	0.19	103	FF	[106]
P8	3	$n_8 = 12$	192	1.049	0.2228		FF	[102]
P8BAL	3	$n_8 = 12$	192	1.167	0.233	149	TB	[107]
P6.8^2	3	$n_8 = 12$	48	2.04	0.488		DFT-LDA	[82]
P688	3	$n_8 = 12$	48		0.5297		FF	[102]
P688	3	$n_8 = 12$	48	1.999	0.523	273	TB	[107]
P7	3	$n_7 = 24$	216	1.02	0.20	75	FF	[66, 106]
P216	3	$n_7 = 24$	216	1.11	0.17	101	FF	[106]
P7	3	$n_7 = 24$	216		0.4711		FF	[102]
P7PAR	3	$n_7 = 24$	216	1.033	0.260	126	TB	[107]
D8	9	$n_8 = 48$	768					[67]
D8	3	$n_8 = 12$	192	1.096	0.283		FF	[102]
D8BAL	3	$n_8 = 12$	96	1.201	0.276	179	TB	[107]
D6.8^2	2	$n_8 = 6$	24	2.19	0.208		DFT-LDA	[82]
D688	2	$n_8 = 6$	24		0.0449		FF	[102]
D688	2	$n_8 = 6$	24	2.145	0.259	325	TB	[107]
D7	3	$n_7 = 24$	216	1.15	0.18	94	FF	[66, 106]
D216	3	$n_7 = 24$	216	1.16	0.16	107	FF	[106]
C_{168}	3	$n_7 = 24$	168	1.28	0.22	115	FF	[106]
C_{168}	3	$n_7 = 24$	168	1.297	0.11	110	FF	[108]
G8	3	$n_8 = 12$	192	1.129	0.2494		FF	[102]
G8BAL	5	$n_8 = 24$	384	1.238	0.266	159	TB	[107]
G688	3	$n_8 = 12$	192		0.2359		FF	[102]
G688	3	$n_8 = 12$	96	2.217	0.479	293	TB	[107]
G7	3	$n_7 = 24$	192		0.1760		FF	[102]
G216	3	$n_7 = 24$	216	1.18	0.17	109	FF	[106]
I-WP	7	$n_5 = 24$ $n_8 = 48$	744	1.061	0.5701		FF	[102]
Schwarzite 1248	12	$n_5 = 38$ $n_7 = 155$ $n_8 = 12$ $n_9 = 1$	1248	1.26	0.23	49	FF	[106]
C_{60}	0	$n_5 = 12$	60		0.6658		FF	[102]
C_{60}	0	$n_5 = 12$	60		0.67		FF	[108]
FCC-C_{60}	0	$n_5 = 12$	60	1.70	0.438		DFT-LDA	[82]
FCC-C_{60}	0	$n_5 = 12$	60	1.71	0.42	14	FF	[66, 106]
Graphite	1	0	4	2.32	0	27.8	DFT-LDA	[78]
Graphite	1	0	4	1.80	0	2.4	DFT-GGA	[78]

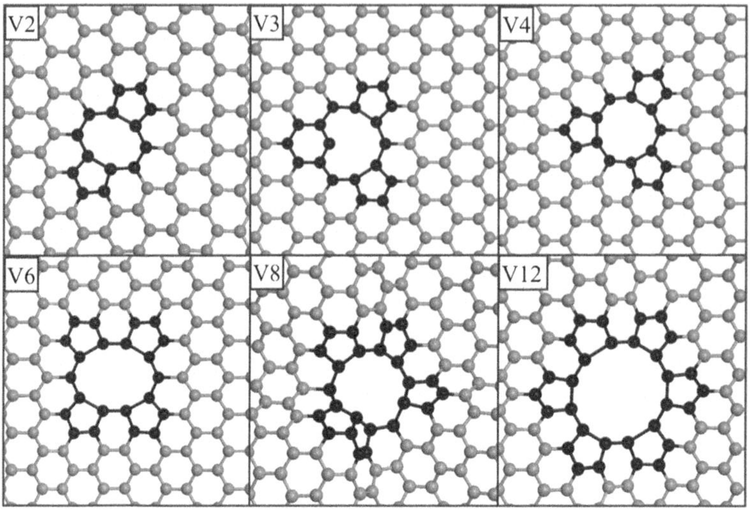

Fig. 4.14 The atomic structure of defect motifs in graphene. The hexagonal C-atoms are *grey* and the C-atoms that belong to non-hexagon rings are highlighted as *black*. A double vacancy, V2 forms an eight-atom ring. The Trivacancy V3, has an undercoordinated, zigzag atom. The Quadrovacancy V4, the Hexavacancy V6, Octavacancy V8 and the Twelvevacancy V12 have 9, 10, 11 and 12 atoms in the central rings that surround the void. The V8 motif becomes more curved than the other motifs as expected from the Kroto's rule, since two pentagons sit next to each other

characterized as multi-atom vacancies as shown in Fig. 4.14. The ab initio atomistic thermodynamics (AIATD) method will now be outlined to calculate the formation energy and estimate the concentration of these defect motifs.

4.3.5.1 Ab Initio Atomistic Thermodynamics

The formation energy ΔG_{form}, concentration and contribution to the cohesive energy of NPC due to such defect motifs is not just a function of the internal energy, but depends also on the temperature and chemical potential of the carbon atoms in the system, μ_{C}. The chemical potential is the derivative of the Gibbs free energy G with respect to the number of atoms, n_{C}. This implies that it is necessary to calculate the *free energy* not just the total energy. Free energy calculations, such as thermodynamic integration or meta dynamics are very demanding and the ab initio atomistic thermodynamics method is a convenient method to obtain the free energy for solids and gases. This method was introduced to estimate the defect concentration in semiconductors [50,90], but it is a general method that has recently been applied in other fields, such as surface physics and catalysis [86].

The general idea is to combine standard ab initio calculations with statistical mechanics and thermodynamics to calculate the free energy, G. This extends the ab initio calculations at $T = 0$ K to more realistic conditions and it enables to study the influence of temperature and pressure on individual reaction steps, such as adsorption or desorption or the exchange of particles with a reservoir to model defect formation. The calculations of the free energy for a solid is described in this section while methods to derive the free energy and chemical potential for a gas at finite temperature and pressure will be discussed in Section 4.5.3.1 in connection with the oxidation of NPC. The calculation of the free energy for a solid

$$G = U + pV - TS \tag{4.7}$$

is divided into two parts. The internal energy U is obtain by a static ab initio calculation, such as density-functional theory (DFT) [39, 52], at $T = 0$K. An overview of how to perform DFT calculations can be found in reviews such as Ref. [83]. The pV-term is negligible except at extreme pressures, such that the thermal contributions originate from phonons and configurational entropy [63]. The phonon contribution to the free energy, F^{phonons}, is calculated from the partition function for the phonons:

$$F^{\text{phonons}}(T) = k_B T \Sigma_{\alpha}^{3N} \int ln\left(1 - e^{-\frac{\hbar\omega}{k_B T}}\right) g_{\alpha}(\omega) V d\omega \tag{4.8}$$

where $g_{\alpha}(\omega)$ is the density of states for the α phonon mode. Methods to calculate phonon spectra are described in Section 4.4.2.

A defect free crystal has a well-defined non-degenerate ground state structure, such that the configurational entropy contribution is absent. However, defects may be created and distributed in different ways, so a defective crystal has a configurational entropy contribution due to the defects. Adding up the contributions gives the AIATD free energy for the solid:

$$G^{\text{solid}}(T) \approx E^{\text{DFT}} + F^{\text{phonons}}(T) + TS^{\text{config}} \tag{4.9}$$

The focus is now turned to the formation energy for the defect motifs. The formation energy is the difference in free energy between the system with a defect motif and the sum of the chemical potential of the constituents.

$$\Delta G_{\text{form}} = G^{\text{Defect}} - \Sigma_i n_i \mu_i \tag{4.10}$$

It can, in an abstract sense, be seen as exchanging n_{C} carbon atoms with a reservoir of carbon atoms that has the chemical potential μ_{C}. The reservoir can correspond to free atoms, μ_{C}^{free}, in the case of a sputtering experiment or graphene, $\mu_{\text{C}}^{\text{G}}$, under growth conditions. It is therefore convenient to treat the chemical potential of the reservoir as a variable $\mu_{\text{C}} = \mu_{\text{C}}^{0} + \Delta\mu_{\text{C}}$ defined with respect to a reference state, μ_{C}^{0}, for instance the free atom at $T = 0$K. In such case $\Delta\mu_{\text{C}}$ corresponds to the cohesive energy of the carbon material that is used as the reservoir. Another choice is the smallest stable carbon compound, i.e. the carbon dimer as is used in Fig. 4.15. The introduction of defects in the lattice leads to phonon softening, which gives a significant change in the vibrational free energy, $\Delta F^{\text{phonons}}(T)$ compared to the perfect lattice. However, the vibrational contribution has comparable amplitude for

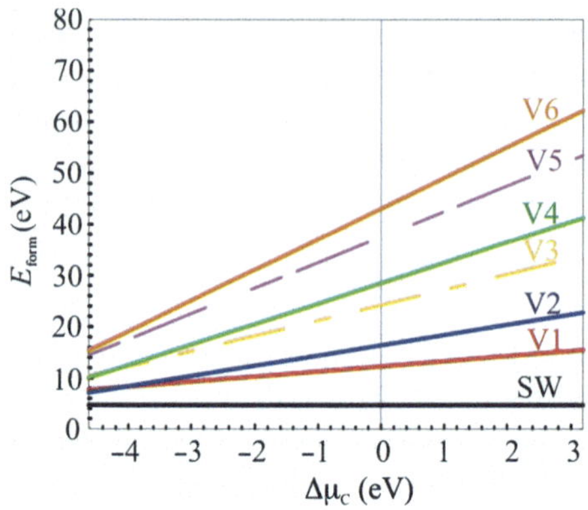

Fig. 4.15 Formation energies, ΔG_{form}, for defects in a graphene sheet as function of $\Delta\mu_{\text{C}} = \mu_{\text{C}} - 1/2E_{\text{C}_2}$. The right limit is the free C-atom at $T = 0$ K and the left limit is μ_{C} of graphene at $T = 0$ K. The *vertical line* at $\Delta\mu_{\text{C}} = 0$ corresponds to the energy of a carbon atom in a C_2 molecule at zero temperature. The labels refer to the Stone–Wales (SW) defect, the vacancy (V1) and vacancy clusters (from V2 to V6). Adopted from Ref. [11]. Copyright (2006) by the American Physical Society

defects of similar type, such that the vibrational contribution does often not change the energetic ordering between the defects [29]. The vibrational energy contribution is therefore often neglected if only the relative formation energies are important to determine the most important defect type. Small point defects, like few atom vacancies, only have a limited number of configurations, such that the configurational entropy is similarly often neglected when studying clean point defects. However, the number of configurations increases rapidly if the vacancies are exposed to molecule adsorption as will be discussed in Section 4.5.3.

In addition, the defect motif can be charged if electrons have been attracted or donated to the surrounding graphene lattice. This depends on the energetic position of the defect states with respect to the chemical potential of the electron reservoir, i.e. the Fermi level $\mu_e = \epsilon_{\text{F}}$ of the graphene lattice. The character of the defect states will be discussed in Section 4.3.5.2. The resulting formation energy of a clean structural motif can be calculated by [11]:

$$\Delta G_{\text{form}}(\mu_{\text{C}}, q, \epsilon_{\text{F}}) = \left\{E^{\text{motif}}(n_{\text{C}}, q) + n_{\text{C}}\mu_{\text{C}}\right\} - E^{\text{G}} + q\epsilon_{\text{F}} \tag{4.11}$$

where E^{G} and E^{motif} refer to the total energy of a graphene sheet before and after the defect is created, n_{C} denotes the number of C-atoms that has been exchanged with the reservoir, μ_{C}.

4.3.5.2 Energetics of Defect Motifs

Calculating the formation energies for the different defect motifs that can appear in NPC, using the formalism introduced in the previous section, reveals that different types of motifs are preferred depending on the external conditions. The right hand side of Fig. 4.15 shows that small defects, such as the single vacancy (V1), are most favorable when the carbon atoms are removed as individual atoms and evaporated into the gas phase. This would correspond to the case when the motifs are formed due to the impact of a projectile, like in a sputtering experiment. During growth of NPC in contrast, the carbon atoms are added and occasionally removed in equilibrium and the chemical potential is close to value in graphene $\mu_{\mathrm{C}} \approx \mu_{\mathrm{C}}^{\mathrm{G}}$. This corresponds to the left hand side of Fig. 4.15, where it can be observed that the vacancy motifs where $(2m-1)$ and $2m$ ($m=$ integer) atoms are missing have approximately the same formation energy. This indicates that the undercoordinated, zigzag atom at the odd vacancy motifs is unfavorable under these conditions. This originates from the fact that the zigzag atom has a dangling bond. It is for instance more favorable to move the zigzag atom to another odd vacancy to generate two even vacancies as two odd vacancies with $(2m-1)$ missing atoms have a higher formation energy than two corresponding even vacancies with $(2m-2)$ and $2m$ missing atoms. Another, more surprising point is that the formation energies are so large that the equilibrium formation and annihilation of vacancies in a perfect graphene sheet is completely negligible. This is also supported by Monte Carlo simulations for graphene sheets at finite temperature, where no spontaneous defect formation was observed [22]. These results support the hypothesis that the pores in NPC are frozen into the structure during the growth phase, forming the porous, metastable structure of NPC.

Another method to estimate the pore concentration in NPC is to calculate the heat of formation, ΔH_{NPC}, as function of the pore concentration by a cluster expansion approach. In this case, the NPC is modeled as a graphene sheet with a random distribution of motifs of the type shown in Fig. 4.14. Each motif has a certain concentration and the heat of formation can be calculated by summing the contributions from the different motifs. The first order contribution or the "on site" term is the formation energy of the motif and higher order terms are the motif–motif interactions. These interactions are mechanical strain interactions, which are quite long ranged in graphene due to the stiffness of the lattice, and charged motifs also have electrostatic interactions. Figure 4.16 indicates that the motif–motif interaction is quite moderate up to a defect concentration of 1%, such that the expansion can be truncated already after the on-site term giving the first order expression for the heat of formation of NPC, H_{NPC} as [11, 12]

$$H_{\mathrm{NPC}}(C_i) = H_{\mathrm{G}} + \frac{1}{1-\sum_j C_j N_{\mathrm{C}}^j} \sum_i C_i E_{\mathrm{form}}^i(\mu_{\mathrm{C}}^{\mathrm{G}}) \tag{4.12}$$

where H_{G} is the heat of formation per atom for graphene and $E_{\mathrm{form}}^i(\mu_{\mathrm{C}}^{\mathrm{G}})$ is the formation energy of the defect in graphene. The defect concentration C_i is defined as the number of defects per lattice site. Figure 4.16 shows that these NPC models with a concentration of motifs up to 1% yield H_{NPC} comparable to single wall nanotubes. It is also comparable to many of the Schwarzite models in Table 4.1, such as the P8Bal structure. This indicates that a graphene sheet can have a moderate amount of

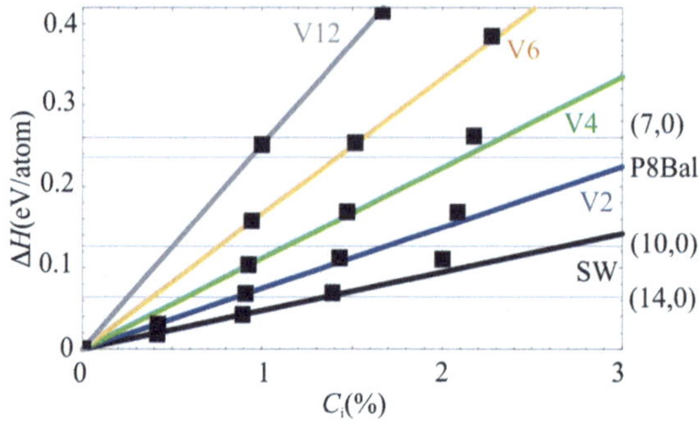

Fig. 4.16 The heat of formation/atom of NPC compared to graphene: $\Delta H = H_{\mathrm{NPC}} - H_{\mathrm{G}}$ as function of the defect concentration for the SW-defect, and the multiple vacancies. The *dots* indicate DFT calculated values and the *lines* show the values calculated from Eq. 4.12. The *horizontal lines* indicate ΔH for three zigzag nanotubes of different diameters (7,0), (10,0) and (14,0) and the P8Bal model of NPC [107]. Adopted from Ref. [11]. Copyright (2006) by the American Physical Society

voids, such as the motifs in Fig. 4.14, and still be metastable as long as the barriers towards graphitization are large enough. This is in agreement with the conclusions from TEM [9, 34] and neutron diffraction experiments [84] that NPC derived from hydrocarbons has the form of defective graphene sheets.

4.4 Properties of NPC

After establishing different structure models for nanoporous carbon, the focus now turns to the properties of nanoporous carbon. This section covers mechanical, electronic and vibrational properties derived from ab initio calculation for NPC generated via supersonic cluster beam deposition, the Schwarzites and the defect motif model for NPC.

4.4.1 Mechanical Properties of NPC

The mechanical properties of carbon materials span the whole spectrum from the extreme hardness of diamond to the soft interlayer binding in graphite. Diamond is able to scratch any material, while it is so easy to remove a few graphene layers from graphite that we use graphite in our pencils. In this context it is interesting to note that the *in-plane* sp^2 bond in graphite is actually $4/3$ stronger than the sp^3-bond in diamond. Only three bonds for every carbon atom provide the whole cohesive energy of graphite, while four bonds contribute to the cohesive energy of diamond.

The mechanical properties of NPC can be calculated once an appropriate structure model has been determined. The general method to calculate the elastic properties starts by expressing the variation of the free energy of the material under deformation $F(V, T)$, in terms of the strain tensors $\epsilon_{\alpha\beta}$ [64]:

$$F(V, T) = \frac{1}{2} C_{\alpha\beta\gamma\delta} \epsilon_{\alpha\beta} \epsilon_{\gamma\delta} \tag{4.13}$$

the tensor $C_{\alpha\beta\gamma\delta}$ is called the elastic module tensor and it is defined as the second derivative of the free energy:

$$C_{\alpha\beta\gamma\delta} = \frac{1}{V} \left(\frac{\partial^2 F(V, T)}{\partial \epsilon_{\alpha\beta} \partial \epsilon_{\gamma\delta}} \right) \tag{4.14}$$

The elements of $C_{\alpha\beta\gamma\delta}$ are called the elastic constants. The indexes are grouped together pairwise to form a combined index giving six combinations, such that the elastic module tensor has 36 elements in total [4]. The tensor is symmetric, such that are 21 individual elements and the symmetry of the crystal determines how many of the elements that are really independent. For hexagonal systems such as graphite only five elastic constant are independent and for cubic systems, like diamond, it reduces to three elastic constant, C_{11}, C_{12} and C_{44} that are necessary to completely describe the elastic properties.

The standard procedure to calculate the elastic constants by ab initio calculations proceeds by deforming the supercell in the appropriate directions and calculate the energy differences as function of the displacement. The entropic contributions are negligible compared to the changes in internal energy for small deformation at moderate temperatures, such that the free energy differences can be replaced by total energy differences. The elastic constants are finally obtained by a quadratic fit of the calculated energy differences. Some of the elastic constants can be related to the macroscopic properties such as the bulk modulus, B, but the bulk modulus is easier to determine by Murnaghan's universal binding curve[79]. In Murnaghan's approach, the total energy is calculated as function of variation of the cell volume and the values are fitted to the binding curve:

$$E(V) = E_0 + \frac{BV}{B'(B'-1)} \left[B' \left(1 - \frac{V_0}{V} \right) - 1 + \left(\frac{V_0}{V} \right)^{B'} \right] \tag{4.15}$$

The parameters in the Murnaghan's binding curve are the Bulk modulus B, the pressure derivative of the Bulk modulus $B' = \frac{\partial B}{\partial P}$, the equilibrium volume V_0 and the ground state energy E_0.

Considering the wide span in elastic properties for different carbon materials, it is expected that the properties of nanoporous carbon are somewhere in between the two extremes of diamond and graphite. Indeed, Donadio et al. found that this is really the case for NPC grown on a diamond substrate by the SCBD method discussed in Sections 4.2.4 and 4.3.3. They discovered that the elastic constants for the NPC films could be described by a power law [19]:

$$C_{ij} = b < N >^{-a} \tag{4.16}$$

where N is the size of the clusters in the molecular beam and a and b are fitting parameters. A diamond film provides one extreme as the carbon material that would be grown on the diamond substrate by single carbon atoms in the molecular beam. The NPC films grown by larger carbon cluster gets successively softer as the size of the clusters in the beam increases. This was explained by the fact that the larger clusters produce NPC films with increasing porosity and consequently decreasing stiffness. This suggests that the elastic properties of NPC films grown by the SCMB methods can be tailored by careful selection of the distribution of the cluster sizes in the molecular beam.

The perfectly crystalline Schwarzites provide another interesting model as they are the first example of a truly three-dimensional sp^2 bonded carbon material, where the van-der Waal's bonding does not play an important role. All other sp^2-bonded carbon materials, such as graphite, the C_{60} crystal or multiwall nanotube bundles are held together by van-der-Waals bonds. The stiffness of such crystals is extremely non-isotropic as they are very stiff along the sp^2-bonds and soft in the van-der-Waals bonded directions. Table 4.2 shows that the elastic constants of the different

Table 4.2 Summary of the mechanical properties of Schwarzites: bulk modulus B, the tetragonal shear modulus $C_{11} - C_{12}$ and the trigonal shear modulus C_{44}. The method used in the calculations is indicated as FF for force field methods, TB for tight binding and DFT for density-functional theory. The names are the same as in Table 4.1

Name	B (GPa)	$C_{11} - C_{12}$ (GPa)	C_{44} (GPa)	Method	Refs.
P192	103		17	FF	[106]
P8BAL	149	61	82	TB	[107]
P688	273	101	79	TB	[107]
P7	75		15	FF	[106]
P216	101		19	FF	[106]
P7PAR	126	69	54	TB	[107]
D8BAL	179	148	76	TB	[107]
D688	325	161	186	TB	[107]
D7	94		16	FF	[106]
D216	107		16	FF	[106]
C_{168}	115		20	FF	[106]
C_{168}	110	260	80	FF	[108]
G8BAL	159	115	78	TB	[107]
G688	293	123	124	TB	[107]
G216	109		18	FF	[106]
Schwarzite 1248	49		14	FF	[106]
Diamond	443		576	FF	[106]
Diamond	484	938	607	TB	[107]
Diamond	440	950	580	FF	[108]
Diamond	432	945	562	DFT-GGA	[78]
Graphite	27.8	883	4.5	DFT-LDA	[78]
Graphite	2.4	862	3.9	DFT-GGA	[78]

Schwarzites models are more isotropic and they have values that are intermediate between the values for diamond and graphite. The bulk modulus of the periodic Schwarzites for instance is roughly just a factor of two to four lower than the value for diamond [66, 107]. The random Schwarzite is softer than the periodic counterparts and approaches values similar to graphite. Another important result is that the moduli shown in Table 4.2 are positive for all the Schwarzite structures. This confirms that the Schwarzite structures are local energy minima and hence mechanically stable even under small deformations [107]. The issue of the stability of NPC and the graphitization process will be further discussed in Section 4.5.1.

4.4.2 Vibrational Properties of NPC

The large variation in the elastic properties for different models of NPC discussed above in Section 4.4.1 indicates that the vibrational properties could also be quite diverse. Vibrational spectroscopy would in such case provide a method to distinguish between different types of NPC. Unfortunately, the Schwarzites are the only models of NPC, where the vibrational properties have been thoroughly investigated by theoretical methods, such there is no complete map of the vibrational properties for different types of NPC available yet.

In order to calculate the vibrational properties of a material, it is necessary to solve Newton's equations for the atoms in the solid. In the harmonic approximation this equation can be written as follows [4]

$$D_{\kappa\kappa'}(\mathbf{q})u_{\kappa'}(\mathbf{q}) = \omega^2(\mathbf{q})u_\kappa(\mathbf{q}) \tag{4.17}$$

where $D_{\kappa\kappa'}(\mathbf{q})$ is the dynamical matrix of the material, ω is the phonon frequency and $u_\kappa(\mathbf{q})$ is the atomic displacement. The dynamical matrix is calculated by a fourier transform of the force constant matrix, $D_{\alpha\beta\gamma\delta}(\mathbf{R}-\mathbf{R}')$. The force constant matrix is defined as the second derivative of the total energy with respect to the atomic displacements

$$D_{\alpha\beta\gamma\delta}(\mathbf{R}-\mathbf{R}') = \frac{\partial^2 E}{\partial u_{\alpha\beta}(\mathbf{R})\partial u_{\gamma\delta}(\mathbf{R}')} \tag{4.18}$$

The force constant matrix is derived by calculating the interatomic force constants by numerical differentiation in a similar way as the elastic constants. However, it is computationally more expensive to calculate the phonon spectrum since all the force constants for every atom in the unit cell are needed. To obtain these force constants by numerical differentiation every atom needs to be displaced in all of the principal directions of the crystal. This requires $3N_{atoms}$ force evaluations or $6N_{atoms}$ total energy evaluations to determine the atomic force constants, where N_{atoms} is the number of atoms in the unit cell. Once the dynamical matrix has been derived by Fourier transforming the ab initio force constant matrix, the solution to Eq. 4.17 is obtained by diagonalizing the dynamical matrix, which finally gives the eigenfrequencies and eigenvectors of the phonons in the crystal. These eigenfrequencies and eigenvectors refer to the Γ-point where $\mathbf{q} = 0$. This is sufficient to compare to experimental IR- or Raman spectra, but to obtain the full dispersion curves and the vibrational density of states (v-DOS) it is necessary to extend the calculations to included finite wave vectors. There are two methods available to extend the vibrational calculations to finite wave vectors, the frozen phonon method [60] and the Density functional perturbation theory (DFPT) [6, 26].

Graphene has two atoms in the unit cell and consequently six phonon modes. Neutron diffraction experiments [72] and ab initio calculations [55, 78, 110] have shown that the vibrational density of states has an almost bimodal distribution, since the acoustic and optical modes are well separated, which can be seen in Fig. 4.17. The low frequencies region up to 630 cm^{-1} is dominated by the two acoustic modes, the in plane transversal, TA-mode and out of plane ZA-mode. The out of plane optical mode, ZO-mode has a small dispersion in the range 630–880 cm^{-1}. These three modes give a strong peak in the v-DOS between 400–800 cm^{-1}. The other two optical modes, the longitudinal LO and transversal TO-modes, give a similar strong peak between 1,310 and 1,600 cm^{-1}. The LO-TO mode is Raman active at the Γ-point giving rise to the characteristic G-peak in the Raman spectra at around 1,580 cm^{-1}. Diamond has a more compact v-DOS as can be seen in Fig. 4.17.

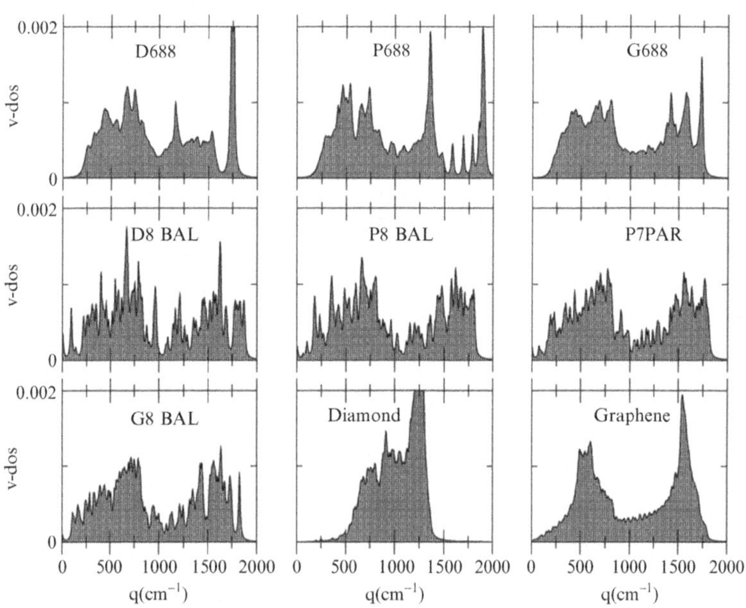

Fig. 4.17 Vibrational density of states (v-DOS) for seven Schwarzite models for NPC together with diamond and graphite calculated by the tight-binding method. Note the narrow, high frequency peak around 1,750 cm^{-1} for D688 and G688, and around 1,900 cm^{-1} for P688. Reprinted with permission from Ref. [107]. Copyright 2003 IOP Publishing

A Raman active mode is located around 1,330 cm^{-1} giving rise to the D-mode in the Raman spectra of diamond. These G- and D-modes are frequently used to distinguish between sp^2 and sp^3-bonded carbons.

The Schwarzites show a similar tendency to a bimodal distribution of the v-DOS as graphene as can be seen in Fig. 4.17. This suggests that a similar division into low frequency acoustic and high frequency optical modes may be possible. However, the division is less pronounced for the Schwarzites indicating that a larger variety of phonon modes are allowed in Schwarzite structures. An interesting feature in the v-DOS for some of the models is the sharp peaks at frequencies that lay at the same region or even above the LO- and TO-modes in graphene. The small dispersion suggests that they can be "defect modes" localized to particular regions of the Schwarzites structure.

4.4.3 Electronic Structure of NPC

The electronic structure of nanoporous carbon has been approached in two ways. One route is the bottom up, where the defect motifs are taken as models for the pores and their influence is seen as a perturbation to the electronic structure of the underlying graphene lattice. The other approach is to directly assess the electronic structure of the Schwarzite models.

4.4.3.1 Electronic Structure of the Defect Motifs

To examine the role of the non-hexagonal rings for the electronic structure of nanoporous carbon, a graphene sheet is taken as the starting point and defect motifs are introduced to determine the effects on the electronic structure. The extraordinary band structure of graphene is very well known as the "Quantum Electrodynamics in a pencil trace" [80]. Wallace was first to investigate graphene by a tight-binding approach and discovered that graphene is a zero-gap semiconductor [109]. The valence band is derived from the p_z-electrons that form the π and π^* bands. They have a linear dispersion around the Fermi level and they meet exactly at the K-point of the Brillouin zone. The electrons in perfect graphene therefore behave as massless, relativistic particles, but move with the Fermi velocity. However, these peculiar properties are very sensitive to quality of the hexagonal lattice. Recent investigations of graphene grown on SiC have shown that even small perturbations of the lattice can disrupt the linear dispersion and open up a band gap at the Fermi level [81]. This suggests that the pores in NPC would perturb the π-band, leading to the opening of a band gap. Based on these observations NPC could naively be expected to be semiconductors, where the size of the band gap would be proportional to the amount of disorder and the density of voids in the structure.

However, STM investigations of sputtered graphite have shown an increased signal around vacancies indicating that the density of states around voids in the lattice

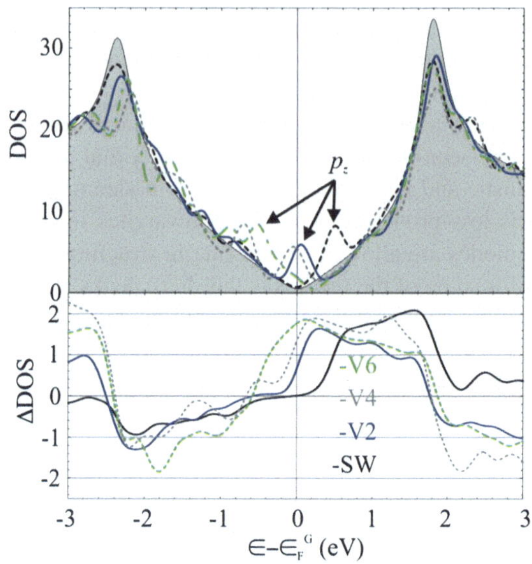

Fig. 4.18 *Upper panel*: Density of states (DOS) for a graphene sheet with a SW-defect (*black line*), divacancy V2, (*blue line*), four atom vacancy V4, (*grey dashed line*) and six atom vacancy V6, (*green dashed-dot line*). The *gray shaded regions* show the DOS for perfect graphene and the energy scale is set to zero at ϵ_F^G, which is indicated by the *vertical line*. The *arrow marked* p_z shows the position of the p_z-state for the vacancies. *Lower panel*: the integrated difference DOS (ΔDOS) for the same defects

is actually higher than on the defect-free regions [30]. This has been supported by DFT calculations that have shown that the defects in graphene lead to additional localized defect states [11]. These defect states are visible as narrow peaks around the Fermi level in the density of states (DOS) in Fig. 4.18. These defect states are the result of the perturbation of the π-band by the voids in the lattice. The p_z-orbitals located on the atoms belonging to the defect motifs are displaced compared to the hexagonal lattice, leading to new combinations of the p_z-orbitals. This breaks up the π-band, such that a local band gap occur and the defect states, indicated by the arrow marked p_z in Fig. 4.18, are formed having an energy in this band gap. The fact that the energy of the defect state occurs in the band gap also implies that it is forbidden to enter the normal lattice, such that it is semi-localized around the vacancies.

The occurrence of these defect states can also be explained at a more qualitative level by an electronic redistribution between the different rings that constitute the defect motif. The pentagons act as acceptors in the hexagonal lattice as they have a propensity to attract one electron to fulfill Hückel's aromatic rule that states that rings containing $4n + 2$ electrons are particularly stable. This has been confirmed by electronic structure calculations for a nanocone. These DFT calculations showed that the isolated pentagon at the apex of the nanocone induces an acceptor states just below the Fermi level [15]. Heptagons can similarly be classified as donors as they have one electron in excess of the prescription of the aromatic rule. The defect state is the result of a partial electron transfer of the excess electrons at the large ring to the surrounding pentagons. This leads to an increased electron density at the common bond between the large ring and the pentagon.

These defect states can attract or donate electrons to the surrounding graphene lattice, such that the defect may become charged. This can be investigated by calculating the difference between the DOS in the defect system and a defect-free graphene sheet. The difference DOS (ΔDOS) can furthermore be integrated to count how many additional states are occupied in the defect system. Such an integrated ΔDOS is shown in the lower panel of Fig. 4.18 and it reveals that the number of occupied states is increasing in the graphene sheet containing defect motifs derived from vacancies. Hückel calculations have also shown that an extra π-electron is located at a single vacancy [37]. The charge state of a defect can also be determined by a charge state analysis where the formation energy in Eq. 4.11 is minimized with respect to the local charge q and as function of the chemical potential of the electrons, $\mu_e = \epsilon_{\mathrm{F}}$ i.e. the Fermi level in the system. This requires performing a set of calculations for the defect model with varying number of electrons in the supercell. However, the electrostatic energy diverges if the net charge of the supercell is not exactly zero so it is necessary to compensate the additional electrons by a compensating background. The simplest approach is to smear out the compensation charge as a background jellium, which is perfectly legitimate when studying point defects in bulk semiconductors. The introduction of the background jellium unfortunately leads to unphysical electrostatic contributions to the total energy, due to the interaction between the localized charge at the defect and the background jellium. A correction term has been proposed to calculate the quadrupole contribution due to the point charge immersed in the background jellium [68]. However, this simple isotropic quadrupole correction term is less accurate for low-dimensional systems, such as graphene and even more pronounced for one-dimensional systems such as nanotubes. A charged graphene slab calculated in a supercell with a jellium background is instead better described by a plate capacitor model since the slab resembles a charged plate immersed in a charged dielectric [93].

Such a charge state analysis for the point defects in a NPC sheet is shown in Fig. 4.19 and it reveals that the SW-defect would remain neutral, while the vacancies would attract electrons to become negatively charged in agreement with the analysis of the ΔDOS in Fig. 4.18. The charge accumulation at the defect motifs that are incorporated in the NPC sheets suggests that the reactivity towards electro negative molecules such as oxygen would increase. The motifs are also more receptive to adsorption because the strained bonds in the vacancy ring are weaker than the bonds in the basal plane. The reactivity of the defect motifs will be further discussed in Section 4.5.3.

4.4.3.2 Electronic Structure of Schwarzites

The previous section showed that introducing isolated pentagons or genus conserving defects leads to semi-localized defect states that may become charged. This

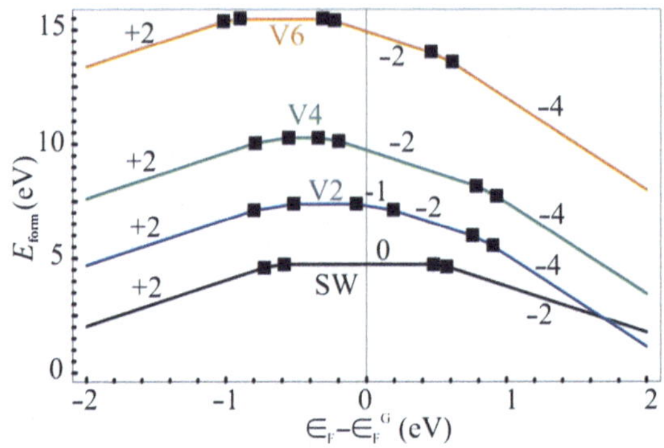

Fig. 4.19 A charge state analysis for the even defects: Formation energy $E_{\text{form}}(\epsilon_F)$ for the even defects as function of the Fermi level ϵ_F in the system, which is given relative to perfect graphene, ϵ_F^G. The numbers indicate the charge state and the *black dots* indicate the position of the charge transfer levels. The intrinsic charge state for the defects is given by the value at ϵ_F^G. Reprinted with permission from Ref [11]. Copyright (2006) by the American Physical Society

suggests that increasing the concentration of the defect motifs to complete coverage could lead to metallization. This has also been observed for the Haeckelites, which have a complete coverage of pentagons and heptagons. The Haeckelites have been predicted by DFT calculations to be metallic [87].

The Schwarzites have larger non-hexagonal rings in the structure that have more donor character. It could be expected that the Schwarzites would be metallized, but tight-binding and DFT calculations have shown that a majority of the studied Schwarzites are semiconductors [107]. In particular the three structures with largest cohesive energies have a significant band gap as can be seen in Fig. 4.20. The authors also noted the structure relaxation enhanced the size of the band gap, indicating that the optimal Schwarzite structures tend to be semiconductors, and that the opening of the band gap was not an artifact of the generation method. A closer look at the density of states (DOS) reveal a more complex spectrum with narrow peaks and deep valleys in between. This indicates that the curvature breaks up the graphene bands into sub bands with less dispersion and leaving band gaps in between the bands.

Comparing the electronic structure of the Schwarzites and the effects of individual defect motifs in NPC suggests that the chemical properties of NPC may have two origins. The molecular sieving ability of NPC may be connected to the defect-free pores that only have one type of non-hexagonal rings. These pores lead to geometrical constrictions that give rise to the mechanical size selectivity towards penetration of gas phase molecules. The chemical activity of NPC may rather be connected to the defect motifs incorporated into the pore walls, since these motifs are able to capture electrons into the defect states leading to local charging. This suggests that the active sites for dissociative molecule adsorption are vacancy induced defect motifs. The chemical properties of NPC will be discussed further in Section 4.5.

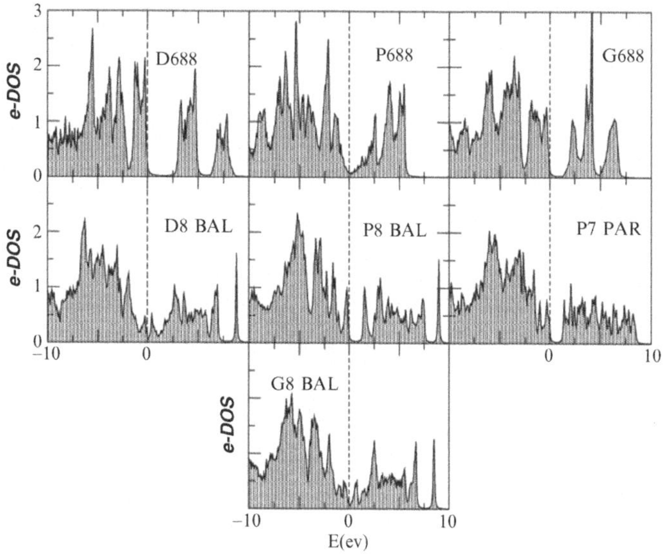

Fig. 4.20 Electronic density of states (DOS) for seven Schwarzite models for NPC calculated by the tight-binding method. The Fermi level is the zero point for the energy scale. Reprinted with permission from Ref. [107]. Copyright 2003 IOP Publishing

4.5 Chemical Properties of NPC

The porosity, large surface area and increased reactivity for chemical reactions makes nanoporous carbons distinctly different from the defect-free, crystalline forms of sp^2-bonded carbon, such as graphite. There are a number of reasons for the increased activity of NPC, such as confinement effects in the pores, strain due to inclusion of non-hexagonal rings in the structure and electronic effects due to defect states. This section will investigate some of these effects by focusing on three important reactions, the graphitization path of a Schwarzite like NPC model, gas absorption in a Schwarzite model and oxidation of defect motifs on a NPC sheet. There has also been theoretical studies of the reactivity of pores and vacancies in graphene for water dissociation [53] and methane decomposition [40].

4.5.1 Graphitization of NPC

The stability of the porous structure is an important question for the use of NPC as molecular sieves or as catalyst. Experiments discussed in Section 4.2.1 indicate that the pore structure is diminishing in NPC, if the pyrolysis temperature is above 1,000°C. However, relatively little is known about the details of this graphitization process. Margine et al. used Tight-binding and DFT calculations together with the Nudged elastic band (NEB) method to follow the graphitization process of a Schwarzite model, nicknamed the "Wormhole" [70]. The Wormhole model consisted of two parallel graphene sheets connected by a pore. The pore contained 12 heptagons along the neck, giving a 5 Å pore width as shown as structure A in Fig. 4.21.

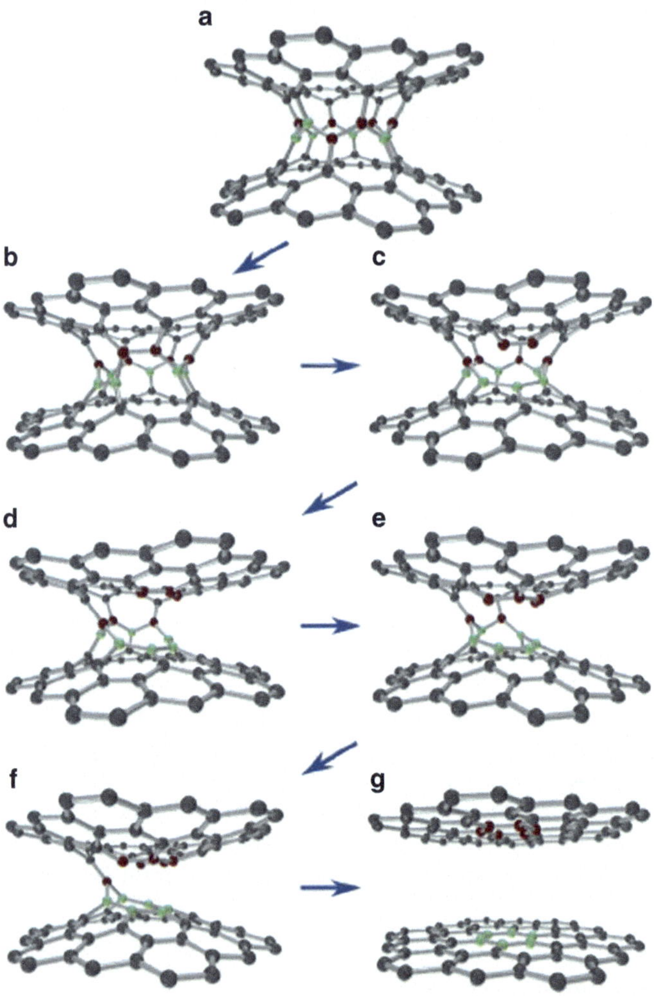

Fig. 4.21 The reaction path for the graphitization process of the Wormhole model of NPC. The starting configuration is the Wormhole (structure A). The first step is a SW-transformation that transforms four heptagons into two octagons and two hexagons (structure B). The next step is breaking of a bond in the octagon that initiates the separation of the sheets. The bond breaking process proceeds until the two sheets are completely decoupled into two parallel graphene sheets (structure G). Reprinted with permission from Ref. [70]. Copyright (2007) by the American Physical Society

The graphitization process was divided into five intermediate steps between the starting configuration of the Wormhole (structure A) and the final structure of two parallel graphene sheets (structure G). The full graphitization process released 19 eV elastic energy and 5 eV of interlayer binding energy as the height of the pore impose to a larger layer separation than normal graphite stacking. The first step along the graphitization path was the SW-transformation at a junction between four heptagons that produces two octagons and two hexagons (structure B). The second step was breaking one of the bonds in one of the octagons to initiate the separation process. The resulting structure C has the largest energy along the graphitization path. These two steps destabilize the pore structure so much that the graphitization process then continues automatically. Calculations of the reaction barriers for the first two steps revealed that the SW-transformation has the largest barrier. This indicates that the Wormhole model would suffer from brittle failure as the first symmetry breaking SW-transformation is the rate-limiting step. The value for the barrier for the SW-transformation was calculated to be 5.73 eV that is much lower then in flat graphene, but still a very large barrier that would explain the non-graphitizing character of NPC.

This also suggests that there might be a fundamental difference between the genus preserving defect motifs derived from vacancies and isolated non-hexagonal rings that induce curvature in the lattice. The genus preserving defects might be easier to anneal out, if atom diffusion is sufficiently fast. The motifs need only to capture the appropriate number of carbon atoms to heal the vacancy. The non-genus preserving defects like isolated non-hexagonal rings on the contrary need to rebuild the surrounding lattice via SW-transformations to change the symmetry.

An interesting experimental observation is that the addition of alkali metal atoms to NPC lowers the graphitization temperature down to 50°C [99]. Margine et al. studied the effect on the barrier for SW-transformation due to alkali atoms located next to a heptagon [70]. The presence of alkali atoms did lower the barrier slightly, but not to the extent required to explain the graphitization of NPC just above room temperature. The authors suggested that a co-initiator would be needed in addition to the alkali metal atoms to explain the low graphitization temperature observed in the experiments. Oxygen is an example of a very strong agent for manipulation of the structure and it would be interesting to investigate the interaction between molecules and NPC, which is the subject of the next two sections.

4.5.2 *Molecular Adsorption in NPC*

Molecular sieves is one of the most mature applications of NPC, where the fine porosity is used to separate out gas molecules from a gas mixture [49]. Jiang and Sandler have used Monte Carlo simulations in the NVT [45] and NPT [46] ensembles to investigate the absorption of oxygen and nitrogen in the C_{168} Schwarzite model for NPC proposed in Ref. [108]. The Monte Carlo protocol involved five type of moves: displacement, rotation, creation, depletion, and identity exchange where one O_2 and N_2 molecule swap places. The acceptance of a move was determined by the Metropolis criterion, where the total energy of the adsorbate system was calculated by summing up the interaction energies between the gas molecules and the NPC described by a Lennard–Jones potential.

Figure 4.22a and b shows how the gas phase molecules fill up the pores. The open regions in the structure have a continuous filling, while the narrow constructions lead to localization of the gas phase molecules into pockets. The simulations furthermore revealed that there are two pressure regimes with different behavior. The adsorption energy determines the adsorption preference below the critical pressure $p_c = 0.5$ MPa. The N_2 molecule has a larger adsorption energy in this regime,

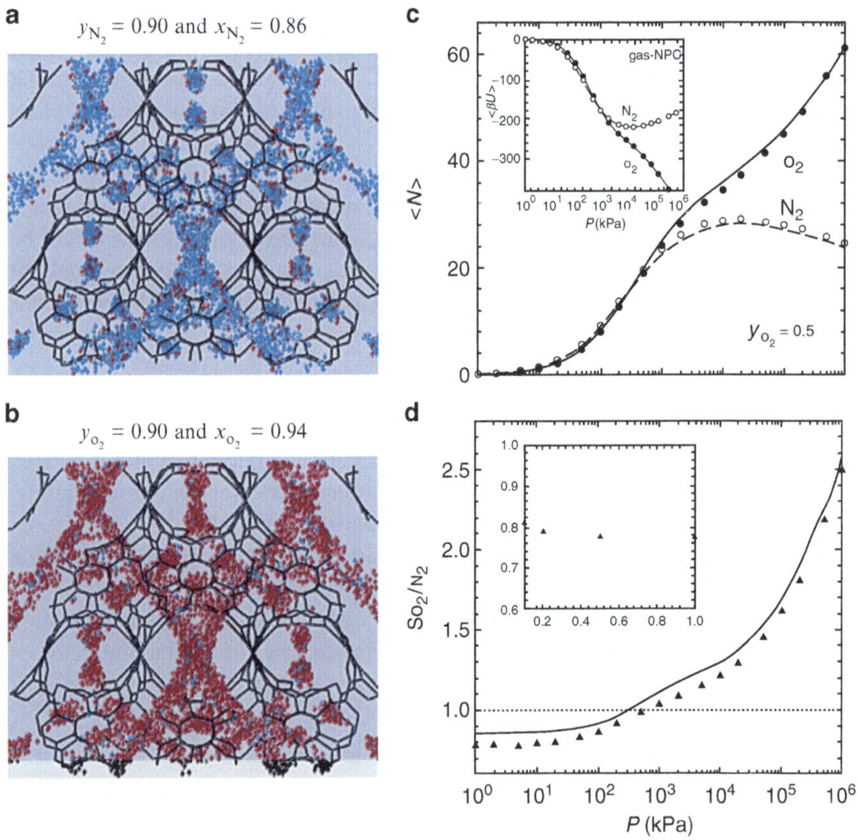

Fig. 4.22 Center of mass density distribution of O_2 (*red dots*) and N_2 (*blue dots*) molecules at $T = 300$ K and $p = 10^5$ Pa in the C_{168} Schwarzite model of NPC. (**a**) Nitrogen to oxygen fraction $y_{N_2} = 0.9$ and $y_{O_2} = 0.1$. (**b**) Nitrogen to oxygen fraction $y_{N_2} = 0.1$ and $y_{O_2} = 0.9$. (**c**) The amount of adsorbed O_2 and N_2 molecules as function of pressure. At low pressures is the adsorption comparable, while the size effect is dominating at high pressure leading to larger adsorption of O_2 than N_2. The *inset* shows the adsorption for O_2 and N_2 in NPC as function of partial pressure. (**d**) shows the selectivity that increases with pressure. Reprinted with permission from Ref. [46]. Copyright 2003 American Chemical Society

such that the adsorbed amount of N_2 molecules is slightly larger than the adsorbed amount of O_2 molecules. The selectivity S_{O_2/N_2} in Fig. 4.22d is consequently less than one. The molecular size on the contrary is the determining factor above the critical pressure. The cross section for the O_2 molecule, $\sigma_{O-O} = 2.99$ Å is smaller than the N_2 $\sigma_{N-N} = 3.32$ Å such that the adsorption of the O_2 molecule increases more strongly than the N_2 molecule as shown in Fig. 4.22c. The selectivity value S_{O_2/N_2} in Fig. 4.22d increases significantly to be more than a factor of two at pressures above 0.1 GPa.

These simulations indicated that the gas separation is controlled by non-dissociative physisorption of the gas molecules in NPC that occurs in between the defect-free pore walls. However, Section 4.3.5 showed that there are defect motifs in the pore walls and Section 4.4.3.1 revealed that these motifs may have a higher reactivity than the pore wall. The motifs can be able to dissociate adsorbed molecules and the process of dissociative adsorption of molecules at defect motifs in NPC is the subject of the next section. The section is focusing on adsorption of oxygen in NPC that leads to oxidation.

4.5.3 Oxidation of NPC

Oxidation is one of the fundamental reactions for carbon materials in general and in particular for NPC. Uncontrolled oxidation lead to combustion, but mild oxidation can add functional groups to the pores that may enhance the properties for particular applications. This indicates that it is very important to control the oxidation process, but it is still not satisfactory understood as it is a very complex process and the experimental results are sometimes contradictory.

Oxidation of graphite has been studied in great detail and scanning tunneling microscopy (STM) investigations have shown that the defect-free regions of the basal plane are quite inert towards molecular oxygen below a critical temperature of 875 K [31]. Molecular oxygen only etches vacancy motifs that are already present on the surface below the critical temperature. Comparing to NPC suggests that the vacancy motifs incorporated into the walls of NPC are much more reactive towards O_2 than the defect-free regions of the pores. Deeper insight into the individual reaction steps at the vacancy motifs is unfortunately very difficult to obtain inside a NPC. It is already difficult to follow the oxidation process at the atomic level on graphite as the STM is unable to distinguish between carbon and oxygen atoms attached to the carbon surface. However, other information about the oxidation process can be obtained by macroscopic methods such as Temperature programmed desorption (TPD) experiments. Such experiments confirm the existence of oxygen (O)-functional groups attached to the carbon surface at a variety of oxidized carbon materials, such as graphite [69], coal char[32], activated carbon [23] and carbon fibers [113]. The O-groups are also identified by other methods such as Fourier transform infrared red (FT-IR) spectroscopy [21] and x-ray photoelectron spectroscopy (XPS) [51]. However, a surprising observation is that the desorption spectrum in the TPD experiments is dominated by CO rather than CO_2. One would expect to observe CO_2 as it is the final product of complete combustion at the relatively low temperatures used in the oxidation experiment, while CO is normally the product of incomplete combustion at high temperatures.

The initial theoretical efforts regarding oxidation of carbon materials focused on the oxidation of the edges of graphene flakes [5, 16, 38, 42, 89, 95]. These investigations led to a model for the oxidation process that might be applicable to incomplete combustion at high temperatures [16]. The model consists of dissociative adsorption of O_2 and direct desorption of CO. However, this model is probably not appropriate to describe the oxidation process at low temperatures. Experiments using different oxygen isotopes in two subsequent oxidation experiments have revealed that the resulting CO_2 contains a complete mixture of the oxygen isotopes [69]. This indicates that the precursor for the formation of a CO_2 molecule on the surface is interacting with the molecules in the gas phase. This suggests that the CO_2 molecules are generated in a sequential process where more than one oxygen molecules take part.

The oxidation of defect motifs in graphene has only recently received significant attention [2, 13, 65]. Recent work presented in the next Section 4.5.3.1 has led to the formulation of a two-step mechanism for low temperature oxidation of defect motifs in graphene that is able to explain the experimental observations of the molecular interaction between the oxygen molecules gas phase and the adsorbed species [13].

4.5.3.1 Ab Initio Atomistic Thermodynamics for Gas Adsorption

The model for low temperature oxidation is based on a study of the elementary processes of adsorption of O_2 and desorption of CO and CO_2 at the defect motifs using the ab initio atomistic thermodynamics method (AIATD) that was introduced in Section 4.3.5.1. The free energy for a solid was discussed in Section 4.3.5.1 and the free energy for a gas molecule is derived here. Small gas molecules like O_2 or CO_2 behave more or less as an ideal gas, such that the fugacity is very close to one. The Gibbs free energy for the molecules (Eq. 4.7) can then be derived from the expressions for an ideal gas with a good accuracy. The derivation contains three steps: The internal energy U is calculated by DFT for the free molecule at T = 0K. The thermal contribution to the free energy at a particular reference pressure p_0, is calculated via the partition function for an ideal gas, $F_{\text{gas}}(T, p_0) = -k_B T ln(Z_{\text{gas}})$. The partition function can be derived from the individual contributions per molecule, $Z_{\text{gas}} = Z_{\text{trans}} Z_{\text{rot}} Z_{\text{vib}} Z_{\text{e}} Z_{\text{n}}$, where Z_{trans} is the translation, Z_{rot} is the rotation and Z_{vib} is the vibrational partition function. The HOMO-LUMO gap is usually larger than 1 eV, such that no electronic excitations occur so the electronic partition function can be taken as unity $Z_e = 1$. The probability for nuclear excitations is also negligible, such that the nuclear partition function is similarly approximated as unity $Z_n = 1$. This gives the three thermal contributions to the free energy as:

$$F^{\text{trans}}(T, p_0) = -k_B T ln\left[\frac{k_B T}{p_0}\left(\frac{m k_B T}{2\pi\hbar}\right)^{\frac{3}{2}}\right] \tag{4.19}$$

$$F^{\mathrm{rot}}(T) = -k_B T ln\left(\frac{2Ik_BT}{\sigma\hbar}\right) \tag{4.20}$$

$$F^{\mathrm{vib}}(T) = -k_B T ln\left(\frac{e^{\frac{\hbar\omega}{2k_BT}}}{1-e^{\frac{\hbar\omega}{k_BT}}}\right) \tag{4.21}$$

The chemical potential is furthermore the free energy per molecule, $\mu(T, p) = \frac{\partial G}{\partial n}$, and the thermal contributions to chemical potential are given as $\mu^{trans}(T, p_0) = F^{trans}(T, p_0)$ and similarly for the rotational and vibrational contributions.

The pressure dependence of the free energy is finally derived from the equation of state for the ideal gas.

$$\left(\frac{\partial G}{\partial p}\right)_T = V = \frac{nk_BT}{p} \tag{4.22}$$

Integrating the differential with respect to pressure gives the free energy change at an arbitrary pressure with respect to the reference pressure, p_0.

$$G(T, p) - G(T, p_0) = \int_{p_0}^{p}\left(\frac{\partial G}{\partial p}\right)_T dp = nk_BT ln\left(\frac{p}{p_0}\right) \tag{4.23}$$

Dividing by the number of molecules, n, gives the pressure contribution to the chemical potential. Summing up the contributions gives the free energy of the molecule. Taking oxygen as example gives the chemical potential for an O-atom in the O_2 molecule:

$$\begin{aligned}\mu_{\mathrm{O}}(T, p) &= \frac{1}{2}\mu_{\mathrm{O}_2}(T, p)\\ &= \frac{1}{2}\Big[E^{DFT}[\mathrm{O}_2] + k_BT ln\left(\frac{p}{p_0}\right)\\ &\quad + \left(\mu_{\mathrm{O}}^{\mathrm{trans}}(T, p_0) + \mu_{\mathrm{O}}^{\mathrm{rot}}(T) + \mu_{\mathrm{O}}^{\mathrm{vib}}(T)\right)\Big]\end{aligned} \tag{4.24}$$

The AIATD method can be used to calculate the free energy of formation ΔG_{form} of defect motifs in equilibrium with a gas phase reservoir of molecular oxygen by inserting the expression for the free energy of the solid in Eq. 4.9 and the gas phase molecules in Eq. 4.24 into Eq. 4.10. The resulting expression for the free energy of formation as a function of T and p:

$$\begin{aligned}\Delta G_{\mathrm{form}}(n_{\mathrm{C}}, n_{\mathrm{O}}, T, p) = E^{\mathrm{motif}}(n_{\mathrm{C}}, n_{\mathrm{O}}) - n_{\mathrm{O}}\mu_{\mathrm{O}}(T, p) - (E^{\mathrm{G}} - n_{\mathrm{C}}\mu_{\mathrm{C}})\\ + \Delta F^{\mathrm{vib}}(T) - k_{\mathrm{B}}T \ln g^{\mathrm{config}}\end{aligned} \tag{4.25}$$

where E^{G} and E^{motif} refer to the total energy of a graphene sheet before and after the oxidized vacancy motif is created, n_{C} denotes the number of C-atoms that have been exchanged with the bulk reservoir μ_{C}, and n_{O} corresponds to the number of O-atoms having the chemical potential μ_O determined by the gas phase molecules. $\Delta F^{\mathrm{vib}}(T)$ is the additional vibrational free energy of the vacancies. In principle, a full phonon calculation is required for each system, but as the change in vibrational free energy at the oxidized motifs ΔF^{vib} is mainly due to the adsorption of oxygen, most other contribution will cancel. It is therefore sufficient to include only the bending and stretch modes of the functional O-groups attached to the vacancy motifs. Finally, g^{config} is the configurational degeneracy of the vacancies, due to different ways that molecules can attach to the vacancy. The number of possible configurations can be obtained by exact enumeration of all possible orientations of the molecules at the different adsorption sites. At typical reaction temperatures above 900 K, these vibrational and configurational contributions amount to several eV, so they are not negligible.

4.5.3.2 Oxidation of Defect Motifs in NPC

Applying the AIATD method to the adsorption of oxygen at the defect motifs [13] showed that the dissociative adsorption of O_2 is favored at the pentagons in the defect motifs (Route I in Fig. 4.23) compared to on the basal plane of the graphene (Route G in Fig. 4.23). The O-atoms saturate the pentagons forming one ether (C-O-C) group per pentagon. The adsorption energy is strongly exothermal indicating that the process is favored and would quickly proceed to full saturation of the available defect motifs. The strong reactivity of the bare defect motifs can be explained by the fact that the introduction of the non-hexagonal rings induces strain as they do not fit perfectly into the lattice. The defect motifs also lead to defect states as mentioned in Section 4.4.3. These defect states may become charged which increases the electron density locally. This can provide additional electrons that are available to interact with adsorbing molecules. The ether groups formed by oxygen adsorption at the defect motifs are very stable. The calculated desorption energy for breaking up an ether group to release a CO molecule is of the order 6 eV, which indicates that temperatures above 1,500 K are needed before the ether groups are decomposing [13]. The ether groups can be expected to remain at the defect motifs once formed and the saturation of the defect motifs by ether group is defined as the first step in the oxidation of NPC.

The defect motifs saturated by ether groups may appear to be inert, but the pentagons with an ether group provide an additional adsorption site for O_2 (Route II in Fig. 4.23) [13]. The upper panel in Fig. 4.23 shows that this adsorption site is less favorable than the clean pentagon, but significantly more attractive compared to the basal plane. The adsorption of O_2 next to the ether group generates a lactone (C-O-C=0) group. Subsequent adsorption of an additional O_2 molecule next to the lactone group forms an anhydride (O=C-O-C=O) group. The lactone and anhydride groups can decompose into CO_2 and the desorption energy is much less compared to the CO desorption from the ether group. This indicates that the low temperature oxidation has a second step where the ether groups get activated by an additional O_2 adsorption. This forms the lactone and even anhydride groups that are decomposing by releasing CO_2.

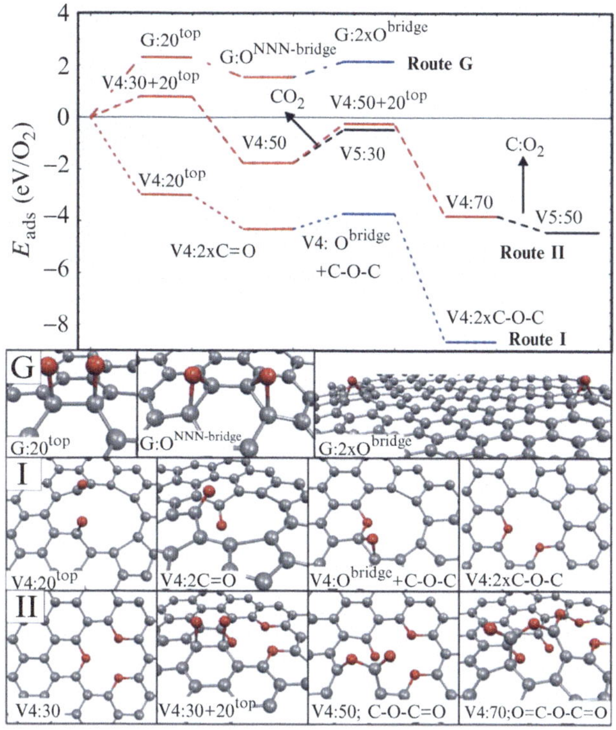

Fig. 4.23 *Upper panel*: The adsorption energy for dissociative O_2 adsorption on the perfect basal plane of graphene (route G), at a bare V4-vacancy (route I) and at an oxygen saturated V4-vacancy (route II). *Red lines* indicate O_2 adsorption, *blue lines* O-atom diffusion and *black lines* CO_2 desorption. *Lower panels*: The corresponding geometries of the stable intermediate adsorption structures for O_2 dissociation on the perfect basal plane of graphene, route G, at a bare V4-vacancy (route I) and at an oxygen saturated V4-vacancy (route II). Adopted from Ref. [13]. Copyright (2009) by the American Physical Society

The chemical potential of oxygen μ_O, can also be explicitly connected to a temperature and pressure scale using Eq. 4.24. This provides the possibility to generate a phase diagram for an oxidized vacancy motif in NPC in by minimizing the formation energy calculated by Eq. 4.25 as function of temperature and pressure. Such a phase diagram for a four atom vacancy V4, is shown in Fig. 4.24. This phase diagram confirms the conclusions from the adsorption process in Fig. 4.23 and it indicates that the V4 vacancy would remain without any O-groups only at extremely low O-pressures and high temperatures. Any higher O-pressure and/or lower temperature lead to a progressive saturation of the vacancy motif until the V4:3O configuration, characterized by three pentagons filled by one ether group each, is formed. Correlating the formation energy of the oxygen functional groups with an analysis of Fig. 4.24 reveals that increasing the oxygen concentration beyond the V4:3O results in the formation of oxygen complexes containing anhydride and ether groups.

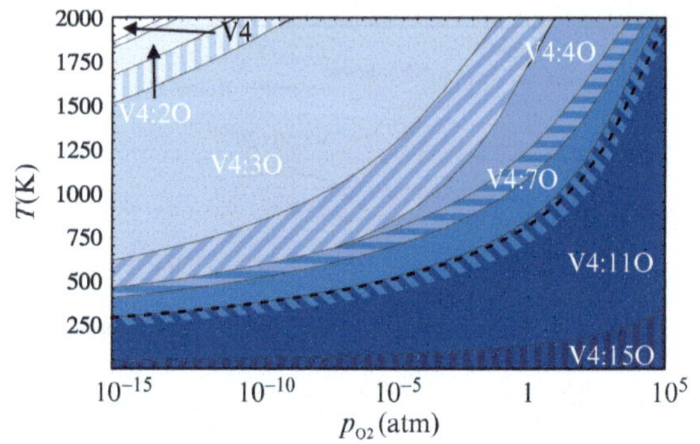

Fig. 4.24 A phase diagram for a four atom vacancy, V4 in graphene, showing how the oxygen concentration changes as a function of temperature, T, and pressure, p. V4:nO denotes how many O-atoms are adsorbed at the vacancy. ΔG_{form} is negative below the *dotted black line*. The hashed regions indicate coexistence regions, where the energy difference between two configurations is below the numerical resolution. Reprinted with permission from Ref. [13]. Copyright (2009) by the American Physical Society

Adding up the results for the adsorption and desorption processes suggests that the low temperature oxidation of NPC sheets proceed by a two-step mechanism. The first step is a fast saturation of the bare defect motifs by attachment of O-atoms to the pentagons that form ether groups. The ether groups are very stable, such that they are not able to desorb at the temperatures used for low temperature oxidation. These ether groups may also explain why NPC is non-graphitizing according to classification of Franklin [24]. The ether groups prevent movable C-atoms to fill in the vacancies in the NPC sheets to form perfect graphite. The combustion of NPC is instead activated in a second step by the adsorption of O_2 at the ether groups that form lactone and anhydride groups that subsequently desorb as CO_2.

4.6 Summary of the Structural and Chemical Properties of Nanoporous Carbon

Summarizing the characterization of nanoporous carbon presented in Sections 4.2 and 4.3 reveals a number of common features in spite of the diversity of starting materials and the variety of methods used to synthesize NPC. There is a general consensus that nanoporous carbon is predominantly sp^2-bonded, such that it forms sheet-like structures similar to graphene. However, there is a varying amount of non-hexagonal rings in the structure, which sets NPC apart from graphite. The origin of these non-hexagonal rings appears to be two-fold. The synthesis simulations in Section 4.3.3 suggests that the isolated non-hexagonal rings are the result of growth mismatch, when smaller graphene fragments merge into larger sheets. The larger, pore-like genus preserving combinations of non-hexagonal rings appear due to multi-atom vacancies according to the defect motif investigations in Section 4.3.5. The exact relationship between the non-hexagonal rings and the porosity and chemical reactivity remains debated. It is clear that isolated pentagons lead to positive curvature and that there are also larger rings such as heptagons that induce negative curvature. Perfect ordering of the non-hexagonal rings would form Schwarzites discussed in Section 4.3.4, but such structures have not yet been observed experimentally. The templating method discussed in Section 4.2.2 appears to be the most promising method to generate Schwarzite like structures, but the Schwarzites are still idealized models until the templating procedure succeeds in generating them. The NPC materials synthesized up to now by pyrolysis (Section 4.2.1) or the arc discharge method (Section 4.2.3) instead appear to consist of more or less randomly oriented defective and buckled graphene sheets. The individual sheets have been characterized as fullerene fragments [34, 35], where the number of pentagons dominated, but negatively curved sheets have also been observed [9]. The buckling of the sheets furthermore prevents an ordered stacking of the sheets like in graphite, such that the radial correlation function is loosing any structure beyond a few nearest neighbor distances. The typical (0002) peak in graphite is absent in NPC and the first "interlayer peak" is often measured around 5 Å a value similar to measured pore widths. This suggests that the porosity measured via methods like BET, reflects the distance between parallel walls of NPC sheets.

Nanoporous carbon is surprisingly stable even though it is a metastable structure. The NPC materials can be heated to temperatures above 2,500°C without obtaining a perfect graphite structure [34], such that NPC can be classified as a non-graphitizing carbon material. Some insight into the resistance to graphitization has been obtained by theoretical investigations of Schwarzites. Sections 4.3.4 and 4.4.1 showed that Schwarzites have cohesive energies comparable to fullerene crystals and positive shear modulus, such that they are local minima that are stable with respect to small distortions indicating that they are metastable structures. A vital step in the graphitization process is to remove the non-hexagonal rings in the structure. Ab initio calculations of the kinetics of the graphitization process showed that the rate limiting step for the process is the SW-transformation that reduces the size of one of the non-hexagonal rings at the neck of the pore [70]. The barrier for the SW-transformation is still very high in Schwarzites, which can explain the resistance towards graphitization. Section 4.5.3 suggested that also foreign atoms, such as oxygen bonded into ether groups, can be an additional obstacle for graphitization. The ether groups are very stable so they remain at defect motifs even when NPC is heated to high temperatures and the ether groups prevent the recombination of movable carbon atoms into vacancies.

The variety of chemical properties of nanoporous carbon can finally be explained by the structural diversity inside the NPC material. Section 4.5.2 indicated that the pores formed between adjacent NPC layers are responsible for the molecular sieving effects as they provide constrictions for penetrating molecules. These pores do adsorb the molecules, but they do not dissociate the molecules. There are instead strong indications that molecule dissociation only takes place at the defect motifs, not just at arbitrarily positions in between the pore walls. Section 4.5.3 showed that the chemical activity is much higher at the defect motifs that are incorporated into the walls of the NPC sheets. This is due to the strain at the defect motifs and the charge accumulation into the defect states that provide a much higher reactivity compared to the defect-free regions of the NPC sheets. It seems that the varying facets of the structure of NPC have different functions, which explains the span of applications that is provided by this fascinating material.

Acknowledgments The author would like to thank M. Scheffler, R. Schlögl, F. Hanke, S. Linic, D. Rosenthal and D. Su for fruitful discussions about different aspects of carbon materials during my employment at the Fritz-Haber-Institut.

References

1. Acharya M, Strano MS, Mathews JP, Billinge SJL, Petkov V, Subramoney S, Foley HC (1999) Phil Mag B 79:1499
2. Allouche A, Ferro Y (2006) Carbon 44:3320
3. Amorim RG, Fazzio A, Antonelli A, Novaes FD, and da Silva AJR (2007) Nano Lett 7:2459
4. Ashcroft NW, Mermin ND (1976) Solid state physics. Saunders College Publishing, Forth Worth
5. Barco G, Maranzana A, Ghico G, Causa M, Tonachini G (2006) J Chem Phys 125:194706
6. Baroni S, Gianozzi P, Testa A (1987) Phys Rev Lett 58:1861
7. Beck RD, Rockenberger J, Weis P, Kappes MM (1995) J Chem Phys 104:3638
8. Belz T, Schlögl R (1996) Synth Met 77:223
9. Bourgeois LN, Bursill LA (1997) Phil Mag A 76:753.
10. Brunauer S, Emmett RH, Teller EJ (1938) J Am Chem Soc 60:309
11. Carlsson JM, Scheffler M (2006) Phys Rev Lett 96:046806
12. Carlsson JM, Phys Stat Sol B (2006) 243:3452
13. Carlsson JM, Hanke F, Linic S, Scheffler M (2009) Phys Rev Lett 102:166104
14. Cerius^2 Modeling Environment. Molecular simulations, San Diego, USA
15. Charlier J-C, Rignanese G-M (2001) Phys Rev Lett 86:5970
16. Chen N, Yang RT (1998) J Phys Chem A 102:6348
17. Diederich L, Barborini E, Piseri P, Podest A, Milani P, Schneuwly A, Gallay R (1999) Appl Phys Lett 75:2662
18. Donadio D, Colombo L, Milani P, Benedek G (1999) Phys Rev Lett 83:776
19. Donadio D, Colombo L, Benedek G (2004) Phys Rev B 70:195419
20. Ebbesen TW, Ajayen PM (1992) Nature 358:220
21. Fanning PE, Vannice MA (1993) Carbon 31:721
22. Fasolino A, Los J, Katsnelson (2007) M Nat Mat 6:858
23. Figueiredo JL, Pereira MFR, Freitas MMA, Orfao JJM (1999) Carbon 37:1379
24. Franklin RE (1951) Proc Royal Soc A 209:196
25. Frenkel D, Smit B (2002) Understanding molecular simulation, 2nd edn. Academic, San Diego
26. Gianozzi P, de Gironcoli S, Pavone P, Baroni S (1991) Phys Rev B 43:72321
27. Grünbaum B, Shephard GC (1986) Tilings and patterns. W. H. Freeman and Company, New York
28. Haeckel E (2005) Art forms from the Ocean: the radiolarian atlas of 1862. Prestel, München, ISBN 3-7913-3327-5
29. Haffenden G (2009) First principles calculations of the spectroscopic and thermal properties of graphitic materials, PhD thesis, University of Sussex, Falmer, UK
30. Hahn JR, Kang H, Song S, Jeon IC (1996) Phys Rev B 53 R1725

31. Hahn JR (2005) Carbon 43:1506
32. Hall PJ, Calo JM (1989) Energy Fuel 3:379
33. Han B-H, Zhou W, Sayari S (2003) J Am Chem Soc 125:3444
34. Harris PJF, Tsang SC (1997) Phil Mag A 76:667
35. Harris PJF, Burian A, Duber S (2000) Phil Mag Lett 80:381
36. de Heer W, Urgate D (1993) Chem Phys Lett 207:480
37. Hjort M, Stafström S (2000) Phys Rev B 61:14089
38. Hod O, Barone V, Peralta JE, Scuseria GE (2007) Nano Lett 7:2295
39. Hohenberg P, Kohn W (1964) Phys Rev 136:B 864
40. Huang L, Santiso EE, Nardelli MB, Gubbins KE (2008) J Chem Phys 128:214702
41. Iijima S (1991) Nature 354:56
42. Incze A, Pasturel A, Chatillon C (2003) Surf Sci 537:55
43. Janes A, Kurig H, Lust E (2007) Carbon 45:1226
44. Janes A, Thomberg T, Lust E (2007) Carbon 45:2717
45. Jiang J, Klaudia JB, Sandler SI (2003) Langmuir 19:3512
46. Jiang J, Sandler SI (2003) Langmuir 19:5936
47. Joo SH, Choi SJ, Oh I, Kwak J, Liu Z, Terasaki O, Ryoo R (2001) Nature 412:169
48. Jun S, Chol SJ, Ryoo R, Kruk M, Jaroniec M, Liu Z, Ohsuna T, Terasaki O (2000) J Am Chem Soc 122:10712
49. Kane MS, Kao LC, Mariwala RK, Hilscher DF, Foley H (1996) Ind Eng Chem Res 35:3319
50. Kaxiras E, Bar-Yam Y, Joannopolous JD, Pandey KC (1987) Phys Rev B 35:9625
51. Kelemen SR, Freund H (1988) Energy Fuels 2:111
52. Kohn W, Sham LJ (1965) Phys Rev 140:A 1133
53. Kostov MK, Santiso EE, George AM, Gubbins KE, Nardelli MB (2005) Phys Rev Lett 95:136105
54. Kravchik AE, Kukushkina JA, Sokolov VV, Tereshchenko GF (2006) Carbon 44:3263
55. Kresse G, Furthmüller J, Hafner J (1995) Eur Phys Lett 32:729
56. Krishnan A, Dujardin E, Treacy MMJ, Hugdahl J, Lynum S, Ebbesen TW (1997) Nature 388:451
57. Kroto HW, Heath JR, O'Brien SC, Curl RF, Smalley RE (1985) Nature 318:162
58. Kroto HW, Nature (1987) 329:529
59. Kumar A, Lobo RF, Wagner NJ (2005) Carbon 43:3099
60. Kunc K, Martin RM (1982) Phys Rev Lett 48:406
61. Kyotani T, Nagani T, Inoue S, Tomita A (1997) Chem Mater 9:609
62. Kyutt RN, Smorgonskaya EA, Danishevskii AM, Gordeev SK, Grechinskaya AV (1999) Phys Sol Stat 41:808
63. Landau LD, Lifschitz EM (1980) Statistical physics, Part 1, 3rd edn. Elsevier-Butterworth and Heinemann, Oxford, UK
64. Landau LD, Lifschitz EM (1986) Theory of elasticity, 3rd edn. Elsevier-Butterworth and Heinemann, Oxford, UK
65. Lee SM, Lee YH, Hwang YG, Hahn JR, Kang H (1999) Phys Rev Lett 82:217
66. Lenosky T, Gonze X, Teter M, Elser V (1992) Nature 355:333
67. Mackay AL, Terrones H (1991) Nature 352:762
68. Makov G, Payne MC (1995) Phys Rev B 51:4014
69. Marchon B, Carazza J, Heinemann H, Samorjai G (1988) Carbon 26:507
70. Margine ER, Kolmogorov AN, Stojkovic D, Sofo JO, Crespi VH (2007) Phys Rev B 76:115436
71. Martin CR (1994) Science 266:1961
72. Maultzsch J, Reich S, Thomsen C, Requardt H, Ordejon P (2004) Phys Rev Lett 92:075501
73. Mayer ST, Pekola RW, Kaschmitter JL (1993) J Electrochem Soc 140:446
74. McGreevy RL, Putszai RL (1988) Mol Simul 1:359
75. Mestl G, Maksimova NI, Keller N, Roddatis VV, Schlögl R (2001) Angew Chem Int Ed 40:2066
76. Meyer JC, Geim AK, Katsnelson MI, Novoselov KS, Booth TJ, Roth S (2007) Nature 446:60

77. Milani P, Ferretti M, Piseri P, Bottani CE, Ferrari A, Bassi AL, Guizetti G, Patrini M (1997) J Appl Phys 82:5793
78. Mounet N, Marzari N (2005) Phys Rev B 71:205214
79. Murnaghan FD (1944) Proc Natl Acad Sci USA 30:244
80. Novoselov KS, Geim AK, Morozov SV, Jiang D, Katsnelson MI, Grigorieva IV, Dubonos SV, Firsov AA (2005) Nature 438:197
81. Ohta T, Brunswick A, Seyller T, Horn K, Rothenberg E (2006) Science 313:951
82. O'Keefe M, Adams GB, Sankey O (1992) Phys Rev Lett 68:2325
83. Payne MC, Teter MP, Allan DC, Arias TA, Joannopoulos JD (1992) Rev Mod Phys 64:1045
84. Petkov V, DiFrancesco RG, Billinge SJL, Acharaya M, Foley HC (1999) Phil Mag B 79:1519
85. Pikunic J, Clinard C, Cohaut N, Gubbins KE, Guet J-M, Pellenq RJ-M, Rannau I, Rouzaud J-N (2003) Langmuir 19:8565
86. Reuter K, Scheffler M (2001) Phys Rev B 65:035406
87. Rocquefelte X, Rignanese G-M, Meunier V, Terrones H, Terrones M, Charlier J-C (2004) Nano Lett 4:805
88. Ryoo R, Joo SH, Jun S (1999) J Phys Chem B 103:7743
89. Sanchez A, Mondragon F (2007) J Phys Chem C 111:612
90. Scheffler M, Dabrowski J (1988) Phil Mag A 58:107
91. von Schnering HG, Nesper R (1987) Angew Chem Int Ed 26:1059
92. Schoen AH (1970) Infinite periodic minimal surfaces without self-intersections, NASA Technical Note TN D-5541
93. Schwarz G (2002) Untersuchungen zu Defekten auf und nahe der (110)-Oberfläche von GaAs und weiteren III-V-Halbleitern, PhD thesis, Techniche Universität Berlin, Germany
94. Schwarz HA (1890) Gesammelte Mathematische Abhandlungen. Springer, Berlin
95. Sendt K, Heynes BS (2007) J Phys Chem C 111:5465
96. Siegal MP, Yelton WG, Overmyer DL, Provencio PP (2004) Langmuir 20:1194
97. Smith MA, Foley HC, Lobo RF (2004) Carbon 42:2041
98. Smorgonskaya EA, Kyutt RN, Danishevskii AM, Gordeev SK (2004) Carbon 42:405
99. Stevens MG, Subramoney S, Foley HC (1998) Chem Phys Lett 292:352
100. Stone AJ, Wales DJ (1986) Chem Phys Lett 128:501
101. Terrones H, Mackay AL (1992) Carbon 30:1251
102. Terrones H, Mackay AL (1993) Chem Phys Lett 207:45
103. Terrones H, Terrones M, Hernandez E, Grobert N, Charlier J-C, Ajayan PM (2000) Phys Rev Lett 84:1716
104. Tersoff J (1989) Phys Rev B 39:5566
105. Thomson KT, Gubbins KE (2000) Langmuir 16:5761
106. Townsend SJ, Lenosky TJ, Muller DA, Nichols CS, Elser V (1992) Phys Rev Lett 69:921
107. Valencia F, Romero AH, Hernandez E, Terrones M, Terrones H (2003) New J Phys 5:123
108. Vanderbildt D, Tersoff J (1992) Phys Rev Lett 68:511
109. Wallace PR (1947) Phys Rev 71:622
110. Wirtz L, Rubio A (2004) Sol Stat Comm 131:141
111. Yushin G, Dash R, Jagiello J, Fischer JE, Gogotsi Y (2006) Adv Func Mater 16:2288
112. Zhao XB, Xiao B, Fletcher AJ, Thomas KM (2005) J Phys Chem 109:8880
113. Zhou J-H, Shi Z-J, Zhu J, Li P, Chen D, Dai Y-C, Yuan W-K (2007) Carbon 45:785

Chapter 5
Amorphous Carbon and Related Materials

Nigel A. Marks[1]

Abstract Amorphous carbons have attracted enormous experimental, theoretical and computational interest over the past two decades. This chapter explains how computer simulation has contributed to the understanding of these important materials. The principle themes are (i) the necessity of accurate interaction potentials, (ii) the ability of computer simulation to discern between competing theoretical models, and (iii) the need to quantify timescales when comparing experiment and simulation. Common methodologies such as liquid quenching and thin-film deposition are explained, and major successes in reproducing experimental behaviour are discussed. Connections are also established between amorphous carbons and related materials such as glassy carbon and hydrogenated amorphous carbon. Finally, unsolved problems and opportunities for future research are identified.

5.1 Introduction

Disordered phases of carbon have always proved difficult to characterize. The absence of long-range order (and sometimes medium-range order too) predicates against standard structural determination methods based on diffraction or coordination. Even with a wealth of intuition it can be virtually impossible to intuit local structure in the absence of a computer model. Non-graphitizing carbons (studied and named by Rosalind Franklin in the 1950s [1]) typify the confusion: an enormous amount of experimental data for this important class of material is available, and yet the many structural models [2–5] proposed over the intervening decades are, with hindsight, rather incomplete. It is only over the past half decade a suite of complementary computer simulation methods have converged on a unified picture based on fullerene-like fragments [6–9].

A similar story from more recent times followed McKenzie et al.'s landmark 1991 paper [10] reporting the synthesis of a diamond-like amorphous phase now

[1]Nanochemistry Research Institute, Curtin University of Technology, Perth WA 6845, Australia
e-mail: N.Marks@curtin.edu.au

L. Colombo and A. Fasolino (eds.), *Computer-Based Modeling of Novel Carbon Systems and Their Properties*, Carbon Materials: Chemistry and Physics 3,
DOI 10.1007/978-1-4020-9718-8_5,

referred to as tetrahedral amorphous carbon (*ta*-C). Here the foundational questions such as the nature of microstructure and the mechanism responsible for the high sp^3 fraction again required computer simulation as an important investigative tool. The critical difference between *ta*-C and non-graphitizing carbon was timing, as significant theoretical advances were occurring in parallel with the experimental discoveries and hence cutting edge computational techniques were quickly applied. The story of amorphous carbon is thus best told with an eye to history, particularly regarding the explosion of computational techniques from 1985 onwards.

The 1991 paper in Physical Review Letters (PRL) by McKenzie, Muller and Pailthorpe (then all based at Sydney University) marks a useful reference point for discussing amorphous carbons. The field of course has a history preceding this date, including the first ion beam experiments of dense carbon films in 1971 by Aisenberg and Chabot [11], and the 1978 cathodic arc deposition experiments by Aksenov et al. [12,13], enabling high deposition rates. What makes the 1991 paper noteworthy is its impact. The McKenzie PRL has 623 citations, and would later spawn (commencing in 1998 in Cambridge) the Specialist Meeting for Amorphous Carbon. The McKenzie paper also defined what came to be common parlance, introducing the acronym *ta*-C to distinguish the high density, predominantly sp^3-bonded phase from its lower density sp^2-rich counterparts. Within the amorphous carbon community this lead to a somewhat strict definition of nomenclature, with *ta*-C referring to the high-density material (nominally above 50% sp^3 bonding) and *a*-C referring to the low-density phase. In contrast, the broader materials community used the more generic descriptor diamond-like-carbon (DLC) for hard carbons involving hydrogen and/or dopants (for details see reviews in Refs. [14,15]). Within this chapter, the stricter definition is applied and unless specified, all materials are pure carbon phases. To further codify shades of gray, *a*-C and *ta*-C will be used when referring a regime where a particular hybridization is predominant, while the term *amorphous carbon* will refer to the entire spectrum, that is, *a*-C and *ta*-C inclusive.

The purpose of this chapter is threefold: firstly, it sets out to provide a definitive summary of the application of computer simulation techniques in the amorphous carbons; secondly, it identifies topics of interest which remain unanswered; and thirdly, it seeks to address the standard questions which an individual typically asks when simulating an amorphous carbon system for the first time:

- With what methodology (e.g. molecular dynamics or monte carlo, liquid quenching or film deposition) should I simulate my amorphous carbon system?
- What kind of interatomic potential or level of theory (i.e. empirical, tight-binding or ab initio) should I apply?
- What properties can be calculated from the simulation and how do these generally compare to experiment?

Finally, this chapter establishes the relationship between amorphous carbons and related materials involving hydrogen, dopants and carbon forms with sp^2-ordering, many of which offer productive future directions for research.

5.2 Experimental Essentials and Conceptual Models

When performing simulations of amorphous carbon it is essential for the simulator to be well-acquainted with typical experimental conditions. This requirement reflects the metastable nature of amorphous carbons; the materials are the product of non-equilibrium preparation techniques and thus the computational approach must reflect the experimental synthesis. In this regard, one should note that metastability is not synonymous with instability; *ta*-C thin films, for example, are thermally stable in air up to 700 K, and as a consequence *ta*-C is extremely robust at room temperature, finding applications in protective coatings for blades and as a ultra-thin corrosion barrier in hard-disks. The key point is that amorphous carbon materials are analogous to quenched liquids, and thus for the simulations to have relevance it is crucial to maintain correspondence with experiment as close as possible.

Atomic beams are the common element amongst all of the various deposition techniques suitable for synthesizing amorphous carbons. The basic principle is that carbon species, typically ions, are deposited onto a substrate with a prescribed energy. The primary experimental control parameter is thus the substrate bias which accelerates ions to the desired energy. The substrate temperature and deposition rate also affect the film properties. The most prominent techniques used to produce dense amorphous carbons (i.e. *ta*-C) are filtered cathodic vacuum arc (FCVA) [10, 17] and mass-selected ion beam (MSIB) [18, 19]. Data for both techniques from a variety of authors is summarized in Fig. 5.1. Other methods include magnetron sputtering [20]

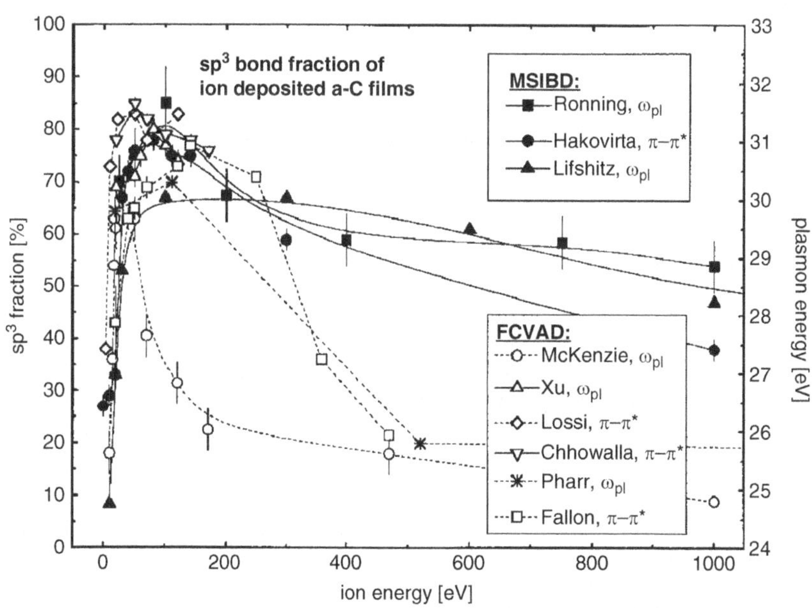

Fig. 5.1 Amorphous carbon films deposited using energetic ion beams exhibit characteristic behaviour in which predominantly sp^3-bonded films are produced within an energy window centered around 30–100 eV. Figure from Ref. [16]: reprinted with kind permission from Springer Science+Business Media

and laser ablation of graphite [21]. The general trend is independent of the apparatus, with *a*-C produced at low and high energies and *ta*-C at intermediate energies. This "energy window" effect is not specific to the sp^3 fraction; the density, compressive stress and optical gap all exhibit a similar energy dependence [17].

Amongst the various apparatus suitable for depositing amorphous carbons, the most important characteristic is the absence of carbon macroparticles. In the FCVA approach, for example, magnetic fields around curved ducts are used to isolate a beam of charged carbon ions, while in the MSIB methodology, the mass-selection process itself extracts only single carbon atoms. Due to this requirement for high-purity beams, the deposition rate is small, around 1–10 Å per second, and consequently most amorphous carbons are prepared as thin films. Another factor limiting the thickness of *ta*-C films is the presence of large biaxial compressive stresses. Above a critical film thickness (ca. 100 nm) the stress exceeds the adhesion strength of the interface and the film delaminates.

Considerable experimental and theoretical attention has been devoted to understanding the origin of compressive stress and sp^3 bonding in amorphous carbon films. These discussions have generated sharp debate, and in some sense the collective jury is still out. Various competing models for the high sp^3 fraction in *ta*-C are summarized schematically in Fig. 5.2. The oldest and most commonly discussed theory is that of subplantation as promoted by Lifshitz [23–25]. Here an analogy is drawn with ion implantation theory, with the model proposing that

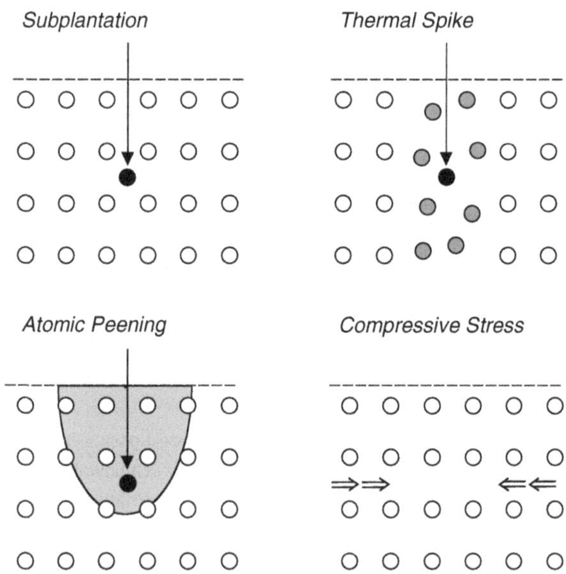

Fig. 5.2 Schematic of proposed models for sp^3 formation and/or compressive stress generation in *ta*-C thin-film deposition. The illustrations emphasize the predominant physical argument: subsurface densification for the subplantation model, impact induced melting for the thermal spike model, a pressure pulse (followed by relaxation) for the atomic peening model, and biaxial stress for the compressive stress model. Figure from Ref. [22]: reprinted with permission from Elsevier

shallow implantation into the sub-surface of the film is responsible for densification, compressive stress and sp^3 bonding. Similar explanations were provided by Robertson [26,27] and Davis [28] in the context of the sp^3 fraction and compressive stress, respectively. An alternative explanation based on a transient pressure pulse followed by relaxation was put forward by Kopenen et al. [29], while Hofsäss et al. [16] proposed a somewhat different relaxation-based model in which cylindrical thermal spikes are responsible for the *generation* of highly tetrahedral phases. Finally, the early work by McKenzie et al. [10] suggested a link between compressive stress and the sp^3 phase, in analogy with crystalline carbon where the Berman–Simon line defines the pressure–temperature phase boundary between graphite and diamond.

In this author's opinion, the merits of the subplantation model have been overplayed. As noted in Section 5.5, simulations show that penetration of the surface is not essential to generate either high stresses or large sp^3 fractions. Furthermore, the language of subplantation is itself misleading; the idea of interstitials and densification is at odds with the non-crystalline nature of an amorphous material. From an experimental point of view, recent work [30] showed conclusively that the stress-sp^3 relationship is non-linear, confirming the original suggestion by McKenzie that sp^3 carbon is synthesized above a critical stress. Once the *ta*-C phase has been created, the stress can be eliminated by either post-deposition annealing or removal of the substrate entirely [31]. Due to the large elastic moduli of amorphous carbons, comparatively small relaxations in density will relieve large biaxial stresses. The precise stress-relief mechanism, however, remains extremely contentious [32].

The discussion of amorphous carbon deposition thus far has implicitly assumed room temperature conditions. However, when the substrate temperature is increased by 100 K or so, a sharp transition occurs in which *ta*-C is no longer observed [33–36]. As seen in Fig. 5.3, above a critical temperature (T_c) the various

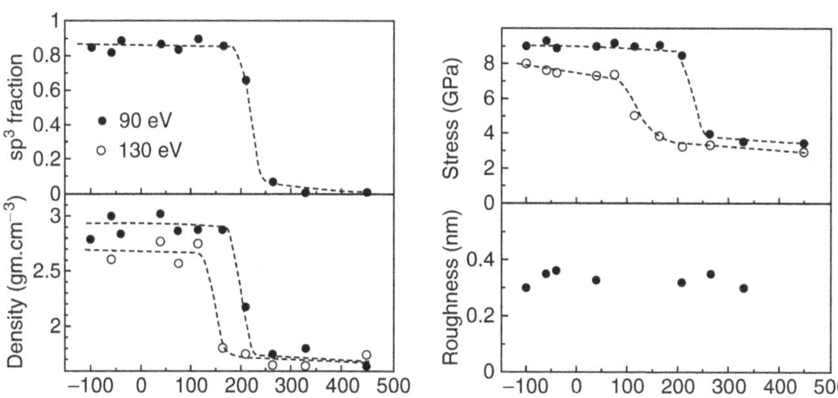

Fig. 5.3 Experimental FCVA data showing the variation of four important film properties as a function of the substrate temperature at two incident energies. The sharp decrease in the properties at the critical temperature T_c correlates to a transition from *ta*-C to a sp^2-phase. Figure adapted from Ref. [33]: reprinted with permission from the American Institute of Physics

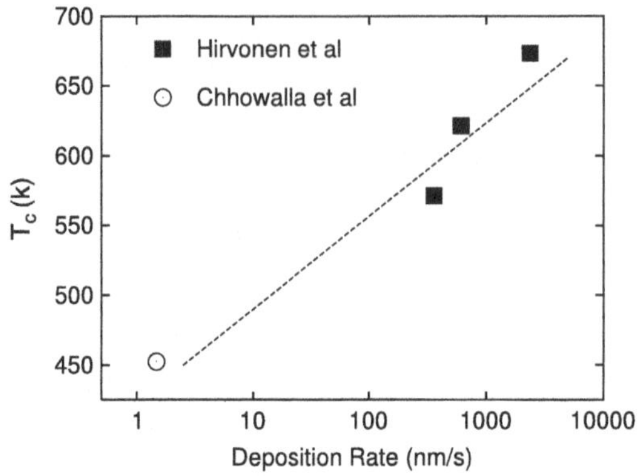

Fig. 5.4 Experimental data showing the dependence of the critical substrate temperature (T_c) on deposition rate. T_c is defined as the temperature above which *ta*-C is not observed (Fig. 5.3). The *dotted line* is to guide the eye, highlighting the activated nature of the process. Original figure generated using data extracted from Chhowalla et al. [33] and Hirvonen et al. [34]

properties associated with *ta*-C universally disappear; the only factor unchanged is the surface roughness which maintains its characteristically smooth value. Above this critical temperature the deposited films are almost entirely sp^2 bonded, possessing an orientation order in which the *c*-axis of graphite-like regions lies in the biaxial plane. As explained in Section 5.5.2, the origin of this effect was only recently revealed via computer simulation. Adding a further complication to the analysis of the experimental data is the effect of deposition rate. Figure 5.4 shows that increasing the deposition rate markedly increases the value of the critical temperature T_c. The other important temperature-related property of *ta*-C is its thermal stability under post-deposition annealing. Whereas 450 K is sufficient to trigger an sp^2 phase *during* deposition, once the material is formed a much higher temperature of 1,400 K is required to activate a change of state in vacuum [37]. In the presence of air, oxidation reduces this threshold to around 700 K [38].

To conclude this brief overview (see Refs. [39–41] for more details), it is sufficient to note that the properties of amorphous carbon are both rich and varied. The combination of multiple competing hybridizations and the subtle interplay amongst the experimental parameters make for a scientifically fascinating system. The good fortune for amorphous carbon science was the parallel development of computer simulation as a predictive and interpretative tool. Advances in interatomic potentials and their timeline are covered in the following section, while applications to bulk carbon and thin-film deposition are considered in Sections 5.4 and 5.5.

5.3 A Short History of Interaction Potentials for Carbon

The story of computational methods for amorphous carbon simulation begins with the landmark work of Car and Parrinello in 1985 [42]. Here, the foundations were established for a density-functional-theory (DFT) framework in which molecular dynamics (MD) could be performed. The calculations were, however, extremely computationally expensive, and 10–15 years would pass until amorphous carbons could be generated routinely. Even now, the calculations involve small number of atoms and cannot describe phenomena involving film growth and surfaces. To address these questions, faster and more approximate methods are required.

Prior to ground-breaking work by Tersoff in the late 1980s, analytic potentials for carbon were effectively non-existent. Useful models were available for ionic solids, metals and noble gases, but the complexity of covalent bonding had hindered progress. Drawing on Abell's bond-order formalism [43] (in effect, a $1/\sqrt{Z}$ dependence), Tersoff released a potential for silicon [44] in 1986, followed by a version for carbon [45] in 1988. Both papers had enormous effect, generating todate 686 and 751 citations, respectively. The carbon potential was particularly significant, as this was the first time that a computationally efficient method was available which treated multiple hybridization states. For a full discussion of the Tersoff methodology, see the extended description in Ref. [46].

Drawing on Tersoff's advance, Brenner soon released the Reactive Bond Order (REBO) potential, adding hydrogen interactions and improving the description of radicals [47, 48]. Twelve years later, a revised version of the Brenner potential was released [49]. For insight into the philosophy of these potentials, Brenner's 2000 article [50] titled "The Art and Science of an Analytic Potential" is highly recommended. As for Tersoff's work, the REBO potentials had enormous impact, with citations numbering 1,646 and 566 for the 1990 and 2002 papers, respectively. Particular application was found in the burgeoning fields of fullerenes and nanotubes following their respective discoveries in 1985 and 1991.

Despite their achievements, the Tersoff and Brenner potentials did not significantly advance fundamental understanding of amorphous carbons. Early effects to simulate thin-film deposition of *ta*-C were spectacularly unsuccessful; the simulations did not predict high sp^3 fractions or densities, contained unphysical bonding configurations, and did not observe a critical substrate temperature T_c. This unhappy state of affairs is exemplified by 1997 simulations (Fig. 5.5) of thin-film deposition using the Tersoff potential. The simulations exhibit not the slightest hint of the experimental transition around 300°C, and the sp^3 fractions (which correlate closely to the elastic modulus) are underestimated by nearly a factor of two.

The origin of the discrepancies between simulation and experiment can be traced to the interaction potentials themselves. The Tersoff and Brenner methods are short-ranged potentials, employing switching functions to identify nearest neighbours. While this benefits computational efficiency, it inverts the density relationship between graphite and diamond. Experimentally, graphite (2.27 g/cc) is 36% less dense than diamond (3.52 g/cc) but in Tersoff–Brenner carbon the interaction cutoff is 2.1 Å, making graphite 3% *more dense* than diamond. This large error eliminates a crucial thermodynamic driving force (the volume) which often promotes the tetrahedral phase. Other undesirable effects of short cutoffs are found in the kinetics of bond-making and -breaking, as the switching functions are arbitrarily chosen, leading to unphysical reaction pathways when sp^2 atoms convert to sp^3 and vice-versa. In MD simulations this deficiency manifests itself as sharp spikes in the

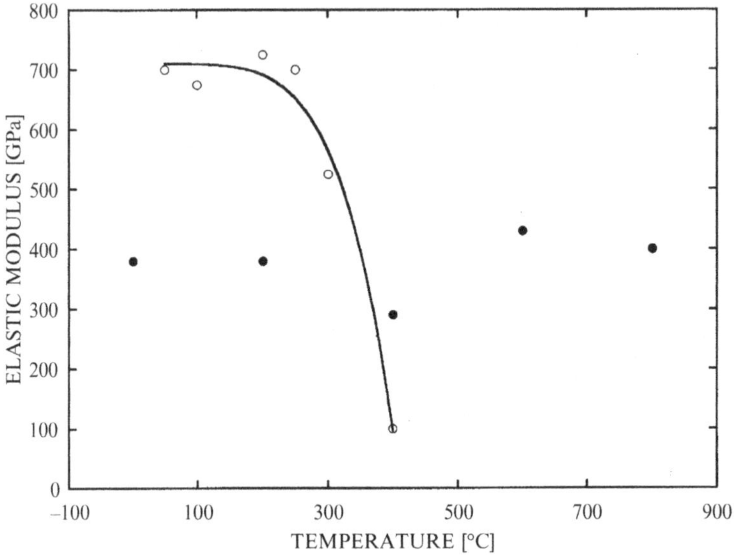

Fig. 5.5 Experimental measurements (*empty circles*) and deposition simulations (*solid circles*) of the Young's modulus of carbon films as a function of substrate temperature. Simulations were performed using the Tersoff potential. Figure from Ref. [34]: reprinted with permission from the American Institute of Physics

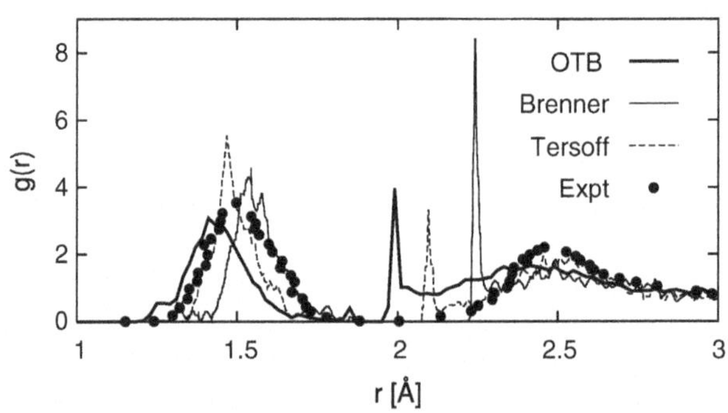

Fig. 5.6 Radial distribution function from simulations of amorphous carbon deposition employing incorrect cutoffs. The simulations exhibit unphysical spikes in the RDF not seen in experiments. Cutoffs for orthogonal tight-binding (OTB) [51], Tersoff [52] and Brenner [52] simulations were 2.0, 2.1 and 2.25 Å, respectively. Figure from Ref. [22]: reprinted with permission from Elsevier

radial distribution function (RDF), indicative of atoms trapped in false intermediate positions. Typical amorphous carbons have broad first and second neighbour peaks centered around 1.5 and 2.5 Å, respectively, but short-cutoff simulations exhibit an additional peak which is unphysical in both its narrowness and location. Figure 5.6 shows examples of this behaviour in MD simulations of thin-film deposition. All three potentials employed short cutoffs, and the values of the cutoff (see the caption for details) need barely be specified, so clear is its effect on the RDF.

The way forward lay in the inclusion of long-range interactions, prototypically, those between graphitic layers. Nordlund et al. [53] proposed in 1996 a long-range extension to the Tersoff potential but the method was never applied to amorphous carbon. The potential of Broughton and Mehl [54] appears to be suitable for amorphous carbon but has been overlooked by the literature. Yet another potential was recently proposed by Kumagai et al. [55]. The most commonly used long-range potentials were released in quick succession; two potentials in 2000 (AIREBO and EDIP), ReaxFF in 2001 and LCBOP in 2003 (version II followed two years later). Virtues and shortcomings of these approaches are summarized below.

- AIREBO – The Adaptive Interaction REBO potential by Stuart et al. [56] is an extension of Brenner's potential: long-ranged interactions between sp^2 sheets are described through a Lennard–Jones (LJ) interaction: the $1/r^{12}$ term provides the missing repulsion, while the $1/r^6$ term captures van der Waals attraction. The LJ term in AIREBO is grafted onto the REBO formalism, employing switching functions to (de)activate the LJ interactions on the basis of distance and bond-order. The strength of this approach is that AIREBO reduces to REBO expressions for isolated fragments. Its weakness, however, stems from the same source. Due to the two-part nature of the derivation, the reaction pathway for bond formation and breaking is not described with a single framework. The potential is thus vulnerable to incorrect kinetics; when an atom is changing its hybridization the switching functions to (de)activative the LJ potential must marry perfectly with the corresponding cutoffs within REBO. In practice this is very difficult to achieve, and is likely part of the explanation for the poor performance of AIREBO in recent liquid quench simulations of amorphous carbon [57]. In 2008 a modified AIREBO potential with improved hydrocarbon properties was released [58], but its suitability for amorphous carbon is not yet known.
- EDIP – The Environment Dependent Interaction Potential [59,60] was developed by this author with an explicit focus on *ta*-C and thin-film deposition. In contrast to most approaches, the Abell-Tersoff formalism was not used as the starting point. Instead, an atom-centered bond-order was employed, drawing on an earlier silicon EDIP method [61,62]. Generalization of silicon EDIP to carbon involved three main steps: non-bonded terms to increase the graphite c-spacing, dihedral rotation penalties for π-bonded atoms, and variable-range pair interactions. Robust numerical fitting was achieved using high-symmetry configurations and reaction pathways. Particular attention was paid to the transformation between diamond and rhombohedral graphite as this reaction changes the hybridization from sp^3 to sp^2 – by tailoring the cutoff functions to density-functional-theory data for this pathway an accurate description of liquid and amorphous carbon is obtained. The main shortcomings of carbon EDIP are the atom-centred bond-order (which does not penalise isolated sp^2 atoms) and the absence of long-range attraction (only a minimum c-spacing is predicted). On the positive side, EDIP is fast and robust, and in a wide variety of situations it has made a significant contribution to the study of amorphous carbons.
- ReaxFF – The Reactive Force Field by van Duin et al. [63] offers a very different approach to modeling carbon. Whereas most potentials are designed with chemical intuition as a key ingredient (see Brenner's Art/Science article, for

example), the ReaxFF framework is instead constructed to be as general as possible, capturing all conceivable interactions spanning covalent (i.e. bond-order) terms, coulomb interactions (include charge equilibration) as well as dispersion and other non-bonding forces. Parameterization of this generic formalism proceeds via a large training set of candidate structures against which a multi-variate fitting procedure is performed. We recently performed ReaxFF simulations of amorphous and liquid carbon [64], and the overall performance is impressive. The main shortcoming is the liquid state, which is not described with the same accuracy seen with EDIP and LCBOP-II. However, given the philosophy used to develop the potential, this deficiency is neither crippling nor surprising. Unusually for an analytic carbon potential, ReaxFF has been extended to describe many other chemical species, and parameter sets are available for systems as diverse as oxides [65], hydrides [66] and metals. The latter, for example, enabled MD studies of Co, Cu and Ni catalysts in carbon nanotube growth [67].

- LCBOP (I and II) – The Long-range Carbon Bond Order Potential by Los et al. [68, 69] is perhaps the best potential for pure carbon systems, drawing on many of the features of the preceding methods. At first glance LCBOP appears similar to the AIREBO potential, offering a Brenner-type bond-order-based description and a Morse-like potential to describe long-range effects. However, the parameter fitting and development of the functional form is much closer to the spirit to EDIP, employing an fully integrated approach for all components. As a consequence, errors associated with a two-step approach are avoided. The payoff for this extra effort is that energetics of bond-making and bond-breaking are well-described, particularly with the second incarnation of the potential (LCBOP-II) (Fig. 5.7 and [68, 70]) which provides improved reaction energetics (amongst other things) with respect to the both REBO and the original LCBOP. The long-range of the potential (6 Å) ensures that interlayer graphitic attraction is

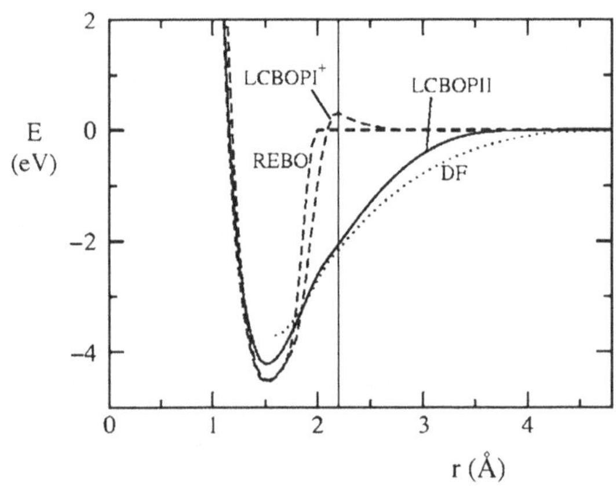

Fig. 5.7 Bonding energy for the single carbon–carbon bond in $(CH_3)_3C–C(CH_3)_3$ as a function of separation. Comparison shown between REBO, DFT and versions I and II of the LCBOP potential. The inclusion of middle-range interactions (1.7–4 Å) in version II of LCBOP markedly improves the description of bond dissociation. Figure from Ref. [68]: reprinted with permission from the American Physical Society

well-described (capturing both attraction and repulsion), and related properties such as elastic constants compare well to experiment. An extended potential incorporating hydrogen is currently in development, further increasing the attractiveness of the LCBOP approach.

Beyond empirical potentials lie a multitude of electronic structure techniques. DFT is the highest level of theory for which MD is practicable, and the agreement with experiment is impressive [71–73]. However, there are many situations where full DFT is not required, and thus approximate treatments of electronic structure are extremely useful. One of the earliest proposed methods was the local basis density functional (LBDF) approach [74] employing a minimal basis. Used primarily by Drabold and co-workers, this methodology enabled calculations comparable to full DFT at reduced cost. At the tight-binding level of theory one of the dominant techniques for carbon is the density-functional tight-binding (DFTB) method of Porezag et al. which uses long-range non-orthogonal orbitals [75, 76]. Another common TB method is the environment-dependent TB method [77] which goes beyond the traditional two-centre approximation used in early TB models for carbon [78]. Yet another contributor to tight-binding simulations is the Bond-Order-Potential (BOP) developed by Pettifor. Employing a ring-counting approach, this methodology is conceptually attractive due to analytic approximations known as ABOPs [79–81]. One key contribution of the Pettifor approach is the demonstration that the Tersoff angular functional form has an explicit quantum-mechanical connection when the moments expansion extends to rings of length four. However, despite their sound theoretical grounds (and their promise to contribute high-quality potentials), in practice analytic BOPs are yet to make significant impact.

Concluding this overview, Fig. 5.8 presents a summary of liquid-quench simulations (Section 5.4) summarizing the transferability of various approaches. Two

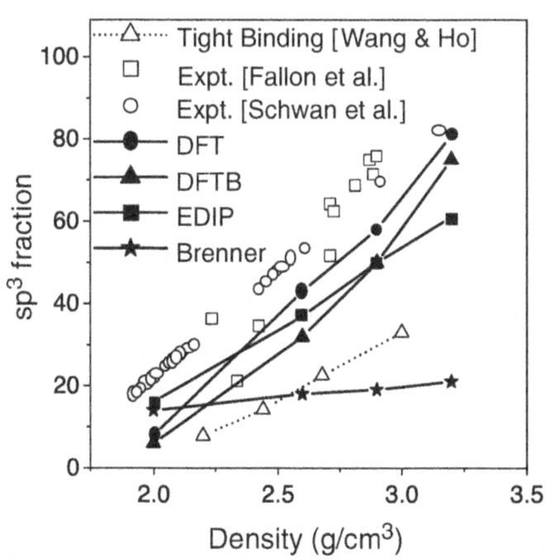

Fig. 5.8 Percentage of sp^3 bonding as a function of density in amorphous carbons generated by liquid quenching. Simulations using DFT agree closely with experimental data (*empty circles and squares*). DFTB and EDIP simulations also reproduce the overall trend, with DFTB providing superior agreement. Simulations using the Brenner potential, orthogonal tight-binding [82] and Tersoff potential (not shown) [83] perform poorly. Figure from Ref. [84]: reprinted with permission of the American Physical Society

trends are apparent, the most obvious being that higher levels of theory correlate more closely with experiment. Amongst the data lies a second lesson – the essential elements for an empirical/analytic potential are long-range interactions and good reaction kinetics. Liquid quenching provides an excellent rule-of-thumb, and provided that a potential passes this test, it can generally be used with confidence.

5.4 The Simulation of Bulk Amorphous Carbon

Computer simulations of amorphous carbon broadly divide into two categories; one class of investigation considers the amorphous network itself (in effect, bulk properties), while the second explores the process of thin-film deposition from energetic species. Computational speaking, the latter is a far more difficult problem, requiring careful simulation design and significant computer time. The order of presentation thus mirrors historical developments: first we will discuss bulk techniques, with a particular emphasis on liquid quenching via molecular dynamics. In Section 5.5 we return to the more complex situation of energetic beams and the techniques required to accurately describe experimental conditions.

Mechanical models of disorder go back to Bernal who studied structure in liquids using rubber balls connected by rods of varying lengths. He undertook this exercise in his office, exploiting regular interruptions to maximize randomness. Suffice to say, computers greatly simplified such studies. A major theoretical breakthrough in the generation of tetrahedral amorphous networks was made in 1985 by Wooten, Winer and Weaire (WWW) [85] who proposed a stochastic bond-switching algorithm. The WWW method generates amorphous silicon and germanium structures in close agreement with experiment; see Drabold's 2009 review [86] for background and other simulation techniques. Unfortunately, it cannot be used for amorphous carbons as the networks contain a minimum of 10–15% threefold (sp^2) bonding.

The first full simulations of amorphous carbon networks used the Tersoff potential and a Monte Carlo (MC) framework in which liquids were rapidly quenched to form a solid [45, 87]. From a transferability point of view, the simulations were hampered by the deficiencies noted in Section 5.3; due to the incorrect density of graphite very high pressures were required to access the sp^3-rich phase. Aside from the choice of potential, there is a fundamental question as to whether MC is an appropriate methodology when the temperature is altered quickly. Here molecular dynamics is at an advantage, describing an explicit reaction trajectory during rapidly changing conditions. Whether in implicit recognition of this situation, or perhaps simply because of history, most amorphous carbon simulations have used molecular dynamics along the lines discussed in the following section.

5.4.1 Liquid Quenching – Justification

Molecular dynamics simulations of amorphous carbon were first performed in 1989 by Galli and co-workers [88, 89]. Paradoxically, the simulations used the highest level of theory presently available; at that time there were no other alternatives, save the Tersoff potential released the year prior. Using the Car–Parrinello method, a rather small 54-atom system at 2.0 g/cc was simulated by generating a liquid sample which was rapid cooled (i.e. quenched) to form an amorphous network. This was a landmark paper, demonstrating the power of computer simulation to explore disordered carbon systems. With considerable motivation provided by the 1991 *ta*-C paper by McKenzie et al., the same liquid-quenching methodology was shortly afterwards applied to dense carbons (circa 3.0 g/cc), achieving considerable success in reproducing experimentally observed coordination fractions, radial distribution functions and electronic properties [71, 90].

Despite the empirical success of the liquid quenching method, there remained an unspoken distrust of the theoretical fundamentals. Computational practicalities of ab initio MD dictate cooling times of around one picosecond, corresponding to a cooling rate in the region of 10^{16} K/s. In many materials systems such quenching rates are unphysically fast by many orders of magnitude [92, 93]. Amorphous carbons, however, are a special case, and the physical basis for liquid quenching was established in a 1997 paper [91] by this author. Using a combination of molecular dynamics and analytic methods, it was shown that experimental deposition conditions for amorphous carbon result in quenching rates very similar to the timescales accessible via MD. This fortunate state of affairs reflects the atom-by-atom nature of the deposition process, in which energetic species induce short-lived local melting in a process known as a thermal spike. Only a small number of atoms attain a molten configuration, and the high thermal conductivity of carbon removes heat from the impact site efficiently and quickly.

The spatial and temporal scales of amorphous carbon thermal spikes are summarized in Fig. 5.9. The right panel provides a conceptual picture in which the incident ion creates a liquid-like state due to collisions with the surface area; the

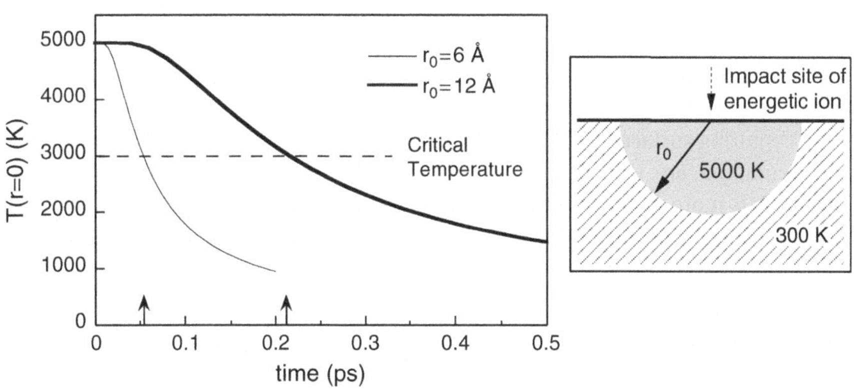

Fig. 5.9 *Left*: Temperature at the centre of two thermal spikes as a function of time. The *dotted line* shows the 3,000 K lower bound used to define the quenching time. *Right*: Schematic diagram showing initial conditions for the analytic model of the spike. Figures from Ref. [91]: reprinted with permission from the American Physical Society

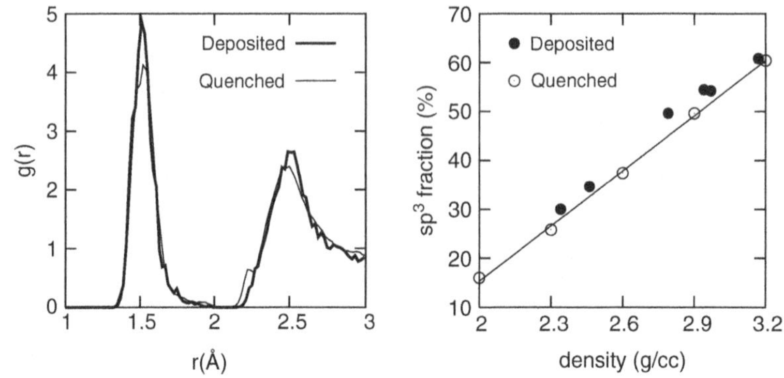

Fig. 5.10 Amorphous carbon structures generated by liquid quenching and thin-film deposition using EDIP. *Left*: radial distribution function g(r) for *ta*-C at 3.2 g/cc (deposition performed at 40 eV); *Right*: Relationship between sp^3 fraction and density for quenched and deposited carbons. Depositions carried out for energies between 1 and 100 eV. Figures adapted from Ref. [84]: reprinted with permission from the American Physical Society

left panel shows analytic solutions of the heat-diffusion equation for two typical radii. Due to the nanometre-scale of the spike, heat diffuses away very quickly, with sub-picosecond thermal spikes being typical. In the original 1997 paper this conceptual picture and timescale was confirmed by two-dimensional MD simulations of thermal spike evolution. Several groups have since performed three-dimensional simulations of ion impact/thermal spikes, but the basic story remains unchanged. Today it is utterly uncontroversial to use picosecond-scale quenching as a robust and physically motivated tool for generating amorphous carbons.

As a illustration of the reasonableness of liquid quenching, Fig. 5.10 shows in its left panel the radial distribution function for 3.2 g/cc *ta*-C simulated using liquid quenching and thin-film deposition. The right panel shows the relationship between the sp^3 fraction and density for amorphous carbons of varying density. In both cases there is considerable consistency between properties computed using the two approaches. Although the deposition simulations are necessarily more complete (capturing surface effects, energy impact processes, density dependence, etc.), the calculations are vastly more computationally expensive. Consequently, liquid quenching is more than justified if a bulk amorphous carbon structure is the primary goal; the only caveat is that the density of the system must be prescribed in advance.

5.4.2 Liquid Quenching – Practicalities

Having established the physical basis of liquid quenching, we now turn to the practical question of how to generate the initial molten configuration. Ideally the equilibrated liquid should be attained very quickly, particularly when using high levels of theory such as DFT where computational resources are at a premium.

This precludes methods commencing with stable crystalline structures where nucleation and superheating will be problematic. In the authors experience, the most reliable procedure for generating liquid carbon uses a slightly randomized simple cubic lattice as the starting structure. This configuration spontaneously melts, conveniently generating a temperature of around 6,000 K in the process (see Fig. 4 in Ref. [59]). The melting process is assisted by a imaginary phonon mode for tetragonal distortions in simple cubic carbon as noted by Schultz et al. [94]. By slightly randomizing the coordinates (a maximum amplitude of 0.25 Å is a good choice) multiple different modes are accessed, enabling prompt decoherence from the starting structure and rapid emergence of a liquid state. A typical starting configuration is shown in Fig. 5.11a.

An appealing aspect of this methodology is that minimal manipulation of temperatures via thermostats is required. Atomic motion alone destroys the initial lattice, and the process is extremely fast. In effect, the structure undergoes a process of self-nucleated melting. After just 10–20 femtoseconds (equivalent to a couple of lattice

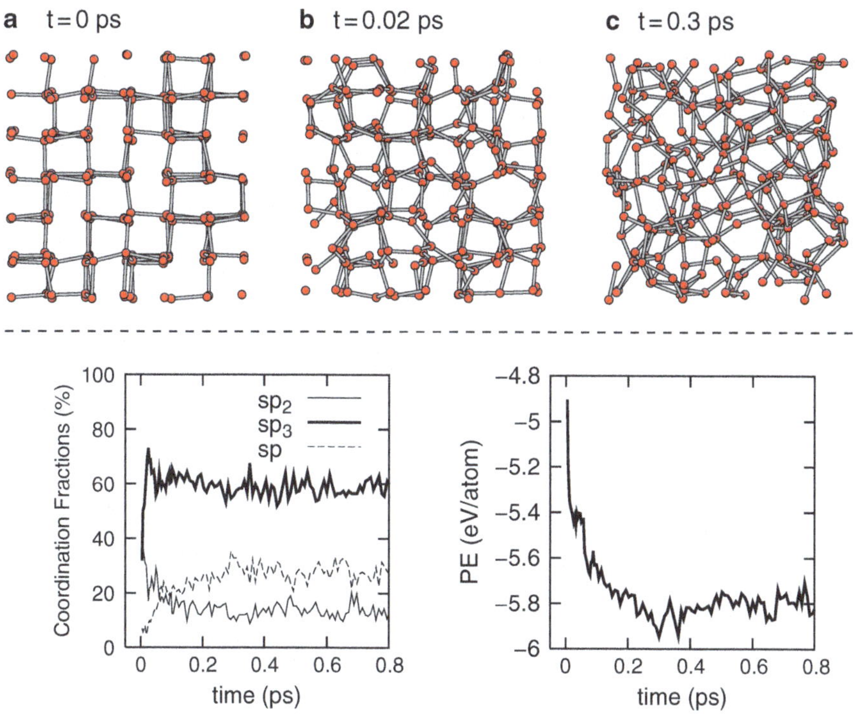

Fig. 5.11 Generation of liquid carbon via spontaneous melting of a slightly randomized simple cubic lattice. Ball-and-stick representations show the structure at various critical stages: (**a**) the initial configuration, (**b**) initial onset of melting at 6,000 K, and (**c**) upon equilibration of several parameters. *Lower left panel* shows coordination fractions determined using a cutoff of 1.85 Å; *lower right panel* shown the potential energy per atom. Calculations performed using EDIP with 216 atoms in an NVE ensemble at 2.9 g/cc. Original figure by the author

oscillations, assuming a vibrational frequency of 10^{14} Hz) the temperature rises to around 6,000 K, well above the melting point. This convenient final temperature reflects another coincidence; the potential energy difference between the liquid and the initial configuration corresponds almost exactly to the kinetic energy of the liquid state. While only a minor detail, and by no means essential, it does further simplify generation of the liquid.

After less than a picosecond (0.3 ps in Fig. 5.11) the coordination fractions and potential energy stabilize, and consistent linear evolution of the mean-square-displacement is observed. At this point, thermostats can be activated to equilibrate the liquid at the desired temperature; typical values might fall in the range 5,000–6,000 K. Note that this methodology is somewhat specific to carbon, and would not be appropriate, for example, in silicon where six-coordinate configurations are stable at high pressure. The liquid state is another point of difference between the group IV cousins; silicon increases its coordination number upon melting (to around 6.4), whereas melting of diamond lowers the average coordination.

Once the liquid is fully equilibrated (by applying thermostats for another picosecond or so), the amorphous state is generated by rapidly removing the kinetic energy. This is most simply achieved via velocity rescaling; NVT thermostats of the Nosé-type are inappropriate as the differential equation describing the heat bath will gain excessive amounts of energy. The choice of temperature profile during the quenching process is somewhat arbitrary; two common choices are exponential and linear. In fact, the shape of the profile is largely irrelevant – the most important quantity is the amount of time available for structural rearrangement during the solidification process. By correlating temperature profiles with the mean-square-displacement, this author determined 3,000 K as an approximate lower limit for diffusion [91]. Accordingly, a *quenching time* was defined as the time over which the liquid temperature reduces to 3,000 K. Typical quenching times are very short, of order 0.1–1 ps; the former corresponds to barely ten lattice vibrations.

As a final reflection on the liquid quenching process, Fig. 5.12 highlights the importance of statistical effects. This analysis shows that otherwise identical

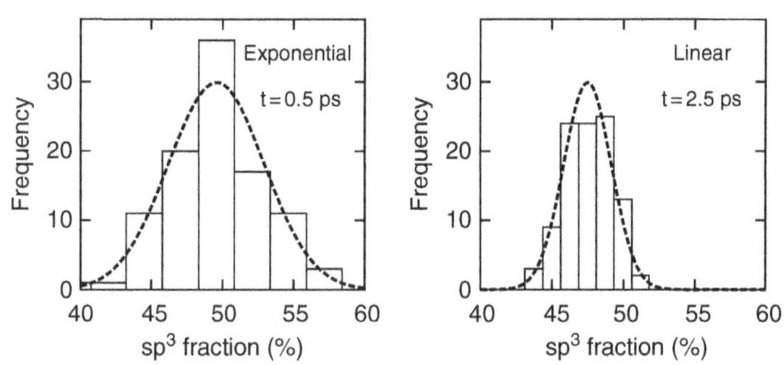

Fig. 5.12 Statistical distribution of sp^3 fractions from a large number of 2.9 g/cc liquid quench simulations performed using EDIP. Cooling curves for the 0.5 and 2.5 ps quenches were exponential and linear, respectively. *Dotted lines* indicate the normal distribution of the data sets. Figures adapted from Ref. [84]: reprinted with permission from the American Physical Society

simulations can generate a quite broad range of final structures. The left-hand panel shows the statistical distribution of one-hundred simulations exponentially quenched from 5,000 to 300 K in 0.5 ps. Here the quenching time as defined above is 0.1 ps. The configurations at the commencement of each quench were indistinguishable, save for an additional 0.1 ps of motion between data sets. Strikingly, the sp^3 fraction typically varies across a range of 10%, helping to explain variability seen in simulations using high-levels of theory where system sizes are necessarily small. The right-hand panel shows that slower quenches narrow this spread, although not enormously so. In this data, linear cooling over 2.5 ps is used, corresponding to a quenching time of 1.1 ps. Even though the quenching rate is an order of magnitude slower, there remains a considerable statistical spread. This is an important factor to keep in mind when generating amorphous carbons; only by going to very large systems will the sp^3 fraction converge to a single value, and even in that case there will likely be local variations in the form of spatial inhomogeneities throughout the cell volume.

5.4.3 Properties of Bulk Amorphous Carbon

The relatively straightforward nature of liquid quenching, and its close correspondence to experiment, has been of great benefit to the computational study of bulk properties in amorphous carbons. A few examples from the literature are summarized here to convey a sense of the breadth of possible studies. Early calculations emphasized the unusual microstructure of amorphous carbons, in particular the topology arising from disorder in combination with multiple hybdrization states. Car–Parrinello simulations shown in Fig. 5.13 emphasize the value of simulation. Neither of these pictures of the atomic bonding could be inferred from experiment alone; the absence of long-range order permits only the crudest assessment of structural relationships between nearby neighbours. The simulations show that a complex ring structure is present, which in the case of *ta*-C includes sp^3-bonded triangles and quadrilaterals. Initially these were thought to be an artifact and unlikely to exist on strain considerations, but the agreement between the simulations and neutron diffraction data was excellent. Furthermore, other calculations using lower levels of theory were found to contain approximations which artificially penalised small ring structures. Today these small rings are an accepted component of the *ta*-C microstructure, and an excellent example of the predictive power of DFT.

The importance of understanding microstructure and related behaviour has led to a raft of analysis techniques to categorize the networks. Some quantities, such as ring statistics and sp^2 clustering measures, are very difficult to compare to experiment; their value lies in fundamental insight and for comparison between methodologies. Most quantities, however, have some kind of experimental equivalent, such as the sp^3 vs density relation (Fig. 5.8), the radial distribution function, the (optical) band gap, elastic/stress properties, and so on. Yet another property of interest is fluctuation electron microscopy (FEM) used to study medium range structural

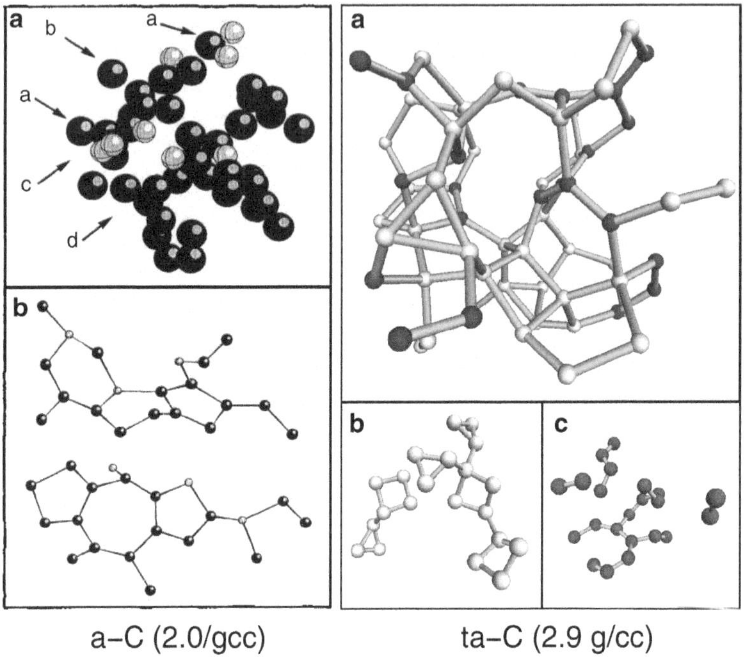

Fig. 5.13 Early simulations of amorphous carbon networks generated by liquid quenching within Car–Parrinello MD. *Left*: *a*-C structure with 54 atoms by Galli et al. *Right*: *ta*-C structure with 64 atoms by Marks et al. Buckled graphite-like layering is evident in the *a*-C structure, while small rings of sp^3-bonded carbon (*light spheres*) are seen in the *ta*-C network. Figures from Refs. [71,88]: reprinted with permission from the American Physical Society

order [95–97]. In FEM the experimental measurement is relatively straightforward but the inversion process needed to interpret the data is non-unique, presenting both a substantial mathematical challenge as well as opportunities for future simulations.

Electronic properties of disordered materials are another challenging area to investigate, and represent something of unfinished business. DFT calculations showed that *a*-C and *ta*-C have deep minima in the electronic density of states at the Fermi level, consistent with the band-gap observed experimentally. However, this comparison is quite approximate, as considerable broadening of the discrete energy levels is required due to the small number of atoms in the simulation. Drabold and co-workers [92, 98] used LBDF calculations to study the relationship between the electronic structure and the local topology of defects (Fig. 5.14). These calculations were notable for their large, state-free, bandgap of 2.5 eV and high sp^3 fraction of 91%. However, the validity of the minimal basis approach in LBDF was questioned by Schultz and Stechel [99] who showed that increasing the quality of the basis set from single-zeta (SZ) to double-zeta-plus-polarization (DZP) dramatically altered the reaction barriers (see Fig. 4 in Ref. [99]). Relaxation of the minimal basis structures with a higher quality (DZP) basis set removed most of the small rings and tripled the number of sp^2-bonded atoms.

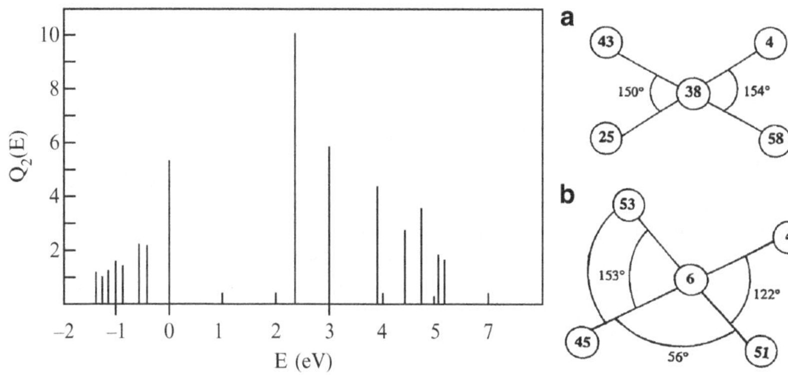

Fig. 5.14 Correlations between electronic states and the physical microstructure in LBDF calculations of *ta*-C. The *left panel* shows the localization metric $Q_2(E)$ for states near the gap. The *right panel* shows two defects with highly strained bonds. While the number of neighbours suggests sp^3 hybridisation, these defects have a significant spectral weight in localized states either side of the gap. Figures from Ref. [92]: reprinted with permission from the American Physical Society

A significant contribution to the spectroscopic measurement of sp^3 fractions was made in 2004 by Titantah and Lamoen [100]. Using all-electron DFT, they applied a microscopic scheme to compute the sp^3 fraction from the unoccupied states. This approach mimics electron-energy-loss-spectroscopy (EELS) commonly used to determine the sp^3 fraction in experimental measurements. They found that at high densities the standard technique of integrating under the π^* peak is correct, but at low and moderate densities this approach can underestimate the coordination number. More recently, the same authors applied their excited-state technique to characterize simulations of annealed *ta*-C [101]. Unfortunately, the MC structural modeling employed the Tersoff potential and thus the physical basis of the calculations is highly suspect. The most glaring shortcoming concerns the sp^3 to sp^2 transformation which is artificially aided by the incorrect density of graphite. In reality, this process is inhibited by the large volume change, but in Tersoff calculations the reaction proceeds without the energetic penalty (and structural rearrangements) associated with repulsion between graphitic elements.

It is surprising that the Tersoff potential is still being used to generate amorphous carbon networks; new researchers are advised to use one of the long-range potentials outlined in Section 5.3. That said, in the early days of amorphous carbon simulation it was very understandable that the Tersoff potential was applied, particularly given the absence of competing methods at the time. The most extensive use of the Tersoff potential was made by Kelires who performed detailed analysis of elastic properties and the microscopic origin of stress [87, 102, 103]. By analyzing the statistical distribution of atomic stresses, he showed that most of the sp^3 sites in *ta*-C are under compressive stress, with sp^2 sites mostly under tensile stress. Kelires also constructed purely sp^3 models (so-called amorphous diamond, or *a*-D) using the WWW method, showing that softening of the bulk elastic properties is primarily due to random orientation of sp^3 sites and the accompanying lowering of density.

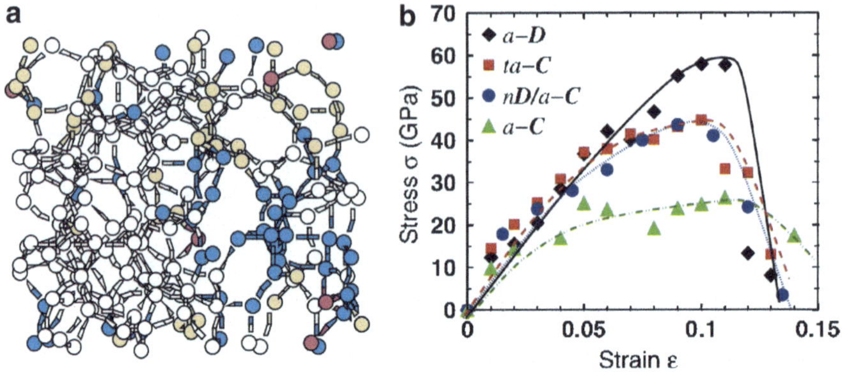

Fig. 5.15 *Left*: fracture in single-phase *ta*-C simulated using tight-binding MD. Beige and empty spheres denote sp^2 and sp^3 sites with no broken bonds. *Pink and blue spheres* denote broken sp^2 and sp^3 sites. *Right*: stress–strain curves for shear load on four amorphous carbon materials. Figures from Ref. [104]: reprinted with permission from the American Physical Society

In the last several years the scope of amorphous carbon simulation has expanded to consider fracture, a macroscopic property often studied from an engineering perspective. At the crack tip, however, fracture is fundamentally atomistic; cracking involves broken bonds and is driven by the energetics of freshly created free surfaces. In a departure from the Tersoff potential, Kelires and co-workers [104] applied tight-binding molecular dynamics (TBMD) to study stress-strain relationships in *a*-C, *ta*-C, *a*-D and a composite comprising nanodiamond and *a*-C (Fig. 5.15). The hardness H was found to correlate with the sp^3 fractions, with H for *ta*-C being nearly 60% that of diamond. Interestingly, minority sp^2 sites are not the point of weakness in *ta*-C and the composite; instead, it is sp^3 sites in the vicinity of clustered sp^2 atoms which fracture. Under strain these tensile-stressed sp^3 sites convert to sp^2, creating the geometrically favored hybridization state for a free carbon surface [9, 105]. Related calculations were subsequently performed by Lu et al. [106] using EDIP. The failure strain was in close agreement with more computationally demanding simulations employed TBMD and DFT, and far superior to Tersoff and REBO models. Calculations of fracture strength revealed the unusual nature of cracking in *ta*-C: simulations with penny-shaped cracks of varying size exhibited a major departure from the standard Griffith criterion.

In concluding this section we note that computer simulation has made many contributions to the understanding of physical and electronic properties in amorphous carbons. Some of these insights are validated by experiment while others offer interpretation. The clever use of the WWW method by Kelires is an excellent example of the latter; by creating hypothetical structures he was able to provide important insight into a question that otherwise could only be the topic of speculation. Still, there remain areas that are either unresolved or unexplored; we will revisit some of these questions in the closing remarks (Section 5.7).

5.5 The Simulation of Amorphous Carbon Thin Films

Although bulk simulations of amorphous carbon have been undeniably successful, a more complete picture is provided by modeling the structure of thin films. Conceptually, this amounts to simulating the deposition process itself, enabling answers to critical questions such as the nature of the surface, the mechanism for generating high sp^3 fractions in *ta*-C (Fig. 5.2), the origin of compressive stress, etc. The first simulations of amorphous carbon deposition were performed in 1992 by Kaukonen and Nieminen [107]. Using the Tersoff potential, they deposited several hundred atoms onto a 384-atom diamond (001) substrate (Fig. 5.16). Panel (a) shows a porous structure deposited with low-energy (1 eV) atoms, while the denser structure in panel (b) is from a 40 eV atomic beam. The panel on the far right shows the variation in the film density as a function of the ion energy. At first this appears to be a successful reproduction of the energy window effect discussed in Section 5.2; however, closer inspection shows that the maximum sp^3 fraction is just 44%, and thus none of the films resemble *ta*-C. With the benefit of hindsight, this situation is not surprising given the absence of critical interactions in the Tersoff potential. Even so, it was nearly a decade before more successful calculations were performed.

The principles of thin-film deposition simulation can be quite simply stated. First, a substrate material is constructed and equilibrated at the desired temperature. Next, an additional atom is introduced a short distance above the upper surface with a prescribed energy. MD is then applied for around one picosecond to describe the impact between the added atom and the surface. For high energy impacts (say above 100 eV), the potential should reduce to the ZBL potential [108] to correctly describe short-range interactions. To prevent unphysical accumulation of heat, temperature thermostatting is applied during and/or after the impact has concluded. At the end of the impact, another atom is added and the process repeats. Typically several hundred atoms or more will be deposited, resulting in a film of nanometre thickness.

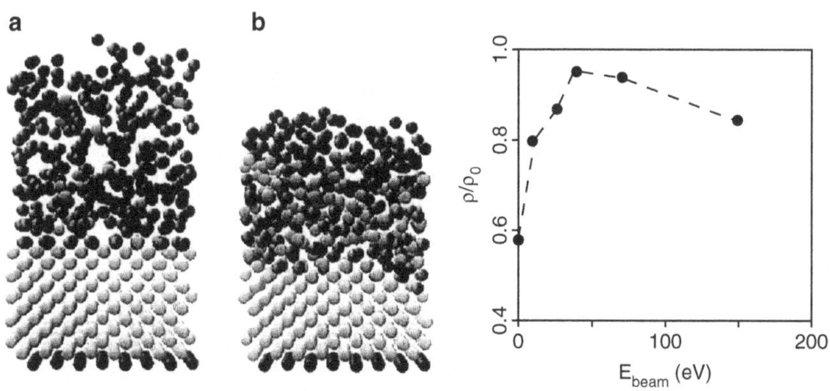

Fig. 5.16 Tersoff simulations of amorphous carbon deposition. *Left*: thin films deposited with (**a**) 1 eV, and (**b**) 40 eV beams. *Right*: Film density relative to diamond as a function of the incident energy. Figures from Ref. [107]: reprinted with permission from the American Physical Society

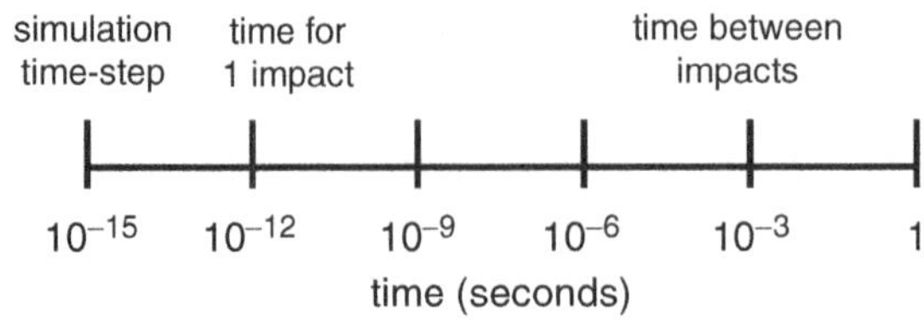

Fig. 5.17 Schematic plot of timescales for energetic impact process in thin-film deposition of amorphous carbons. The thermal spike lifetime (not shown) is $\approx 10^{-14}$ ps. Original figure by author

The most important concept to keep in mind when performing simulations of thin film deposition is time (Fig. 5.17). At the numerical level the integration timestep must be short enough to correctly integrate the high energy processes arising during impact. On the timescale of the thermal spike the atomic motion must be followed for sufficiently long that the system can return to the background temperature of the substrate. From the point of view of film growth the comparative long time between impacts must be taken into account; qualitative inclusion of this effect was a major recent advance. Lastly, the wallclock time of the simulation must also be acceptable to the user. Calculations lasting one month are not atypical and thus a delicate balance is required between transferability and efficiency.

In this section we consider the two dominant concepts in thin-film deposition simulations; the energy window effect (and associated impact processes), and the effect of temperature. Attention is paid to the various theoretical models of film-growth mechanisms which have attracted attention through the years. We also cover aspects related to thin-film deposition, including MD studies of thermal spikes, as well as MC simulations of surface processes.

5.5.1 Energy Effects

The experimental discovery of the energy window effect stimulated major theoretical and computational interest. While the Tersoff simulations by Kaukonen and Nieminen reproduced some aspects of the energy dependence, the low sp^3 fractions casts doubt on the validity of the findings. Another limitation is that only ten particles were followed with full dynamics, the remainder being periodically rescaled to the substrate temperature. Particularly for the highest energies considered (150 eV), this approximation seriously comprises the quality of the simulations.

In recognition of these difficulties, this author performed 2D simulations of thin-film deposition [109]. Although the coordination was by definition fixed (i.e. sp^2-like), the simulations did reproduce the energy window effect for compressive stress (Fig. 5.18). In excellent qualitative agreement with experiment, the simulations show tensile stress at low energies, a transition to compressive stress at intermediate energies, and a reduction in stress at high energies. The simulations also enabled a simple test of the subplantation theory as discussed in Section 5.5.1.1.

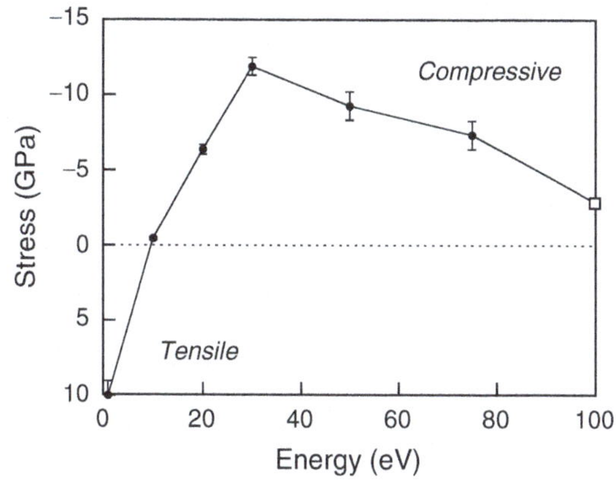

Fig. 5.18 The energy window effect for compressive stress reproduced using 2D simulations of carbon thin-film deposition. Interactions were described using a Stillinger–Weber potential parameterized to graphite. *Solid circles* denote deposited films – the data point at 100 eV was obtained by ion bombardment. Figure from Ref. [109]: reprinted with permission from the American Physical Society

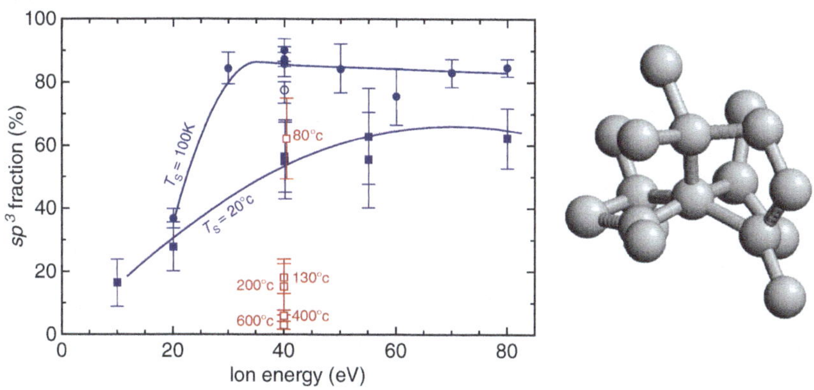

Fig. 5.19 *Left*: Energy and temperature dependence of the sp^3 fraction in thin-film deposition simulations using the Brenner potential with modified cutoffs. *Right*: local environments exhibiting unphysical fivefold coordination. Figures from Refs. [52, 110]: reprinted with permission from the American Institute of Physics and the American Physical Society, respectively

The first 3D simulations of amorphous carbon reporting high sp^3 fractions were performed by Jäger and Albe in 2000 [52]. They performed Tersoff and Brenner potential simulations of thin-film deposition, finding that both methods performed very poorly, consistent with liquid quenching calculations and the early work of Kaukonen and Nieminen [107]. In the case of the Brenner potential, films expected to be *ta*-C-like contained just 1% sp^3 bonding. However, when the R and S cutoff parameters of the Brenner potential were slightly increased, the sp^3 fraction increased to a rather respectable 85%. Furthermore, the simulations exhibit some aspects of the energy window effect as seen in the left panel of Fig. 5.19.

Unfortunately, this improved agreement comes at the expense of highly unphysical aspects in the microstructure. Firstly, the networks contain around 6% fivefold coordinated structures, a typical example of which is shown in the right panel of Fig. 5.19. Carbon atoms cannot form such local arrangements (it is inconsistent with the hybridization), casting a element of doubt on the validity of the simulations. Secondly, the RDF contains a large spike at 2.25 Å (Fig. 5.6), indicating that atoms are artificially trapped in metastable configurations intermediate between the first and second neighbour distances. The presence of these spikes is a warning sign that kinetics will be poorly described. We will return to the question of whether these modified-Brenner simulations can truly capture temperature effects in Section 5.5.2.

Jäger and Albe's work was accompanied by a flurry of other deposition simulations: Kaukonen and Nieminen using the Tersoff potential in 2000 [111], Yastrebov and Smith using the Brenner potential in 2001 [112], Kohary and Kugler using orthogonal tight-binding (OTB) in 2001 [113, 114], Cooper et al. using OTB in 2002 [51], and this author using EDIP in 2002 [60, 84] and subsequently [22, 115]. The Tersoff simulations reinforced the general conclusions from eight years prior [107], while the OTB simulations of Kohary and Kugler were restricted by the computational expense of tight-binding; they considered *a*-C grown at low energies (1–5 eV) using relatively small numbers of deposited atoms. In work involving this author, Cooper et al. applied OTB to thin-film deposition at the much higher deposition energy of 100 eV. These simulations employed a short cutoff to accelerate the performance, with hindsight a poor choice as evidenced by the sharp spike in the RDF in Fig. 5.6. Yastrebov and Smith did not reproduce high sp^3 fractions, but they were able to simulate the drop in sp^3 fraction at very high energies (500 eV).

In 2005 Zheng et al. [116] performed OTB simulations using a 2.6 Å cutoff and reproduced many aspects of the energy window effect. While not having a sharp spike, the second neighbour peak of the RDF compares quite poorly with experiment (see Fig. 2 in Ref. [116]). More recently Ma and co-workers [117] performed a detailed examination of deposition using the Brenner potential (see Section 5.5.1.1). This was followed in 2008 by Halac and Reinoso's Tersoff-based study [118] of multilayer thin films, work that complements similar MC simulations performed by Kelires and coworkers [119]. Another Tersoff simulation of deposition was reported in 2008 by Kim et al. [120] who associate the "satellite peak" in the RDF at 2.1 Å with residual stresses in the film. This interpretation is physically incorrect, given that the peak arises from deficiencies in the potential.

At the risk of appearing partisan, the best simulations of *ta*-C deposition todate have used EDIP. These simulations do not have unphysical spikes in the pair distribution function, and the RDF itself is in good agreement with neutron diffraction data (see Fig. 3 in Ref. [115]). EDIP also reproduces the energy window effect for the density, sp^3 fraction and compressive stress (Fig. 5.20). This success is qualified by two aspects which motivate future work: firstly, the maximum sp^3 fraction is not quite as high as observed experimentally, and secondly, the data at high incident energies is generated using impacts rather than explicit growth.

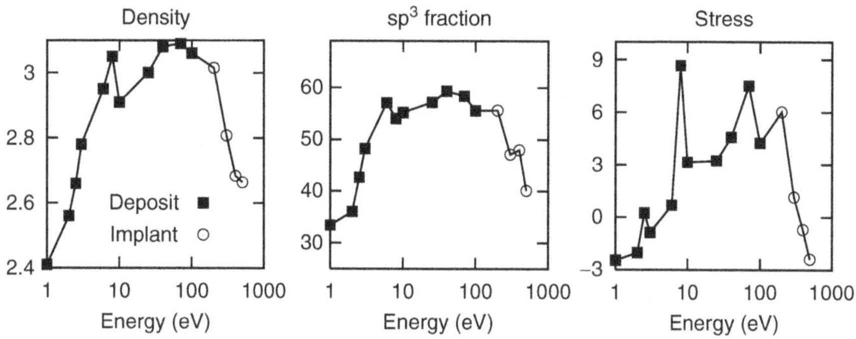

Fig. 5.20 The energy window effect reproduced by EDIP simulations of amorphous carbon thin-film deposition. *Solid squares* denote deposited films; *empty circles* denote equilibrium values following repeated high-energy impacts (25–50 atoms) into the 25 eV film. Units for the density, sp^3 fraction and stress are g/cc, percentages, and GPa, respectively. Figure adapted from Ref. [115]: reprinted with permission from Elsevier

5.5.1.1 Surface Effects and Impact Processes

One of the great attractions of deposition simulation is being able to answer vexing questions concerning growth mechanisms, transformation processes and the like. Perhaps the most contentious issue is the subplantation model (see Section 5.3). It is self-evident that at sufficiently high energies (say, 30–40 eV) the impact species will penetrate the surface and implant at varying depths into the film. The real question is whether the onset of subplantation correlates with the *a*-C to *ta*-C transition, and the concomitant densification and high compressive stress.

The first indication that subplantation may not be an essential element of the energy window effect came from the 2D deposition simulations [109] seen in Fig. 5.18. Significantly, it was found that newly arriving atoms do not implant into the subsurface; instead, new atoms deposit onto the surface, creating stress in a process of "energetic burial". Despite claims to the contrary [121], the classification is robust and not easily dismissed, as the 2D geometry accommodates a rigorous distinction between surface and bulk. In contrast, surface/bulk definitions in amorphous 3D systems are difficult to define and can be quite ambiguous.

Further tests of the subplantation model became possible when 3D simulations of film growth were performed. In a controversial result, EDIP simulations showed that the energetic burial concept also applies to sp^3 formation [22,60,115]. Detailed analysis of deposited *ta*-C films showed that newly arriving atoms primarily adopt *sp* and sp^2 configurations, with conversion to a majority sp^3 phase occurring only when further atoms are deposited. Recently, Ma et al. [117] reproduced the same result using the Brenner potential. They showed that newly deposited atoms (Fig. 5.21, left panel) primarily locate in the surface layer. In this region (see right panel of Fig. 5.21 for a structural view) the coordination is typically three or less, as compared to a predominantly tetrahedral phase (60–70%) in the bulk region beneath. Using 20 eV as an example, the Brenner depositions find sp^3 fractions of 70% in

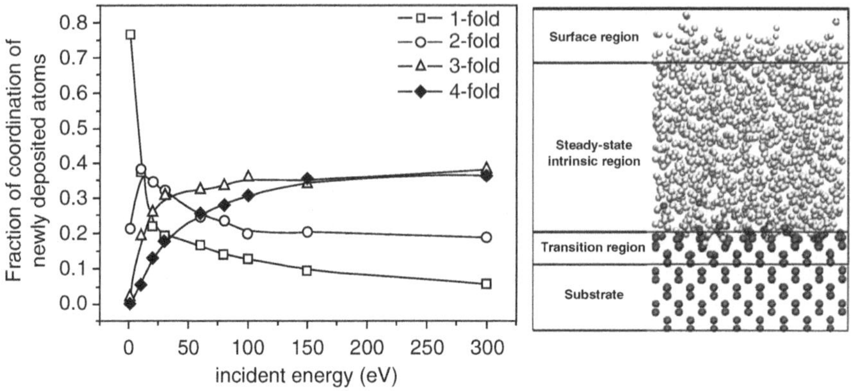

Fig. 5.21 Brenner potential simulations of amorphous carbon deposition. *Left*: Coordination fractions for newly deposited atoms as a function of the incident energy. Note the small amount of sp^3 bonding (*solid diamonds*) at the lowest energies producing *ta*-C (10–20 eV). *Right*: Depth profile of a film deposited with 30 eV ions, highlighting various regions in the microstructure. Figures adapted from Ref. [117]: reprinted with permission of the American Physical Society

the steady state region, and yet the newly deposited atoms (solid diamonds) have less than 15% sp^3 bonding. This behavior is incompatible with the subplantation model, but entirely consistent with the energetic burial process observed in the 2D simulations. Evidently, subplantation is not the critical phenomenon which separates *a*-C from *ta*-C. Finally, we note in Fig. 5.21 that even at high energies the coordination fraction of newly arriving atoms is still only around half of the average coordination of the bulk; EDIP simulations find a similar trend.

On this issue it is relevant to mention 1998 DFTB simulations by Uhlmann et al. [122] in which it is claimed that densification and sp^3 formation occur via subplantation. The critical point is that computational constraints severely restricted the scope of the simulations, effectively rendering the conclusions void. Calculations of successive bombardment considered only 10, 18 and 17 trajectories at energies of 20, 40 and 80 eV, respectively. Such a small number of deposited atoms is far too small to draw conclusions on how amorphous carbon films grow as it precludes the establishment of a steady-state region as seen in Fig. 5.21 and ignores statistical fluctuations associated with multiple hybridizations and the absence of crystalline order. To converge the analysis with respect to these quantities at least several hundred atoms be deposited. With modern computing resources these calculations may now be possible. Given the excellent transferability of DFTB to amorphous carbon (Fig. 5.8), such a calculation would very extremely worthwhile.

Aside from subplantation, another important question in amorphous carbon deposition is the role and nature of the thermal spike. Several detailed studies have been performed, the first being Monte Carlo simulations of dynamic roughening by Casiraghi et al. [123]. Using a discrete (non-atomistic) model of the thermal spike (Fig. 5.22), they explained why the surface of *ta*-C is unusually smooth (Fig. 5.3). The crucial insight is that the thermal spike flattens a small local region, the size of which can be inferred by comparing scaling exponents from the simulation with

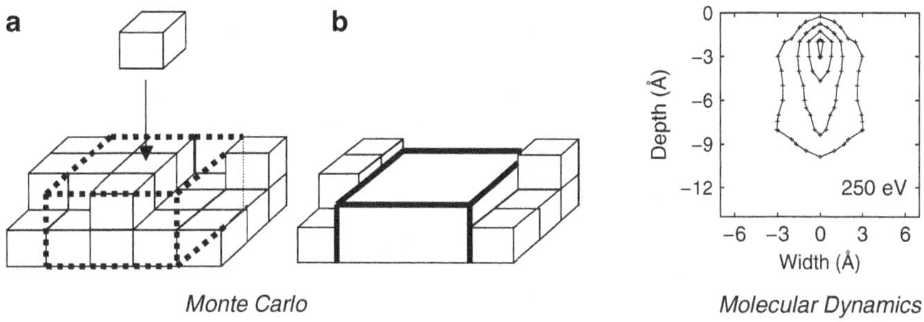

Fig. 5.22 *Left*: Model used in Monte Carlo simulations of dynamic roughening. Schematic in (**a**) illustrates the thermal spike volume (*dotted line*) arising from an energetic impact. The resultant local melting flattens the surface locally as shown in (**b**). *Right*: Contour plot of the spatial distribution of broken bonds in EDIP MD simulations of 250 eV thermal spikes in *ta*-C. Statistics accumulated over 80 impacts using unique impact parameters. Figures from Refs. [123, 124]: reprinted with permission from the American Physical Society and Elsevier, respectively

those obtained from AFM experiments. In further work, Moseler et al. [125] applied multiscale methods to illustrate how downhill currents from thermal spikes lead to ultrasmoothness in *ta*-C. Continuum methods were applied to show how the density profile evolves to fill in valleys, with the coefficients determined from MD simulations of the thermal spike using the Brenner potential. The Brenner simulations (validated in the same work by a smaller DFTB study), show the presence of small impact craters, even for energies as low as 30 eV. Several of these conclusions were independently confirmed by EDIP calculations performed at the same time [124]. The EDIP studies also quantified the shape and energy-dependence of the spike. To a good approximation the spike is cylindrical, with a depth that varies linearly with energy. In contrast, the radius has a near constant value of 3 ± 0.5 Å, directly confirming the spatial extent inferred from the MC simulations of roughening.

5.5.2 Temperature Effects

In the experimental synthesis of amorphous carbon, the substrate temperature is a straightforward control variable. With computer simulation, however, it can be extremely difficult to capture temperature-related effects. The reason is fundamental and easily stated: thermally activated processes occur infrequently over long times (milliseconds, seconds, and higher), while the timescale of MD is very short (typically picoseconds). This enormous difference immediately precludes traditional MD from reproducing many temperature-driven processes. Indeed, one should in fact be quite suspicious of any MD simulation in which direct agreement with experiment is obtained regarding temperature, as this typically means the activation barriers in the simulation are much too small compared with experiment.

For amorphous carbon deposition, the gap in timescale is around nine orders of magnitude (Fig. 5.17), this being the ratio between the time between impacts (circa 1 ms) and the time to simulate a single impact (around 1 ps). In MD simulations of deposition the standard approach is to rethermalize the system after the spike has concluded and then add the next atom. As compared to experiment, this omits any thermally activated process which might occur prior to the arrival of the next atom. By applying the Arrhenius expression,

$$f = A\exp[-E/kT] \tag{5.1}$$

one can estimate the barrier height of reaction processes which are missing in the simulation. As an example, consider an experimental substrate temperature of T = 500 K, a time between impacts of 1 ms, and a typical carbon vibrational frequency A = 10^{14} Hz. For these conditions, reaction processes with a barrier $E = 1.1$ eV will occur at a frequency of 1,000 Hz, that is, once during the inter-impact period of 1 ms. Reaction processes of lower barrier will occur much more rapidly due to the exponential dependence. In the simulation, however, these thermally-activated processes are never seen, since the picosecond timescale of the calculation essentially freezes out these infrequent events.

The solution to this problem was first established by this author in 2006 [126,127] when MD simulations were performed using a inter-impact temperature-pulsing technique to activate the rare events. The entire system was heated for one picosecond between impacts, and a large set of deposition simulations were performed for varying values of the activation temperature T_{act}. Figure 5.23 provides a dramatic

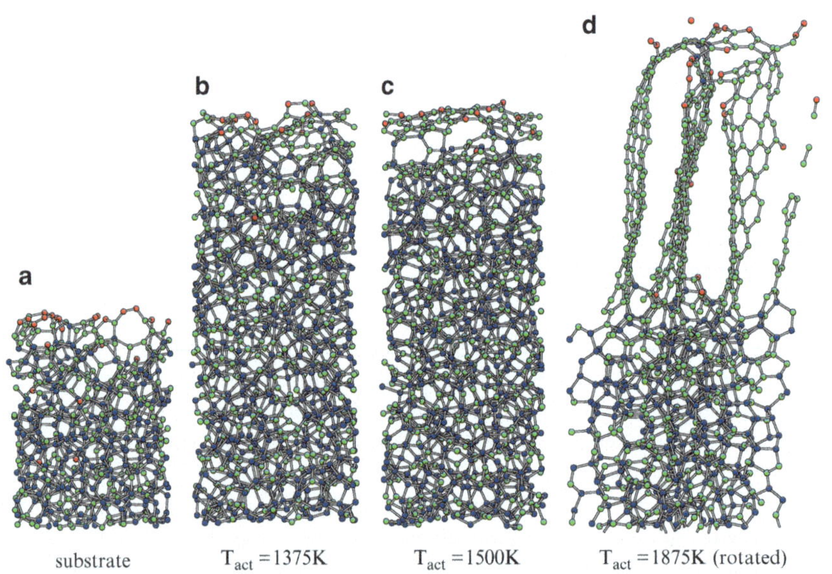

Fig. 5.23 EDIP simulations of thin-film deposition in which inter-impact temperature-pulsing (with varying T_{act}) is applied to activate rare events. (**a**) *ta*-C substrate prior to deposition, (**b**) conventional *ta*-C growth is observed, (**c**) *ta*-C growth plus horizontal layering is observed, (**d**) vertically oriented graphite-like sheets form (structure has been rotated to highlight their relative alignment). Figure from Ref. [126]: reprinted with permission from the Taylor and Francis Group

illustration of the importance of capturing these rare events. For activation temperatures up to 1,500 K the presence of temperature-pulsing has no effect, with *ta*-C produced independent of the temperature used. There is, however, some surface ordering noticeable at 1,500 K which is not seen at 1,375 K. Applying Eq. 5.1 and using $f = 10^{12}$ Hz, this indicates that the energy barrier for surface relaxation into horizontal graphite-like structures is around 0.6 eV. At a slightly higher temperature, a profound transition occurs in which the film no longer deposits as *ta*-C, but instead grows as vertically oriented graphite sheets. This is not a phase transition in the thermodynamic sense, but instead a entirely different growth mode which is accessed due to the activation of the rare events. For further details on the atom-by-atom nature of this process and its experimental implications see Refs. [30, 128].

The agreement between experimental data and the simulations employing temperature-pulsing is excellent. Qualitatively, the simulations reproduce all aspects of the experimental data, including observations that the structure above the transition point is fully sp^2 bonded and that vertical orientation is present. The simulations also reproduce the abrupt nature of the transition (left panel of Fig. 5.24) with the horizontal-axis scaling of the experimental data accounting for the timescale difference. Another significant insight arising from the temperature-pulsing simulations is the explanation of the deposition rate effect. As seen earlier (Fig. 5.4), increasing the deposition rate raises the critical temperature of the transition. The rare-event framework explains why: higher deposition rates reduce the inter-impact time for thermally activated relaxation. Consequently, higher substrate temperatures are required in order to trigger the critical atomic rearrangements. Note, however, that simulations todate have not elucidated actual reaction mechanism. Understanding this process will illuminate a fundamental connection between *ta*-C and graphite, and represents an important area for future study.

Having established the physical processes relating to substrate temperature during deposition, it is appropriate to consider other simulations in the literature.

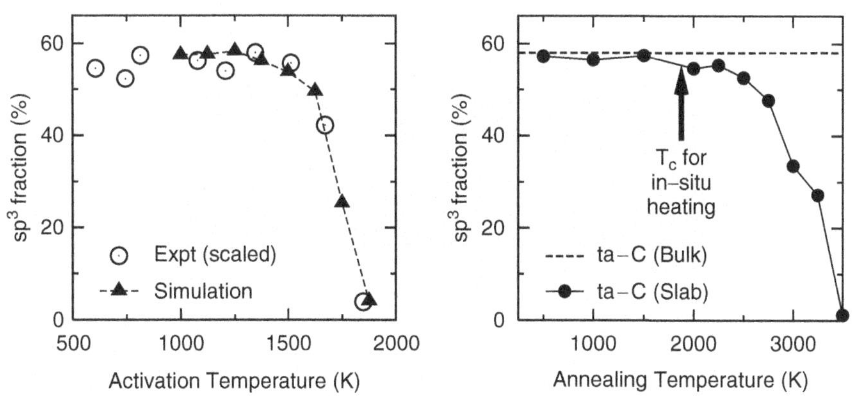

Fig. 5.24 *Left*: EDIP simulations of thin-film deposition at 70 eV using temperature pulsing. *White circles* denote scaled experimental data from Chhowalla et al. [33]. *Right*: EDIP simulations of extended annealing (100 ps) for *ta*-C in slab (*solid line*) and bulk (*thin line*) geometries. Figures adapted from Ref. [126]: reprinted with permission from the Taylor and Francis group

Paradoxically, the early Tersoff simulations (Fig. 5.5 and Ref. [129]) in which no temperature effect was observed are in some sense "correct". One does not of course mean physically correct, but rather that the simulations do not capture the correct timescales, and hence conventional MD simulations performed at the experimental substrate temperature should not see a transition. In ironic contrast, the simulations that at first appear to agree well with experiment (Brenner simulations by Jäger and Belov [110, 130, 131], and the OTB simulations by Zheng et al. [116]) are in fact suspect. Although both of these studies report strong variations with substrate temperature (see, for example, Fig. 5.19 in the chapter and Fig. 5 in Ref. [116]), there are several signs that these transitions are artifacts related to low-barriers. The Brenner calculations, for example, have a sharp peak in the RDF (Fig. 5.6) and five-fold coordinated atoms (Fig. 5.19). With the OTB calculations the second-neighbour peak in the RDF is much too small, and the post-*ta*-C transition involves an *a*-C-like material with 20–40% sp^3, in marked contrast to experiment where the bonding is 100% sp^2. All of these factors are at odds with experimental and DFT benchmarks: when combined with the physical arguments based on timescale one is forced to conclude that these simulations do not accurately describe the situation in the laboratory.

The right-hand panel of Fig. 5.24 summarizes EDIP results for the other critical temperature effect in *ta*-C, namely, post-deposition annealing. Again the limitations with respect to timescale apply; at best MD can simulate nanoseconds, while experiments operate on the timescale of hours (20 or 60 min in the case of Ferrari et al. [37]). Correspondingly, the anticipated computational result is that annealing transition temperatures should significantly exceed the experimental value of 1,400 K. Indeed, this is exactly what was observed in 100 ps EDIP simulations of a *ta*-C slab [126], in which a gradual bulk conversion to sp^2 occurs above 2500 K, accompa-

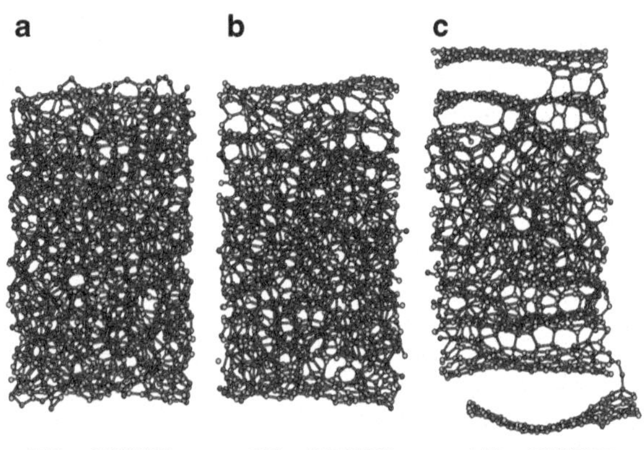

Fig. 5.25 EDIP annealing simulations (t = 100 ps) of a slab of *ta*-C generated by liquid quenching at 3.0 g/cc. (**a**) Below 2,000 K no change was observed. (**b, c**) At higher temperatures a gradual decrease in the bulk density occurs, accompanied by surface exfoliation of graphitic layers. Figure adapted from Ref. [126]: reprinted with permission of the Taylor and Francis group

nied by exfoliation of graphite-like layers (Fig. 5.25). The importance of this free surface was recently shown in EDIP simulations [9] in which bulk *ta*-C samples where subjected to extended annealing. Even at temperatures as high as 3,500 K, the sp^3 fraction remained unchanged, suggesting that *ta*-C annealing is primarily a surface-driven phenomena.

Considering the case of a slab with a free surface, an estimate of the appropriate temperature ratio between the simulation and experiment can be obtained by eliminating E from Eq. 5.1 applied at both temperatures. Reasonable values ($f_{\mathrm{sim}} = 10^{10}$ Hz, $f_{\mathrm{expt}} = 1$ Hz) give a temperature ratio of around 3.5, corresponding to a simulation temperature of $\approx$4,900 K. This value sits well above the melting point, and so it is unsurprising that sp^2 conversion commenced in Fig. 5.24 at lower temperatures. The key point, in fact, is that even longer (i.e. more gentle) simulations are required in order to activate the rare events without melting the material.

These arguments show that computational studies of *ta*-C annealing should not observe transitions at the experimental values. On this matter, the use of elevated temperatures by Kim et al. [132] in their studies of stress reduction is entirely correct. Belov and Jäger also use a elevated temperatures up to 3,000 K in their study of relaxation effects [130], but the explicit correlation with the experimental timescale is not discussed. More questionable is the 100 ps annealing simulation by Zheng et al. [116] in which MD performed with OTB predicts a gradual change, followed by a sharp transition at 1,400 K. On the basis of the arguments presented here, this computational result is unlikely to be correct. It is more difficult to assess the validity of the Tersoff/MC study by Titantah and Lamoen [101], although they do observe a sharp transition at 1,500 K. The question of whether MC is an appropriate framework to study annealing is unanswered; at the very least a systematic study of convergence with respect to the number of MC steps is required. This is in fact a general issue for MC simulations of amorphous carbons which has yet to be addressed. The absence of time as a defined quantity in MC makes all such studies necessarily qualitative, a somewhat disconcerting situation given the metastability of amorphous carbons. In contrast, MD simulations enjoy a strong physical justification and a strict connection with experimental conditions, enabling the robust liquid quenching and deposition methodologies as described above.

5.6 Related Carbon Materials

For clarity and scope this chapter has thus far concentrated on the amorphous carbons (*a*-C and *ta*-C). There remain, however, many other carbon-based systems which are closely related, and for which computer simulation is an attractive tool. Principally, these associated areas fall into three categories: (i) hydrogenated amorphous carbons, in which the H atoms saturate undercoordinated atoms, (ii) doping of *a*-C and *ta*-C with impurities such as boron and nitrogen, and (iii) disordered or glassy carbons which are predominantly sp^2 bonded yet lack the long-range order of graphite. Each of these topics is a field in itself, and so we emphasize only the general trends with respect to amorphous carbon simulation.

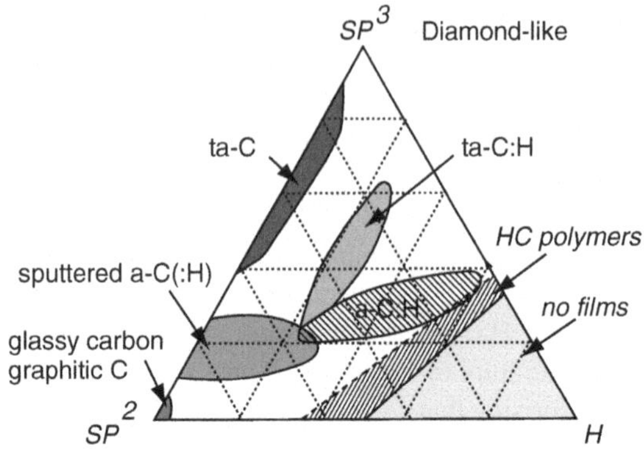

Fig. 5.26 Ternary phase diagram of bonding in amorphous carbon–hydrogen alloys. Amorphous carbons fall along the left-hand edge, with hydrogen-containing structures extending into the interior. The network topology changes markedly with H, trending towards polymeric; above a critical value there is insufficient connectivity to form films. Figure from Ref. [39]: reprinted with permission from Elsevier

The starting point for discussing hydrogenated amorphous carbons is the standard ternary phase diagram (Fig. 5.26) which summarizes the relationship between the various phases according to the carbon hybridization and fractional content of hydrogen. Amorphous carbons are found on the left side of the triangle, accompanied by various ordered sp^2 phases such as graphite and glassy carbon. With increasing H content a wide variety of materials are produced, many of which have useful physical and optical properties. As in hydrogenated amorphous silicon, hydrogen acts to passivate isolated sp^2 atoms, thereby passivating dangling bonds. While electronically attractive, the presence of hydrogen significantly alters the topology, and so *a*-C:H and *ta*-C:H films are necessarily much softer than their pure carbon counterparts. At even higher hydrogen/carbon ratios the films becoming increasingly polymer-like, eventually reaching a hydrogen fraction too great to provide 3D connectivity. Above this limit no films are found.

Initially it is tempting to borrow *carte blanche* the computational machinery of pure amorphous carbons to describe their hydrogenated analogs. Indeed, liquid quenching studies have been performed at multiple levels of theory, using a hydrocarbon Tersoff potential [133], the Brenner potential [134], OTB [135], DFTB [136–138] and DFT [139–141]. These studies have examined the effect of the density and hydrogen fraction on the microstructure and electronic properties, and agreement with experiment is generally reasonable, particularly for the higher levels of theory. The question that has to be asked however, is whether the computational and experimental synthesis are in any way alike. Hydrogenated amorphous carbons are prepared using deposition systems which are rather more complex than the ion beam approaches used for pure amorphous carbons (see Ref. [27] for a discussion of the differences). Since CH_x species are necessarily present, the growth process becomes akin to plasma-enhanced chemical-vapour-deposition, in which both physical and chemical effects act at the surface, accompanied by a rich hydrocarbon chemistry in the plasma [142]. The correspondence with amorphous carbon deposition is closest for *ta*-C:H films produced using highly ionized plasmas [143], where the incident ions ($C_2H_2^+$ species) generate the familiar energy window effect. However, under more conventional conditions appropriate for *a*-C:H formation, any resemblance to amorphous carbon deposition is lost, and there is little justification for borrowing the concepts and principles applying to pure carbon phases.

Due to the chemical complexity of *a*-C:H and *ta*-C:H synthesis, few deposition simulations of hydrogenated amorphous carbons have been performed. Godwin et al. [144] used a specially parameterized TB model for hydrocarbons to model formation of *a*-C:H in the presence of ion bombardment. Their model is particularly notable for the care taken to mimic the corresponding experimental conditions. The most significant contributors to *a*-C:H deposition have been Neyts and Bogaerts and co-workers, who performed an extensive series of REBO simulations [145–148] exploring the effect of hydrogen, the role of internal energy and impact angle, as well as the different chemical reactivity of incident species. Simulations of the initial stages of film growth (50 deposited molecules) were also reported by Mrovec et al. [81] using the ABOP potential. As a final comment on *a*-C:H it is now recognized that significant amounts of carbon "dust" in the interstellar medium is some form of hydrogenated amorphous carbon [149]. This material (known as HAC in the astrophysical community) accounts for $\approx$80% of all carbon, far outweighing the more commonly known PAH's (polycyclic aromatic hydrocarbons). For a overview of carbon in space see the review by Enhrenfreund and Charnley [150].

The presence of a significant optical gap in both *a*-C and *ta*-C immediately suggested doping as a route to controllably modify optical and electronic properties. Consequently, considerable experimental attention was devoted to studying these effects, with boron [151–153], nitrogen [151, 154–156] and phosphorus [151, 155] the common dopants. In contrast, simulations of these systems are much less common, a situation which is not surprising given the complexity of describing even the pure phases. Another restriction comes from the need to describe electronic structure effects, essentially setting tight-binding as the minimum level of theory.

The first simulations of doping in amorphous carbon were performed in 1996 by Sitch et al. [157] who used DFTB to investigate nitrogen and boron substitution in a pure *ta*-C network previously created by liquid quenching. Shortly afterwards, similar studies were applied using the LBDF method [158] (nitrogen doping), and DFT [159] (nitrogen, oxygen and chlorine doping). Although these substitution calculations found general agreement with experiments and early theoretical work by Robertson and Davis [160], the simulations were fundamentally limited by the artificial use of a pure carbon network as the structural template. Post-quenching substitution leads to the breaking of bonds and the creation of dangling-bond defects, whereas in reality, the presence of dopants significantly modifies local structure *during* the quenching of the liquid-like state.

Quenching of doped liquid carbon was first performed in 1999 by Köhler et al. [162] using DFTB. They found two types of nondoping nitrogen impurities, both of which were electronically inactive due to the presence of lone pairs. They

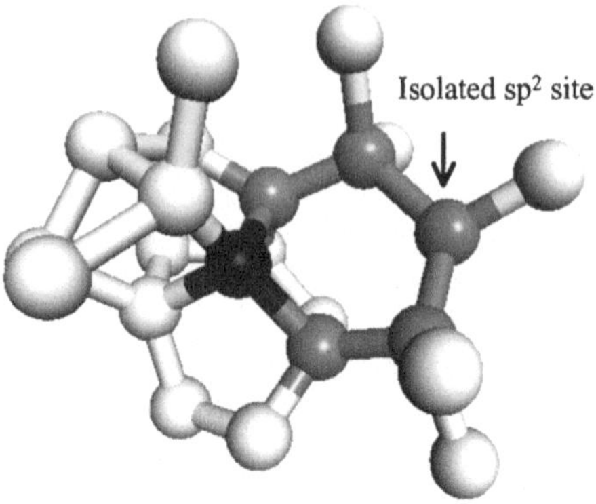

Fig. 5.27 Example of sp^3 doping in *ta*-C:N arising from a fourfold coordinated nitrogen atom (*dark sphere*). All other atoms are carbon, with those in six-membered rings shown as grey. The isolated sp^2 site indicated by the arrow is an unusual, yet not unknown, structural element. Calculations performed using DFT and quenching of doped liquid carbon. Figure from Ref. [161]: reprinted with permission from Elsevier

further found that the most prevalent electronically active dopant was a threefold-coordinated sp^2-bonded nitrogen with one additional π bond to a carbon neighbour. Significantly, this structure had been previously been discarded based on its energetics in post-substitution of pure *ta*-C networks. The other dopant configuration for nitrogen is the sp^3 bonded fourfold site (Fig. 5.27) familiar from conventional semiconductor doping of Si and Ge. Several other DFT calculations of *ta*-C:N have been performed [161, 163, 164], but generally speaking, interest has cooled due to the realisation that *ta*-C is not a particularly useful electronic material.

Other dopants for amorphous carbon have received even less attention than nitrogen. The only calculation for phosphorus doping that the author is aware of has dubious value, involving a 100% sp^3-bonded WWW structure as the model of the *ta*-C network [165]. Slightly more information is available for *ta*-C:B, where, in addition to the original DFTB work [157], both DFT [166] and Tersoff [167] simulations have been performed. The DFT calculations apply quenching to a boron-doped liquid, helping to explain why boron-doped films have lower stress [153]. The Tersoff simulations take an interesting approach, addressing ion implantation as the doping mechanism, but one must question the wisdom of using the Tersoff potential to describe the amorphous carbon component.

The third class of system which relates closely to the amorphous carbons are the semi-ordered sp^2 phases on the lower-left of the ternary phase diagram (Fig. 5.26). These materials are synthesized by pyrolysis of a carbonaceous material, resulting in a highly sp^2-bonded material such as glassy carbon or a carbon char. Carbons generated in this manner contain microstructural pores and have extremely large surface areas, leading to numerous industrial applications in areas such as filtration, gas adsorption and water treatment. As for the amorphous carbons, obtaining atomistic insight into the structure of these materials has been challenging.

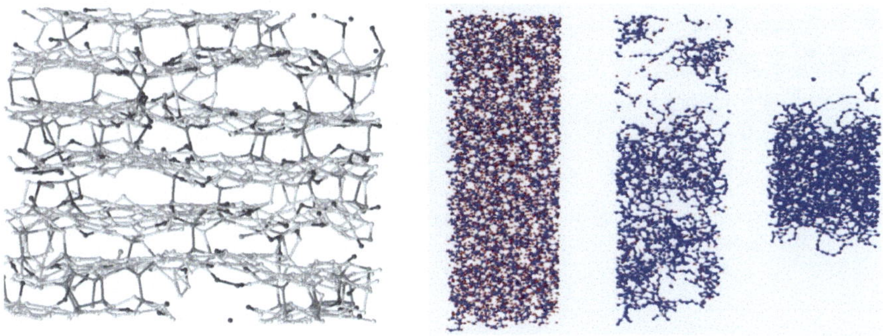

Fig. 5.28 Two approaches for the simulations of structure in a carbon char. *Left*: inversion method employing HRMC with EDIP. *Right*: explicit MD simulation in which hydrogen atoms (shown as *red dots*) are progressively removed to mimic pyrolysis. Figures from Refs. [168, 169]: reprinted with permission from Elsevier and the American Physical Society, respectively

Semi-ordered sp^2 structures are now studied using a variety of techniques employing either inversion or direct simulation. An example of each methodology is shown for a carbon char in Fig. 5.28. Inversion techniques date to 1988 when McGreevy and Putsai [170] proposed the Reverse Monte Carlo (RMC) approach as a general tool for the determination of real-space structure in disordered solids. Their insight was that experimental diffraction data, $S(Q)$, can serve as a target function for a "backwards" Monte Carlo simulation in which random atomic displacements are accepted or rejected on the basis of the difference between the simulated and experimental $S(Q)$. RMC was applied to the structural modeling of *ta*-C by Walters et al. [171], but RMC simulations of highly sp^2 carbons exhibit excessive numbers of three-membered rings [172]. Circumventing this problem, Gubbins et al. [173, 174] performed constrained RMC (applying geometric and hybridization restrictions), while Opletal et al. [175] proposed a hybrid RMC in which the S(Q) target is augmented with energy information from an interatomic potential. The hybrid approach significantly improves structural stability, and HRMC simulations todate have used both EDIP [168, 176, 177] and REBO [178]. Extending the HRMC concept, Nguyen et al. [179] recently proposed an alternative augmentation in which pore information dramatically improved the description of medium range order.

Direct simulation of sp^2 structures follow the conventional MC/MD pathway in which an atomic configuration is evolved on the basis of energies or forces. The difficulty here is time, as the experimental conditions are not amenable to quantitative MD simulation. Nevertheless, Lawson and Srivastava [169] achieved a milestone in their 2008 MD simulation of pyrolysis of a hydrocarbon precursor. Using the REBO potential, they mimicked hydrogen abstraction by eliminating individual H atoms, driving the dramatic structural change evident in Fig. 5.28.

Efforts to understand glassy carbon by direct simulation neglect hydrogen altogether, concentrating instead on the microstructure of the final phase. Recent progress in this area includes MC modeling of purely sp^2 materials (Fig. 5.29),

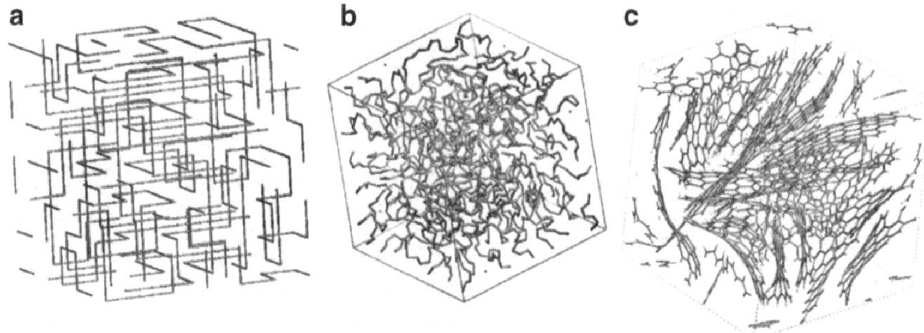

Fig. 5.29 Monte Carlo simulation of an sp^2 bonded disordered carbon. (**a**) polymer backbone on a simple cubic lattice, (**b**) starting polymer structure, (**c**) final structure after 4,000 MC cycles at 800°C and ρ=1.72 g/cc. Figure from Ref. [7]: reprinted with permission from Elsevier

structural assembly of fullerene fragments [6], MD simulation with an sp^2 potential [8], and MD simulation with EDIP [9]. Broadly speaking, all four methods generate structures similar to those in Fig. 5.29. The simulation resolve a 50-year old problem, demonstrating that fullerene-like fragments are an essential element of the glassy carbon microstructure [180]. This insight underpins modern explicit studies of gas adsorption employing Virtual Porous Carbons [181]. As a final comment, the EDIP simulations [9] revealed a remarkable result. Upon extended annealing, amorphous carbons will spontaneously transform into ordered sp^2 structures, such as carbon onions, nanotubes, graphite layers and glassy carbon. The governing factor is free surfaces (external boundaries and internal voids) which control self-assembly and rehybridization. This connection reveals a profound and unexpected relationship between amorphous carbon and other phases exhibiting extended order.

5.7 Summary

Atomistic simulation of amorphous carbons is well summarized by the Einstein maxim, “as simple as possible, but not simpler”. Interatomic potentials and simulation methodologies which do not capture the *essential* chemistry and physics should be put aside and replaced with more accurate versions. In practical language, this means that amorphous carbon simulations should no longer be performed with short-ranged interaction potentials, be they analytic potentials such as the Tersoff and Brenner methods, or tight-binding simulations with aggressive cutoffs (i.e. less than 3 Å). The verdict from simulations of liquid quenching and thin-film deposition is clear; when short cutoffs are used the sp^3 fraction, the microstructure, the radial distribution function and reaction kinetics are seriously compromised. In some instances these deficiencies manifest themselves as unphysical properties, while in other cases experimental trends are falsely reproduced. This leads to great confusion, creating a spectrum of results spanning fully trustworthy to completely wrong. Needless to say, this makes life very difficult for new entrants to the field.

In contrast, when long-range interactions are properly included, analytic potentials such as EDIP, LCBOP and ReaxFF invariably perform well. In liquid quenching simulations the correct experimental trends are reproduced, and the microstructural elements are physically realistic. Even better performance is obtained with higher levels of theory such as environment-dependent TB, DFTB, LBDF and full DFT. The compromise when using electronic structure methods is of course computational cost. Since matrix diagonalisation scales as $O(N^3)$, where N is the number of atoms, analytic potentials will always have an advantage for large systems. And yet, for properly chosen conditions, it is now possible to perform modest thin-film deposition simulations using a tight-binding approach such as DFTB.

On the question of Monte Carlo versus Molecular Dynamics, it is uncontroversial to note that MD has provided the most significant insights into amorphous carbon todate; MD is a natural match for the metastable formation conditions of amorphous carbons. That said, there may well be a number of niche applications for MC, such as the connection between structures in the carbon family, the nature of rare event activation in deposition, and perhaps even in the context of liquid quenching. In the case of the latter, however, detailed comparison between MD and MC is required, employing a common, and physically reasonable, potential.

On the matter of amorphous carbons, there are many questions awaiting future study. The energy-window effect, for example, has yet to be modeled properly at high energies (i.e. without implantation). Post-deposition annealing of stress is not understood, and the rare events which are triggered at the critical temperature T_c are not known. The reason for differences in the surface roughness and energy dependence between FCVA- and MSIB-deposited films are similarly not known, and simulations are yet to consider the finite width of the incident ion energy spectrum. Finally, there remain the many materials related to the amorphous carbons in which hydrogen and/or sp^2 order is present. These fields are even less mature than the amorphous carbons, and once again, simulation has much to offer. Given the amazing flexibility of carbon, it is an offer one is well-advised to accept.

Acknowledgements The author is only in the position of being able to write this account due the scientific leadership and contributions of many different individuals and groups through the years. Personal thanks go to David McKenzie at The University of Sydney, Dougal McCulloch at RMIT University, Mike Finnis (formerly at Queen's University of Belfast), and Michelle Parrinello and Marco Bernasconi (formerly at the Max-Plank-Institute for Solid-state Physics, Stuttgart). Many other colleagues, students and co-workers also assisted in the development of ideas, computer codes and applications. Their contributions and friendship is greatly appreciated.

References

1. Harris PJF (2001) Interdiscipl Sci Rev 26(3):204
2. Franklin RE (1951) Proc Roy Soc A 209(1097):196
3. Ban LL, Crawford D, Marsh H (1975) J Appl Crystallogr 8:415
4. Jenkins GM, Kawamura K (1971) Nature 231:175
5. Crawford D, Johnson DJ (1971) J Microsc 94:51

6. Harris PJF (2004) Phil Mag 84(29):3159
7. Kumar A, Lobo R, Wagner N (2005) Carbon 43(15):3099
8. Shi Y (2008) J Chem Phys 128(23):234707
9. Powles RC, Marks NA, Lau DWM (2009) Phys Rev B 79:075430
10. McKenzie DR, Muller D, Pailthorpe BA (1991) Phys Rev Lett 67(6):773
11. Aisenberg S, Chabot J (1971) Appl Phys Lett 42:2953
12. Askenov II, Belous VA, Padalka VG, Khoroshikh VM (1978) Sov J Plasma Phys 4:425
13. Askenov II, Vakula SI, Padalka VG, Strelnitskii VE, Khoroshikh VM (1980) Sov Phys Tech Phys 25:1164
14. Angus JC, Hayman CC (1988) Science 241(4868):913
15. Grill A (1999) Diam Relat Mater 8(2-5):428
16. Hofsäss H, Feldermann H, Merk R, Sebastian M, Ronning C (1998) Appl Phys A-Mater 66(2):153
17. Fallon PJ, Veerasamy VS, Davis CA, Robertson J, Amaratunga GAJ, Milne WI, Koskinen J (1993) Phys Rev B 48(7):4777
18. Lifshitz Y, Lempert GD, Rotter S, Avigal I, Uzans-aguy C, Kalish R, Kulik J, Marton D, Rabalais JW (1994) Diam Relat Mater 3(4–6):542
19. Grossman E, Lempert GD, Kulik J, Marton D, Rabalais JW, Lifshitz Y (1996) Appl Phys Lett 68(9):1214
20. Schwan J, Ulrich S, Roth H, Ehrhardt H, Silva SRP, Robertson J, Samlenski R, Brenn R (1996) J Appl Phys 79(3):1416
21. Pappas DL, Saenger KL, Bruley J, Krakow W, Cuomo JJ, Gu T, Collins RW (1992) J Appl Phys 71(11):5675
22. Marks NA (2005) Diam Relat Mater 14(8):1223
23. Lifshitz Y, Kasi SR, Rabalais JW (1989) Phys Rev Lett 62(11):1290
24. Lifshitz Y, Kasi SR, Rabalais JW, Eckstein W (1990) Phys Rev B 41(15):10468
25. Lifshitz Y, Lempert GD, Grossman E (1994) Phys Rev Lett 72(17):2753
26. Robertson J (1993) Diam Relat Mater 2:984
27. Robertson J (1994) Diam Relat Mater 3:361
28. Davis CA (1993) Thin Solid Films 226(1):30
29. Koponen I, Hakovirta M, Lappalainen R (1995) J Appl Phys 78(9):5837
30. Lau DWM, McCulloch DG, Taylor MB, Partridge JG, McKenzie DR, Marks NA, Teo EHT, Tay BK (2008) Phys Rev Lett 100(17):4
31. Ferrari AC, Rodil SE, Robertson J, Milne WI (2002) Diam Relat Mater 11(3–6):994
32. Alam TM, Friedmann TA, Schultz PA, Sebastiani D (2003) Phys Rev B 67(24):245309
33. Chhowalla M, Robertson J, Chen CW, Silva SRP, Davis CA, Amaratunga GAJ, Milne WI (1997) J Appl Phys 81:139
34. Hirvonen JP, Koskinen J, Kaukonen M, Nieminen R, Scheibe HJ (1997) J Appl Phys 81:7248
35. Koskinen J, Hirvonen JP, Keränen J (1998) J Appl Phys 84:648
36. Tay BK, Shi X, Liu EJ, Tan HS, Cheah LK (1999) Thin Solid Films 346(1-2):155
37. Ferrari AC, Kleinsorge B, Morrison NA, Hart A, Stolojan V, Robertson J (1999) J Appl Phys 85:7191
38. Tay BK, Sheeja D, Lau SP, Shi X, Seet BC, Yeo YC (2000) Surf Coat Tech 130(2-3):248
39. Robertson J (2002) Mat Sci Eng R 37(4-6):129
40. McKenzie DR (1996) Rep Prog Phys 59(12):1611
41. Silva SRP (ed) (2003) Properties of amorphous carbon (INSPEC, 2003)
42. Car R, Parrinello M (1985) Phys Rev Lett 55(22):2471
43. Abell GC (1985) Phys Rev B 31(10):6184
44. Tersoff J (1986) Phys Rev Lett 56(6):632
45. Tersoff J (1988) Phys Rev Lett 61(25):2879
46. Tersoff J (1988) Phys Rev B 37(12):6991
47. Brenner DW (1990) Phys Rev B 42(15):9458
48. Brenner DW (1992) Phys Rev B 46:1948
49. Brenner DW, Shenderova OA, Harrison JA, Stuart SJ, Ni B, Sinnott SB (2002) J Phys Condens Mat 14(4):783

50. Brenner DW (2000) Phys Status Solidi B 217(1):23
51. Cooper NC, Fagan MS, Goringe CM, Marks NA, McKenzie DR (2002) J Phys Condens Mat 14(4):723
52. Jäger HU, Albe K (2000) J Appl Phys 88:1129
53. Nordlund K, Keinonen J, Mattila T (1996) Phys Rev Lett 77(4):699
54. Broughton JQ, Mehl MJ (1999) Phys Rev B 59(14):9259
55. Kumagai T, Hara S, Choi J, Izumi S, Kato T (2009) J Appl Phys 105(6):064310
56. Stuart S, Tutein A, Harrison J (2000) J Chem Phys
57. Stuart SJ, Knippenberg MT, Kum O, Krstic PS (2006) Phys Scr T124:58
58. Liu A, Stuart SJ (2008) J Comput Chem 29(4):601
59. Marks NA (2000) Phys Rev B 63(3):7
60. Marks N (2002) J Phys Condens Mat 14(11):2901
61. Bazant MZ, Kaxiras E, Justo JF (1997) Phys Rev B 56(14):8542
62. Justo JF, Bazant MZ, Kaxiras E, Bulatov VV, Yip S (1998) Phys Rev B 58(5):2539
63. van Duin ACT, Dasgupta S, F. Lorant, W.A. Goddard (2001) J Phys Chem A 105(41):9396
64. Higginbottom P, Marks NA (2009, unpublished)
65. van Duin ACT, Strachan A, Stewman S, Zhang QS, Xu X, Goddard III WA (2003) J Phys Chem A 107(19):3803
66. Cheung S, Deng WQ, van Duin ACT, Goddard III WA (2005) J Phys Chem A 109(5):851
67. Nielson KD, van Duin ACT, Oxgaard J, Deng WQ, Goddard III WA (2005) J Phys Chem A 109(3):493
68. Los JH, Ghiringhelli LM, Meijer EJ, Fasolino A (2005) Phys Rev B 72(21):14
69. Los JH, Fasolino A (2003) Phys Rev B 68(2):14
70. Los JH, Ghiringhelli LM, Meijer EJ, Fasolino A (2006) Phys Rev B 73(22):1
71. Marks NA, McKenzie DR, Pailthorpe BA, Bernasconi M, Parrinello M (1996) Phys Rev Lett 76(5):768
72. McCulloch DG, McKenzie DR, Goringe CM (2000) Phys Rev B 61(3):2349
73. Han J, Gao W, Zhu J, Meng S (2007) Phys Rev B 75(15):1
74. Sankey OF, Niklewski DJ (1989) Phys Rev B 40(6):3979
75. Porezag D, Frauenheim T, Köhler T, Seifert G, Kaschner R (1995) Phys Rev B 51(19):12947
76. Frauenheim T, Seifert G, Elstner M, Niehaus T, Köhler C, Amkreutz M, Sternberg M, Hajnal Z, Di Carlo A, Suhai S (2002) J Phys Condens Mat 14(11):3015
77. Tang MS, Wang CZ, Chan CT, Ho KM (1996) Phys Rev B 53(3):979
78. Xu CH, Wang CZ, Chan CT, Ho KM (1992) J Phys Condens Mat 4(28):6047
79. Pettifor DG, Oleinik II (1999) Phys Rev B 59(13):8487
80. Oleinik II, Pettifor DG (1999) Phys Rev B 59(13):8500
81. Mrovec M, Moseler M, Elsaesser C, Gumbsch P (2007) Prog Mater Sci 52(2-3):230
82. Wang CZ, Ho KM (1994) Phys Rev B 50(17):12429
83. Stephan U, Haase M (1993) J Phys Condens Mat 5(49):9157
84. Marks NA, Cooper NC, McKenzie DR, McCulloch DG, Bath P, Russo SP (2002) Phys Rev B 65(7):9
85. Wooten F, Winer K, Weaire D (1985) Phys Rev Lett 54(13):1392
86. Drabold DA (2009) Eur Phys J B 68(1):1
87. Kelires PC (1992) Phys Rev Lett 68(12):1854
88. Galli G, Martin RM, Car R, Parrinello M (1989) Phys Rev Lett 62(5):555
89. Galli G, Martin RM, Car R, Parrinello M (1990) Phys Rev B 42(12):7470
90. Marks NA, McKenzie DR, Pailthorpe BA, Bernasconi M, Parrinello M (1996) Phys Rev B 54(14):9703
91. Marks NA (1997) Phys Rev B 56(5):2441
92. Drabold DA, Fedders PA, Stumm P (1994) Phys Rev B 49(23):16415
93. Cobb M, Drabold DA, Cappelletti RL (1996) Phys Rev B 54(17):12162
94. Schultz PA, Leung K, Stechel EB (1999) Phys Rev B 59(2):733
95. Chen X, Sullivan JP, Friedmann TA, Gibson JM (2004) Appl Phys Lett 84(15):2823
96. Johnson JA, Woodford JB, Chen XD, Andersson J, Erdemir A, Fenske GR (2004) J Appl Phys 95(12):7765

97. McBride WE, McKenzie DR, McCulloch DG, Cockayne DJH, Petersen TC (2005) J Non-Cryst Solids 351(5):413
98. Drabold DA, Fedders PA, Grumbach MP (1996) Phys Rev B 54(8):5480
99. Schultz PA, Stechel EB (1998) Phys Rev B 57(6):3295
100. Titantah JT, Lamoen D (2004) Phys Rev B 70(7):1
101. Titantah JT, Lamoen D (2008) J Phys Condens Mat 20(3):035216
102. Kelires PC (1993) Phys Rev B 47(4):1829
103. Kelires PC (1994) Phys Rev Lett 73(18):2460
104. Fyta MG, Remediakis IN, Kelires PC (2006) Phys Rev Lett 96(18):1
105. Davis CA, Amaratunga GAJ, Knowles KM (1998) Phys Rev Lett 80(15):3280
106. Lu Q, Marks N, Schatz GC, Belytschko T (2008) Phys Rev B 77(1):9
107. Kaukonen HP, Nieminen RM (1992) Phys Rev Lett 68(5):620
108. Ziegler JF, Biersack JP, Littmark U (1985) *The Stopping and Range of Ions in Solids*, vol. 1 (Pergmon, New York,)
109. Marks NA, McKenzie DR, Pailthorpe BA (1996) Phys Rev B 53(7):4117
110. Jäger HU, Belov AY (2003) Phys Rev B 68(2):13
111. Kaukonen M, Nieminen RM (2000) Phys Rev B 61(4):2806
112. Yastrebov S, Smith R (2001) Nucl Instrum Meth B 180:145
113. Kohary K, Kugler S (2001) Phys Rev B 63(19):1
114. Kohary K, Kugler S (2002) J Non-Cryst Solids 299:824
115. Marks NA, Bell JM, Pearce GK, McKenzie DR, Bilek MMM (2003) Diam Relat Mater 12(10–11):2003
116. Zheng B, Zheng W, Yu S, Tian H, Meng F, Wang Y, Zhu J, Meng S, He X, Han J (2005) Carbon 43(9):1976
117. Ma T, Hu YZ, Wang H, Li X (2007) Phys Rev B 75(3):1
118. Halac EB, Reinoso M (2008) Phys Rev B 77(22):7
119. Mathioudakis C, Kelires PC, Panagiotatos Y, Patsalas P, Charitidis C, Logothetidis S (2002) Phys Rev B 65:205203
120. Kim KS, Lee SH, Kim YC, Lee SC, Cha PR, Lee KR (2008) Met Mater Int 14(3):347
121. Kelires PC (1998) J Non-Cryst Solids 227:597
122. Uhlmann S, Frauenheim T, Lifshitz Y (1998) Phys Rev Lett 81(3):641
123. Casiraghi C, Ferrari AC, Ohr R, Flewitt AJ, Chu DP, Robertson J (2003) Phys Rev Lett 91(22):226104
124. Pearce GK, Marks NA, McKenzie DR, Bilek MMM (2005) Diam Relat Mater 14(3–7):921
125. Moseler M, Gumbsch P, Casiraghi C, Ferrari AC, Robertson J (2005) Science 309(5740):1545
126. Marks NA, Cover MF, Kocer C (2006) Mol Sim 32(15):1271
127. Marks NA, Cover MF, Kocer C (2006) Appl Phys Lett 89(13):131924
128. Taylor MB, Lau DWM, Partridge JG, McCulloch DG, Marks NA, Teo EHT, McKenzie DR (2009) J Phys Condens Mat 21(22):225003
129. Kaukonen MO, Nieminen RM (1995) Surf Sci 331:975
130. Belov A, Jäger H (2003) Nucl Instr Meth B 202:242
131. Belov A (2003) Comp Mater Sci 27(1–2):16
132. Kim KS, Lee SC, Lee KR, Cha PR (2008) J Kor Phys Soc 52(4):1272
133. Tersoff J (1991) Phys Rev B 44(21):12039
134. Salonen E, Nordlund K, Keinonen J (2001) Phys Rev B 63(19):195415
135. Kopidakis G, Wang CZ, Soukoulis CM, Ho KM (1998) Phys Rev B 58(21):14106
136. Blaudeck P, Frauenheim T, Porezag D, Seifert G, Fromm E (1992) J Phys Condens Mat 4(30):6389
137. Frauenheim T, Blaudeck P, Stephan U, Jungnickel G (1993) Phys Rev B 48(7):4823
138. Frauenheim T, Jungnickel G, Stephan U, Blaudeck P, Deutschmann S, Weiler M, Sattel S, Jung K, Ehrhardt H (1994) Phys Rev B 50(11):7940
139. Iarlori S, Galli G, Martini O (1994) Phys Rev B 49(10):7060
140. Bilek MMM, McKenzie DR, McCulloch DG, Goringe CM (2000) Phys Rev B 62(5):3071
141. Houska J, Klemberg-Sapieha JE, Martinu L (2009) Surf Coat Technol 203(24):3770

142. Weiler M, Sattel S, Giessen T, Jung K, Ehrhardt H, Veerasamy V, Robertson J (1996) Phys Rev B 53(3):1594
143. Sattel S, Robertson J, Ehrhardt H (1997) J Appl Phys 82:4566
144. Godwin PD, Horsfield AP, Stoneham AM, Bull SJ, Ford IJ, Harker AH, Pettifor DG, Sutton AP (1996) Phys Rev B 54(22):15785
145. Neyts E, Bogaerts A, Van De Sanden MCM (2006) Appl Phys Lett 88(14):141922
146. Neyts E, Bogaerts A, Van De Sanden MCM (2006) J Appl Phys 99(1):014902
147. Neyts E, Bogaerts A (2006) Phys Chem Chem Phys 8(17):2066
148. Titantah JT, Lamoen D, Neyts E, Bogaerts A (2006) J Phys Condens Mat 18(48):10803
149. Dartois E, noz Caro GMM, Deboffle D, Montagnac G, d'Hendecourt L (2005) Astron Astrophys 432(3):895
150. Ehrenfreund P, Charnley SB (2000) Annu Rev Astron Astr 38:427
151. Davis CA, Yin Y, McKenzie DR, Hall LE, Kravtchinskaia E, Keast V, Amaratunga GAJ, Veerasamy VS (1994) J Non-Cryst Solids 170:(1), 46
152. Ronning C, Griesmeier U, Gross M, Hofsäss HC, Downing RG, Lamaze GP (1995) Diam Relat Mater 4(5-6):666
153. Chhowalla M, Yin Y, Amaratunga GAJ, McKenzie DR, Frauenheim T (1996) Appl Phys Lett 69(16):2344
154. Veerasamy VS, Yuan J, Amaratunga GAJ, Milne WI, Gilkes KWR, Weiler M, Brown LM (1993) Phys Rev B 48(24):17954
155. Veerasamy VS, Amaratunga GAJ, Davis CA, Timbs AE, Milne WI, McKenzie DR (1993) J Phys Condens Mat 5(13):L169
156. Kleinsorge B, Ferrari AC, Robertson J, Milne WI (2000) J Appl Phys 88:1149
157. Sitch PK, Köhler T, Jungnickel G, Porezag D, Frauenheim T (1996) Solid State Commun 100(8):549
158. Stumm P, Drabold DA, Fedders PA (1997) J Appl Phys 81(3):1289
159. Pöykkö S, Kaukonen M, Puska MJ, Nieminen RM (1998) Comp Mater Sci 10(1–4):351
160. Robertson J, Davis CA (1995) Diam Relat Mater 4(4):441
161. Zheng B, Zheng WT, Zhang K, Wen QB, Zhu JQ, Meng SH, He XD, Han JC (2006) Carbon 44(5):962
162. Köhler T, Jungnickel G, Frauenheim T (1999) Phys Rev B 60(15):10864
163. Kaukonen M, Nieminen RM, Pöykkö S, Seitsonen AP (1999) Phys Rev Lett 83(25):5346
164. Merchant AR, McKenzie DR, McCulloch DG (2001) Phys Rev B 65(2):024208
165. Kádas K, Ferenczy GG (1999) J Mol Struc Theochem 463(1–2):175
166. Gambirasio A, Bernasconi M (1999) Phys Rev B 60(17):12007
167. Dai Y, Yan Y, Wang J, Sun B, He X, Shen H (2005) J Appl Phys 98(1):013525
168. Petersen T, Yarovsky I, Snook I, McCulloch DG, Opletal G (2004) Carbon 42(12–13):2457
169. Lawson JW, Srivastava D (2008) Phys Rev B 77(14):6
170. McGreevey RL, Putsai L (1988) Mol Sim 1:359
171. Walters JK, Gilkes KWR, Wicks JD, Newport RJ (1998) Phys Rev B 58(13):8267
172. O'Malley B, Snook I, McCulloch D (1998) Phys Rev B 57(22):14148
173. Thomson KT, Gubbins KE (2000) Langmuir 16(13):5761
174. Pikunic J, Clinard C, Cohaut N, Gubbins KE, Guet JM, Pellenq RJM, Rannou I, Rouzaud JN (2003) Langmuir 19(20):8565
175. Opletal G, Petersen T, O'Malley B, Snook I, McCulloch DG, Marks NA, Yarovsky I (2002) Mol Sim 28(10):927
176. Opletal G, Petersen TC, McCulloch DG, Snook IK, Yarovsky I (2005) J Phys Condens Mat 17(17):2605
177. Nguyen TX, Bhatia SK, Jain SK, Gubbins KE (2006) Mol Sim 32(7):567
178. Jain SK, Pellenq RJM, Pikunic JP, Gubbins KE (2006) Langmuir 22(24):9942
179. Nguyen TX, Cohaut N, Bae JS, Bhatia SK (2008) Langmuir 24(15):7912
180. Harris PJF (2005) Crit Rev Solid State 30(4):235
181. Terzyk AP, Furmaniak S, Gauden PA, Harris PJF, Włoch J, Kowalczyk P (2007) J Phys Condens Mat 19(40):406208

Chapter 6
Structural, Mechanical, and Superconducting Properties of Clathrates

Xavier Blase[1], Giorgio Benedek[2,3], and Marco Bernasconi[4]

Abstract Clathrates are superhard cagelike structures with sp^3 bonding that have been up to now synthesized for silicon . There is however some evidence of a carbon based counterpart in meteorites. If realized, carbon clathrates could on the one hand provide a material harder than diamond and on the other hand offer the possibility for selective doping and superconducting properties by intercalation. In this chapter all these aspects are thoroughly discussed by means of ab initio calculations based on density functional theory.

6.1 Introduction

Initially introduced in the late 1940s by Powell [1] to describe crystal structures with organic inclusions, the word clathrate was further used in the 50s by Pauling and Marsh [2] for describing chlorine hydrate structures, a member of a family of materials discussed nearly 200 years ago by Davy [3] and Faraday [4]. The clathrate structure is mostly known for the water clathrates [5], or clathrate hydrates, consisting of a network of water molecules enclosing or trapping other gas molecules (CH_4, CO_2, etc.) The discovery that large amounts of methane may be trapped in such hydrates at the ocean floor [6] has been largely publicized. It has been suggested indeed that this may constitute the largest fossil-fuel reservoir on earth, a potential blessing for industry, but also a potential nightmare for global warming in case of uncontrolled degazing in the atmosphere under temperature increase.

[1]Institut Néel, CNRS and University Joseph Fourier, Grenoble, BP 166 cedex, France
e-mail: xavier.blase@grenoble.cnrs.fr

[2]Department of Materials Science, University of Milano-Bicocca, Via R. Cozzi 53, I-20125 Milan, Italy

[3]Donostia International Physics Center (DIPC), Paseo Manuel de Lardizàbal 4, 20018 Donostia/San Sebastian, Spain
e-mail: giorgio.benedek@unimib.it

[4]Department of Materials Science, University of Milano-Bicocca, Via R. Cozzi 53, I-20125 Milan, Italy
e-mail: marco.bernasconi@unimib.it

L. Colombo and A. Fasolino (eds.), *Computer-Based Modeling of Novel Carbon Systems and Their Properties*, Carbon Materials: Chemistry and Physics 3,
DOI 10.1007/978-1-4020-9718-8_6,

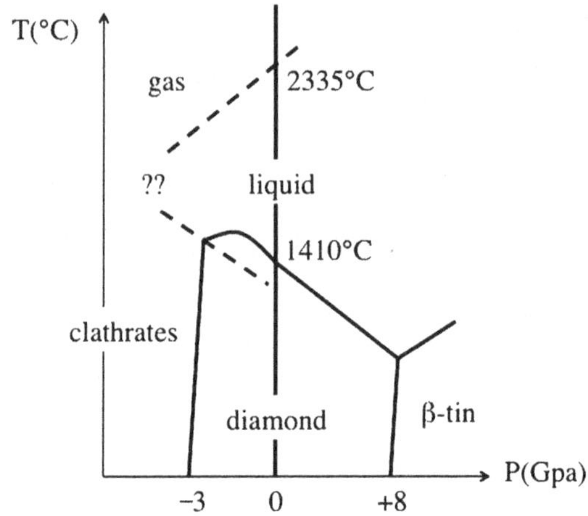

Fig. 6.1 Hypothetical silicon phase diagram, combining existing experimental and theoretical results. Adapted from P.F. McMillan in Ref. [9]

In the mid-1960s [7, 8], sodium intercalated silicon clathrates have been synthesized using thermal decomposition of alkali-metal silicides. As in the case of the hydrates, the silicon cage structure forms around an intercalated guest, sodium atoms in these first synthesis. Built from face-sharing fullerene-like Si_n clusters (n = 20,24 and 28 typically), silicon clathrates are low density phases of the diamond and high-pressure β-tin structures [9, 10] (Fig. 6.1). Since then, various synthesis techniques, such as high temperature–high pressure (HPHT) approaches, various guest intercalated atoms (barium, iodine, etc.), and several host network chemical compositions, including germanium and tin, have been reported. The discovery of unexpected properties, or interesting potential applications, including superconductivity and large Seebeck coefficient for thermoelectric applications, has been at the origin of a large interest from the material science community.

While most of the existing experiments concern the case of silicon or germanium compounds, the calculated thermodynamic metastability of carbon clathrates, the large number of predictions concerning their exceptional properties, and the numerous analogies with existing materials such as polymerized-C_{60}, certainly motivate this chapter devoted to the description of the potentialities of carbon clathrates and cage-like sp^3 carbon materials in general.

6.2 Clathrate Topology

6.2.1 The Simplest Fullerenic Clathrates

An infinite fullerenic clathrate (FC) lattice is a polyhedral filling of the three-dimensional (3D) space obtained through the coalescence of fullerenic cages, so that all atoms are fourfold coordinated, with eclipsed sp^3 bonding to their neighbors, and form exclusively five- or six-membered rings. The polyhedra which can constitute

FC's are the smallest ones, i.e., 12-hedra (C_{20}), 14-hedra (C_{24}), 15-hedra (C_{26}), 16-hedra (C_{28}), and 17-hedra (C_{30}), all considered in their most symmetric isomeric form. The fullerenic $(12 + h)$-hedron is made of twelve five-membered rings and h six-membered rings. Larger polyhedra, like for example a 20-hedron (C_{36}, with eight six-membered rings) or a 32-hedron (C_{60}, with twenty six-membered rings) can in principle enter a clathrate structure, at the cost, however, of introducing small non-fullerenic cages like, e.g., C_{16}, which have also square rings. For example the coalescence of 20-hedra yields the hexagonal clathrate hex-C_{16} with two C_{14} cages containing two square rings each [11]. To better understand the concepts and variety of forms underlying clathrates one may revisit some earlier studies on the topology of polyhedric networks [12, 13] and refer to a previous review on hollow diamonds [14–16]. The basic instrument for such analysis is the Euler's theorem [17]. For any polyhedral network in the 3D space with cyclic boundary conditions it states that

$$\nu - e + f - V = 0 \tag{6.1}$$

where ν is the number of atoms, e the number of bonds, f the number of rings and V the number of cages forming the network. For a FC Eq. 6.1 is easily proved. Since each atom has a fourfold coordination, $e = 2\nu$ and with only five-membered and six-membered rings, in the number of f_5 and f_6 respectively, $f = f_5 + f_6$ and $V = f_5/6$; then, by considering that each atom belongs to six rings, it is $\nu = f_6 + 5\, f_5/6$, and Eq. 6.1 is identically fulfilled.

For periodic FC's one may consider the numbers of atoms N, six-membered rings h_c, and cages V_c belonging to the unit cell, and derive from Euler's theorem the relationship

$$h_c = N - 5V_c \tag{6.2}$$

By introducing the number $N_h = 2h + 20$ of atoms of the $(12 + h)$-hedron and the number V_h of $(12 + h)$-hedra in each unit cell, it is

$$h_c = \frac{1}{2}\sum_h{}' hV_h, \qquad V_c = \sum_h{}' V_h, \qquad N = \frac{1}{4}\sum_h{}' N_h V_h \tag{6.3}$$

where the prime means that the $h = 1$ is excluded since the fullerenic 22-hedron does not exist. Inserting these equations into Eq. 6.3 it is found that the numbers of different polyhedra in a single cell of N atoms must obey the relationship

$$\sum_h{}' (5 + \frac{1}{2}h)V_h = N \tag{6.4}$$

For a given number N of atoms in the unit cell, Eq. 6.4 provides the possible sets of integer numbers V_h of fullerenic cages in the unit cell which can form a periodic FC. However a network which is allowed by topology may not be possible from the point of view of translational symmetry, and what is allowed by topology and translational symmetry may not be possible from the metric point view, since bond lengths and angles are restricted by the requirement of an approximate sp^3 bonding.

Additional constraints arise from point symmetries in much the same way they occur in two dimensions for fullerenes. These symmetry requirements could perhaps be rigorously formulated through the extension of Euler's theorem to symmetry properties which Ceulemans and Fowler have proven for polyhedra [18], but to our knowledge a further extension to 3D networks has not been given yet. In Ref. [14] the matter is thoroughly discussed in terms of *dual lattice* (the network formed by the centres of the cages which are linked through common six-membered rings) and *iso-oriented unsaturated six-membered rings*, through which each unit cell is connected to the adjacent ones – two concepts which greatly restrict the acceptable solutions of Eq. 6.4. In practice for small N one finds a few well known FC's made of either two or three different types of fullerenic cages, corresponding to the combinations:

(I) sc-C_{46} The coalescence of six C_{24} and two C_{20} ($V_2 = 6,\ V_0 = 2$) yields a simple-cubic lattice (space group Pm3n) with $N = 46$ atoms per unit cell. The backbone of this crystal consists of three orthogonal interpenetrating, but not intersecting, square arrays of C_{24} tubes. The residual space outside the tubes forms a body-centred cubic array of C_{20} cages.

(II) fcc-C_{34} The coalescence of two C_{28} and four C_{20} ($V_4 = 2,\ V_0 = 4$) yields a face-centred cubic lattice (space group Fd3m) with $N = 34$ atoms per unit cell; chemists prefer to view this structure as a simple-cubic crystal with 136 atoms per unit cell. The dual lattice of this FC, obtained by connecting the centres of the C_{28} cages, is in turn a diamond lattice, the two C_{28} cages of the unit cell being connected through a six-membered ring in a staggered configuration.

(III) hex-C_{40} This FC, resulting from the coalescence of three types of cages instead of two, actually two C_{26}, two C_{24} and three C_{20} ($V_3 = V_2 = 2,\ V_0 = 3$), has a hexagonal lattice (space group P6/mmm) with $N = 40$ atoms per unit cell. The backbone of this beautiful structure (Fig. 6.2) is a hexagonal array of parallel tubes formed by infinite piles of C_{24} cage pairs. The tubes are held together by rings of C_{26} and C_{20} cages arranged on alternate planes normal to the tubes. This structure is less known that the cubic FC's described above. His silicon version, hex-Si_{40}, has been the object of a first-principle study of its possible photo-luminescence properties, by virtue of the quasi-one dimensional character of the tubes merged into a compact monolithic 3D crystal [18].

There is another possible FC made of seven cages per unit cell with $N = 40$ and a body-centred orthorhombic structure, bco-C_{40}, resulting from the coalescence of two C_{30} and five C_{20}. The model of this structure shows highly distorted bonds, which should make it rather unstable.

For larger values of N Eq. 6.4 allows for an increasing variety of possible FC's. Ref. [14] provides some rules for the construction of plausible FC's with larger numbers of cages per unit cell. Since carbon is apparently an inexhaustible source of complex structures which call for a suitable modeling, it may instead be useful to mention here some topological algorithms which generate series of complex structures from an initial simple clathrate structure.

Fig. 6.2 Perspective view along the c-axis of hex-C_{40} showing the honeycomb array of tubes formed by infinite piles of C_{24} cage pairs

6.2.2 *From Complex Clathrates to Nanodiamonds*

A class of topological transformations useful to generate infinite classes of new structures is the replacement algorithm. It consists in replacing each atom of a clathrate structure with a cluster unit of many atoms which binds to its four neighbours in an eclipsed configuration leaving no dangling bonds. Consider, for example, the series of diamond *tetrahedra* of $N_T = n(n+1)(2n+1)/6$ atoms with n atoms on each edge (Fig. 6.3) [14, 19]. Identical clusters can be covalently bound through the faces in an eclipsed configuration, thus generating from any clathrate of N atoms per unit cell infinite series of diamond-like crystals with NN_T atoms per unit cell belonging to the same space group as the original clathrate. Interatomic bonds are all sp^3, either staggered (inside each tetrahedron) or eclipsed (at the twin boundary between two tetrahedra). The resulting crystals contain a superlattice of cavities corresponding to the fullerenes constituting the original clathrate. For this property these hypothetical structures have been termed *hollow diamonds* [14, 19].

When this topological algorithm is applied to a single $(12 + h)$-hedron of those allowed in a clathrate, a diamond-like cluster (*nanodiamond*) of $2(10 + h)N_T$ atoms and $3(10 + h)$ twin boundaries is obtained with a fullerenic cavity at the centre and the same point symmetry of the original polyhedron. The simplest and probably more stable series follows from $h = 0$ (*icosahedral diamonds*). If identical

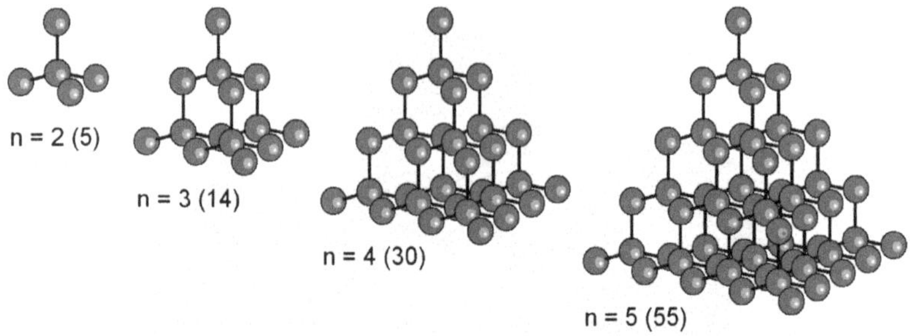

Fig. 6.3 Tetrahedral diamond clusters of $N_T = n(n+1)(2n+1)/6$ atoms where n is the number of atoms along the edge. Identical clusters can be covalently bound through the faces in an eclipsed configuration, thus generating from any clathrate of N_T atoms per unit cell infinite series of diamond-like crystals with N/N_T atoms per unit cell belonging to the same crystallographic class as the original clathrate. Bonds are either staggered (inside each tetrahedron) or eclipsed (at the twin boundary between two tetrahedra). The resulting crystals contain a superlattice of cavities corresponding to the fullerenes constituting the original clathrate (*hollow diamonds*) [14, 19]

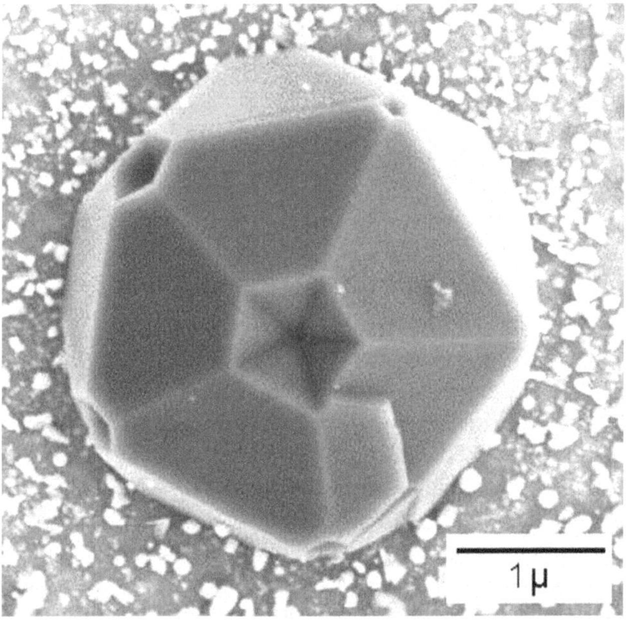

Fig. 6.4 An icosahedral diamond grown by hot-filament chemical vapour deposition on a TiN-coated steel substrate [from J. Breza et al. [20], 2004 Elsevier Ltd.]. The structure is likely to be constituted by multiple twinning of truncated tetrahedral units

truncated tetrahedral diamond clusters are joined a larger cavity is formed at the center and cup-like concavities appear at the external vertices. Actually icosahedral diamonds of this kind have been grown up to the micrometric scale: see, e.g., the beautiful samples obtained by Breza et al. [20] by means of hot-filament chemical vapor deposition on a TiN-coated steel substrate (Fig. 6.4). Although the synthesis of micro-diamonds with fivefold symmetries and multiple twinning can be traced back to the late eighties [21], the interest for the topological algorithms generating families of hollow diamonds from fullerenes [14, 19] has been renewed in recent years [22–24]. The multiple twinning of tetrahedral diamond nanocrystals synthesized by pulsed-laser irradiation, eventually leading to fivefold symmetries which are maintained during growth, has been recently described [25]. The saturation of (111) surface dangling bonds in hollow diamonds, apart from ordinary hydrogen passivation [23], may lead to intriguing fullerene-like reconstructions (*bucky diamond clusters*) [26].

The topological construction of hollow diamonds can be easily extended to the generation of periodic clathrate families by means of any covalent cluster of M atoms which can coordinate four identical clusters in a tetrahedral eclipsed configuration. If one starts from any simple clathrate, a new clathrate is generated through the replacement algorithm where each atom of the initial clathrate is replaced by one cluster with no dangling bond left. For example the 41-atom unit of tetrahedral symmetry depicted in Fig. 6.5 when placed at the nodes of an fcc-C_{34} lattice generates a giant clathrate through the transformation

$$\begin{aligned} fcc - C_{34} &= \frac{1}{4}(2\,C_{28} + 4\,C_{20}) \quad \stackrel{\times 41}{\longrightarrow} \\ fcc - C_{1,394} &= \frac{1}{4}(68\,C_{26} + 36\,C_{30} + 4\,C_{36} \\ &\quad + 4\,C_{60} + 8\,C_{84} + 136\,C_{16}) \end{aligned} \tag{6.5}$$

where are all the fullerenic cages C_n have the structure of their most symmetric isomer. The factor 1/4 reminds that each atom belongs to four different cages, so that $1/4(68\times26+36\times30+4\times36+4\times60+8\times84+136\times16) = 1{,}394$. This clathrate structure is an example of exact inclusion of the well known fullerenes C_{60} and C_{84} in a fully covalent sp^3 crystal. It appears that the clusters C_{60} and C_{84} are centered

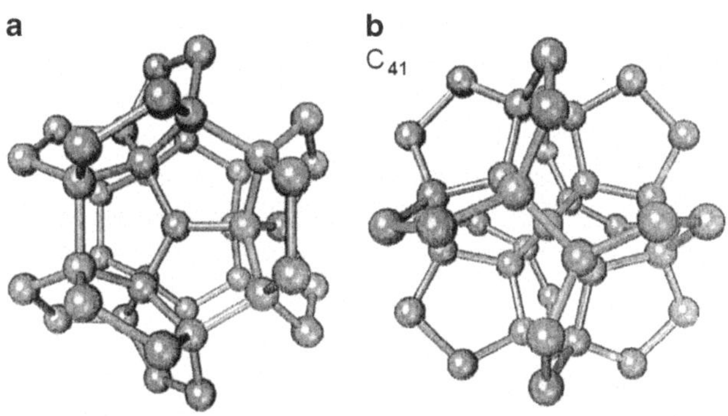

Fig. 6.5 A cluster of 41 carbon atoms organized in a tetrahedral symmetry, as seen along a ternary (**a**) and a binary (**b**) symmetry axis, which works as a building unit of a family of giant clathrates [14]

at the positions of the clusters C_{28} and C_{20} of the original fcc-C_{34} clathrate, respectively. The sp^3 bonds of the atoms on the clusters C_{60} and C_{84} are quite distorted due to their comparatively small curvature. This is however compensated by the insertion of 136 small non-fullerenic cages C_{16}, which are decahedra formed by one hexagon, six pentagons and three squares. The inclusion of four-membered rings, which also occur in stable sp^3-hybridized carbon crystal polymorphs [27] and in C_{60} polymerization by cyclo-addition (see Section 6.2.4), enormously extends the number of families of sp^3-bonded clathrates by allowing the use of larger fullerenic cages. For example the coalescence of C_{36} cages (20-hedra) through their hexagonal faces leads to the clathrate hex-$C_{16} = 1/4(C_{36} + 2C_{14})$ where the small C_{14} cage is a 9-hedron made of six pentagons and three squares. From the pure topological point of view, i.e., disregarding the large sp^3 bond distortions, all bonds of the resulting giant clathrate are eclipsed sp^3 and the replacement algorithm can be repeated leading to a new fcc clathrate with 57,154 atoms in the unit cell, etc.

In general a cluster unit which has an eclipsed bond structure and can coordinate four identical units in the eclipsed configuration with a complete saturation of the dangling bonds can be used in the atom-to-unit replacement process, starting either from a simple clathrate or a small fullerene cage. Depending on the structure of the initial unit the replacement process applied to a simple clathrate can generate at each step either a clathrate with larger and larger fullerenic cages compensated by an increasing number of small cages, as in the case illustrated by Eq. 6.5, or clathrates exclusively made of small fullerenic cages (with $h \leq 4$). In the first case the iteration *ad infinitum* eventually leads to a sort of *fractal* sp^3 *foam*.

The second case, where the new cages generated at each iteration are requested to keep within the set $h \leq 4$, has an interesting realization with the 32-atom tetrahedral cluster shown in Fig. 6.6a. This cluster is made of a C_{28} cage plus four adatoms (A) at the four tri-pentagonal cusps, thus totalling 32 atoms. Since the clusters are joined through the six-membered rings, their atoms count 1/2 thus giving a building unit of 20 atoms. When two such clusters are joined by the coalescence of two six-membered rings a unit dimer is obtained as a replacement of two nearest neighbor nodes (Fig. 6.6b, gray discs) of the original clathrate. By connecting the nearest neighbor adatoms with sp^3 bonds, thus saturating all dangling bonds, new cages are formed which have the same shape and relative positions as the cages of the original clathrate. They are however disjoined by the interposition of the C_{28} cages and rings of C_{20} and C_{24} cages, with numbers of C_{20} and C_{24} cages equal to the number of fivefold and sixfold rings of the previous clathrate, respectively. This describes what we shall call the ×20-replacement iteration. For example starting from fcc-C_{34} the unit cell of the new clathrate will contain $20N$ atoms and is described by the transformation

$$
\begin{aligned}
fcc - C_{34} &= \frac{1}{4}(2\,C_{28} + 4\,C_{20}) \;\stackrel{\times 20}{\longrightarrow} \\
fcc - C_{680} &= \frac{1}{4}(76\,C_{20} + 36\,C_{28} + 8\,C_{24})
\end{aligned}
\qquad (6.6)
$$

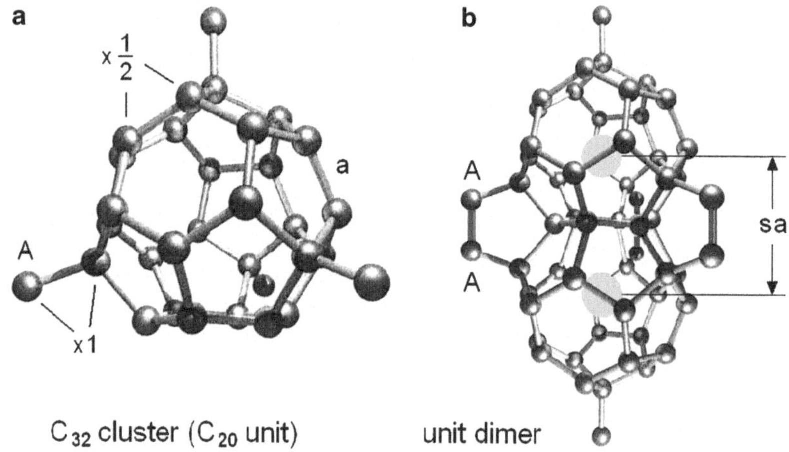

Fig. 6.6 (**a**) The building unit for the generation through the atom-to-unit replacement algorithm of a series of clathrates from a simple clathrate, all made of the smallest fullerenic cages. The cluster is made of a C_{28} cage plus four adatoms (A) at the four tri-pentagonal cusps, thus totalling 32 atoms; since the clusters are joined through the six-membered rings, their atoms count 1/2 thus giving a building unit of 20 atoms. (**b**) When two units are joined by coalescence of two six-membered rings a unit dimer is formed as a replacement of two nearest neighbor nodes (*gray discs*) of the original clathrate. By connecting the nearest neighbor adatoms (A), thus saturating all dangling bonds, new cages are formed which have the same shape and relative positions as the cages of the original clathrate. They are however disjoined by the interposition of the C_{28} cages and rings of C_{20} and C_{24} cages. Assuming equal interatomic distance a, the length scale after the replacement process has been changed by the factor $s = \sqrt{2} + \sqrt{3/2}$

Clearly after n iterations the clathrate will have 34×20^n atoms per unit cell. Again for $n \to \infty$ a fractal sp^3 clathrate is obtained. In the ideal case of equal bond lengths a, the length scale, as illustrated in Fig. 6.6 is changed at each iteration by a factor $s = \sqrt{2} + \sqrt{3/2} = 2.638958....$ After n iterations the unit cell contains $N_n = 20^n N$ and has a volume $V_n = s^{3n} V$, where V is the cell volume of the initial clathrate. Thus the number of atoms

$$N_n = N(V_n/V)^{ln20/3lns} \tag{6.7}$$

does not grow linearly with the volume. The fractal dimension in the present idealized case would be $D \equiv ln20/lns = 3.08716 ...$, which is unphysical since no fractal dimension can exceed the dimensionality of the host space. The atomic volume would in fact go to zero for $n \to \infty$, albeit very slowly. The reason is that in the ideal case the adatom–adatom distance after an iteration (A-A in Fig. 6.6b) turns out to be slightly smaller than a, which slowly accumulates a compressive stress. At each iteration the stress should be released in some way. If the bond lengths of the cages generated by the adatoms would relax to a then D would become slightly smaller than 3 and the atomic volume would slowly increase with n. An interesting compromise is the one where the average atomic volume remains constant, thus stabilizing the fractal dimension at $D = 3$. In the $n \to \infty$ limit the clathrate has no periodicity but preserves an orientational order. This may be regarded as a special case of *quasicrystal* defined as the limit of a fractal structure whose dimension becomes integer. The clathrate of type II has acquired popularity in mathematics in connection with Kelvin's conjecture [28]. Lord Kelvin argued that the division of space with equal volumes and a minimum partitional area can be realized by filling the space with tetrakaidecahedra, i.e., truncated octahedra (well known to solid state physicists as the fcc Brillouin zone) with slightly curved surfaces [29]. More than a century after Kelvin's conjecture Denis Weaire and Robert Phelan proved with a massive computer analysis that a tesselation of space with a type II clathrate, deformed in a way that all cages have the same volume, yields a minimal surface which is 0.3% smaller than Kelvin's solution [30]. The Weaire–Phelan structure, besides decorating the entrance hall of the SNIAM Building at Trinity College in Dublin and shaping the walls and the roof of the new Beijing Water Cube for the 2008 Olympics, has acquired considerable importance in the physics of foams [31] and soft materials in general [32]. A further conjecture about Kelvin's problem is that a fullerenic clathrate solution should have the minimum ratio h/p of hexagon to pentagon numbers [28]. For example for the clathrate of type I $h/p = 1/8$, for type II $h/p = 1/9$, and for type III $h/p = 5/42$. It turns out that at each step of the $\times$20-replacement iteration the ratio h/p transforms as

$$\left(\frac{h}{p}\right)_{n+1} = \frac{5}{54}\frac{1 + 3(h/p)_n}{1 + (h/p)_n} \tag{6.8}$$

It is easily proved that the iteration tends to $(h/p)_\infty = 1/9$. Thus $h/p = 1/9$ is a fixed point, which means that h/p is conserved in the $\times$20-replacement iteration applied to the type I. Should the minimality of h/p be a sufficient condition for the a solution of Kelvin's problem, then the $\times$20-replacement iteration would generate an infinite series of possible solutions.

6.2.3 Lattice vs. Electronic Structure of Fullerenic Clathrates

The ground state properties of the type I and type II carbon clathrates have been calculated with ab initio methods in the early nineties [33–35], whereas density functional (DFT) calculations for the less popular type III clathrate have been first made for hex-Si_{40} in search of a photoluminescent form of silicon [36], and only later have been applied to a study of hex-C_{40} as a possible p- and n-dopable diamond-like semiconductor [37]. Comparative calculations for all the three types of the ground state properties such as the energy per atom, the geometry, the elastic moduli and the phonon structure [19] have been made first on the basis of the phenomenological Tersoff potential [38], whereas the electron densities of states have been obtained with the tight binding (TB) method [14, 15]. Whenever a comparison with DFT results is possible, these calculations proved to be reliable; they are presented in the following paragraphs and in Section 6.3.1, whereas the DFT results for

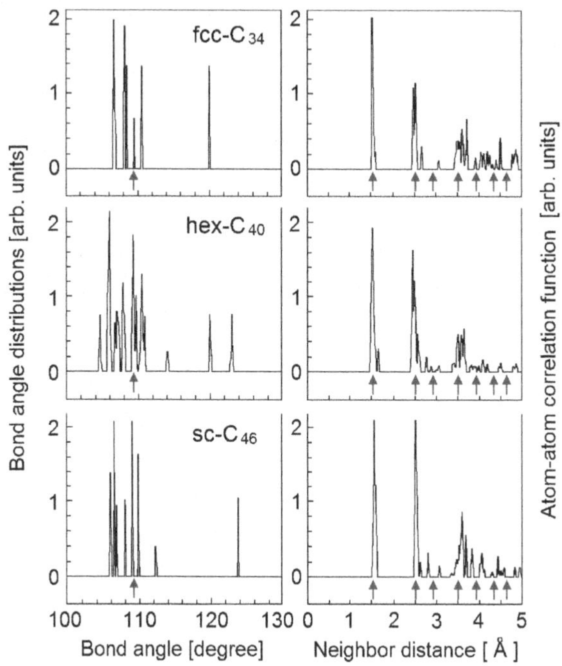

Fig. 6.7 Calculated bond-angle and neighbor-distance distributions at equilibrium for the simple clathrates as compared to diamond (*arrows*)

the mechanical properties, the electronic structure, and the electron–phonon interaction properties of pure and doped clathrates and similar cage structures shall be discussed in the Sections 6.4–6.6.

The equilibrium bond-angle and neighbor-distance distributions for the three simple clathrates, calculated by means of the Tersoff potential and a simulated annealing procedure [19] are shown in Fig. 6.7. It appears that the pentagon angles (240°, 180° and 210° for types I, II and III, respectively) are fairly spread around the regular pentagon angle of 108° or the very close ideal sp^3 angle of cubic diamond (arrows). In fcc-C_{34} the hexagons of the 16-hedra show no distortion by symmetry, their 24 angles being all exactly 120°. In hex-C_{40} the 12 angles of the 14-hedra show no distortion (120°) by symmetry, whereas the three hexagons belonging to the 15-hedra are flattened so as to give 12 angles of 123° and 6 angles of 114°. In sc-C_{46} all the six hexagons of the 14-hedra are distorted in the same way so as to give 24 angles of 124° and 12 angles of 112°. Despite such angular deviations from the ideal sp^3 structure, the nearest and next-nearest neighbor distances are well concentrated around the cubic diamond distances (arrows) and only from the third neighbors the clathrate structures start differentiating substantially from the cubic diamond distribution.

It appears that simple clathrates and their families generated by topological algorithms offer a large variety of structures and models which would allow, in principle,

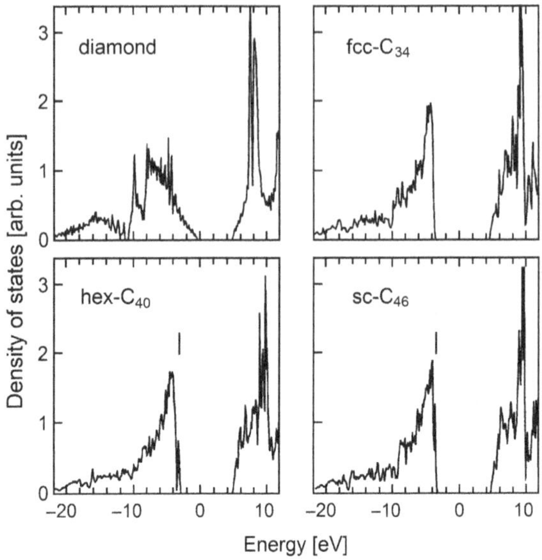

Fig. 6.8 Electronic density of states for the simple carbon clathrates calculated with the tight-binding method [15, 19]. The quasi-localized highest-occupied state (*vertical mark*) occurring in hex-C_{40} and sc-C_{46} is associated with the piles of C_{24} clusters. They form a hexagonal array of parallel tubes in hex-C_{40}, and three square arrays of non-intersecting tubes in the orthogonal directions in sc-C_{46}, which confer in both cases a quasi-1D character to these states

to tailor desired physical properties by designing a suitable structure. A typical example is the band engineering to obtain, e.g., certain photoelectronic properties which the basic material does not have. Tight-binding calculations of the electronic band structure show that simple carbon clathrates are insulators with a large gap of about 8 eV [15, 19], thus considerably larger than that of cubic diamond (Fig. 6.8). This finding disagrees with ab-initio calculations, whereas for Si clathrates TB and ab-initio calculations consistently predict larger gaps than for ordinary silicon (cfr. Sec. 6.4.1). The presence of comparatively large cages as C_{28} means a certain amount of internal surfaces, i.e., some 2D character in the electronic properties which is preserved from the isolated fullerenes. Similarly the tubular structures made of C_{24} piles induces a certain amount of 1D localization. The latter property offers an interesting example of band engineering related to the structure. Both hex-C_{40} and sc-C_{46} are characterized by a highest valence state (HVS) localized along the tubes a contributing a sharp feature at the upper edge of the valence band (Fig. 6.8, vertical marks). This state is also found in hex-Si_{40} with both TB and ab initio calculations [36]. The TB calculated electronic density of states of this system is shown in Fig. 6.9a: the calculated TB gap of about 2.6 eV is quite wider than the ab initio LDA gap even after scissor operator correction (1.6 eV). Both calculations, however, agree in giving a direct gap at the A point of the Brillouin zone which is just a few tens meV wider than the absolute gap, thus allowing in principle for photoluminescence applications (Fig. 6.10). The highest valence state (HVS) and the

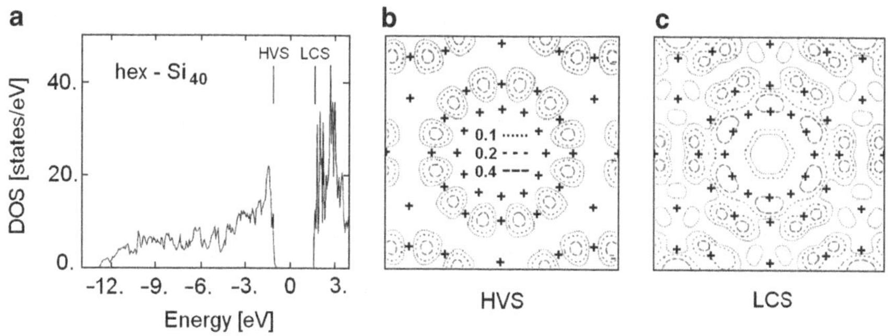

Fig. 6.9 (**a**) Electronic density of states of hex-Si_{40} calculated with DFT [36] ; the gap of about 2.6 eV is substantially larger than in cubic silicon and is direct, thus allowing for photoluminescence. The indicated highest valence state (HVS) and the lowest conduction state (LCS) correspond to states which occur at the A symmetry point of the Brillouin zone and are localized along the C_{24} tubes. The contour plots of the corresponding charge densities (in arbitrary units) are shown in (**b**) and (**c**), respectively, in a (0001)-plane cut perpendicular to the tubes. The crosses are the projections of the atom positions

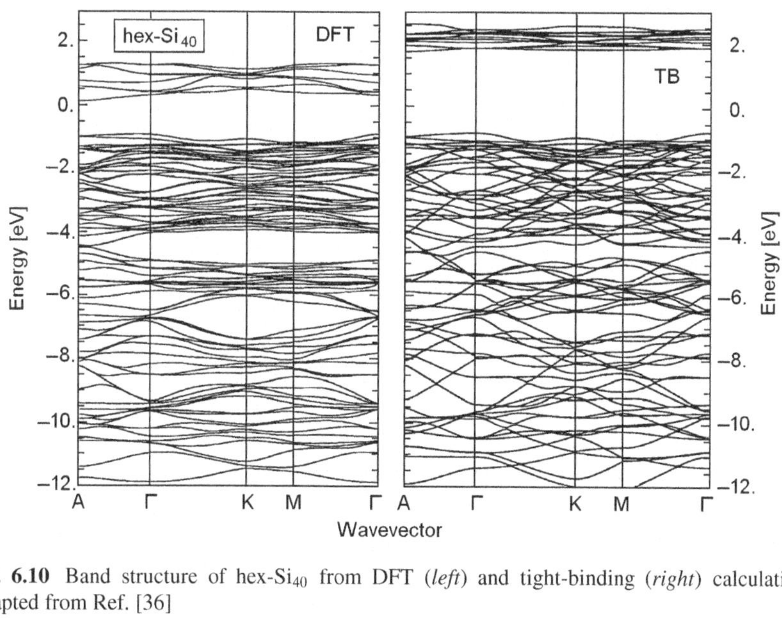

Fig. 6.10 Band structure of hex-Si_{40} from DFT (*left*) and tight-binding (*right*) calculations. Adapted from Ref. [36]

lowest conduction state (LCS) at the A symmetry point, indicated by vertical lines in Fig. 6.9a, are found to be localized along the Si_{24} tubes, just as for the corresponding carbon structure. The contour plots of the charge densities for the HVS and the LCS are shown (in arbitrary units) in Fig. 6.9b and c, respectively, in a (0001)-plane cut perpendicular to the tubes. The direct HVS $\rightarrow$ LCS transition brings the electron tightly bound to the five-membered rings of the Si_{24} cages (Fig. 6.9b) to a state only slightly more expanded around the tube and with some density along the axis of the tube (Fig. 6.9c). This brief discussion aimed at showing an example of structure–property relationship. A more detailed description of ab initio calculations for pure and doped clathrates, as well as for more complex structures originating from the polymerization of fullerenic cages is presented in the next sections.

6.2.4 Three-Dimensional Fullerene Polymers

In the search of possible routes to synthesize carbon clathrates, fullerite crystals have been subjected to high pressure and temperatures. Under high pressure and high temperature conditions ($P = 1 - 8$ GPa and $T = 350 - 450$°C) fullerite molecular crystal transforms into one- or two-dimensional polymeric structure via [2 + 2] cycloaddition reactions between adjacent C_{60} molecules in the (001) (orthorhombic 2D phase) or (111) (rhombohedral 2D phase) planes [39]. Evidence of the synthesis of three-dimensional polymeric forms of C_{60} have also been reported in recent years [40, 41]. Actually, these latter structures are not clathrates according to the definition given in Section 6.2.1 since they display a mixture of sp^3 and sp^2 bonding geometries and the fullerenic-like cages are strongly distorted in a cuboidal-like geometry. Nevertheless these systems are often referred to as clathrate-like forms of carbon. For completeness, we here briefly review the properties of the three dimensional polymers of C_{60} as reported in literature. Marques and coworkers [40] provided experimental evidence of a 3D polymeric phase of C_{60} obtained by compressing a C_{60} powder up to 13 GPa under non hydrostatic condition at 550°C. In the new phase, quenchable at normal conditions, C_{60} cages were suggested to be covalently bonded to the others neighboring cages forming a 3D network, but the small size of the crystallites of the polycrystalline product prevented a compelling refinement of the crystal structure. Later on, different possible 3D polymeric phases have been proposed from the analysis of powder x-ray diffraction data all belonging to the body-center orthorhombic (bco) Immm space group [42] but with different intermolecular bonding. These structures can be seen as generated from the 2D orthorhombic polymer by introducing (3 + 3) bonding between C_{60} in adjacent planes, following the notation introduced in Ref. [42]. Besides [2 + 2] cycloadditions, common four-sided rings have also been proposed for the in-plane polymerization of the 3D phase [40]. Moreover, the hardness of some 3D polymers have been reported to surpass those of diamond [43]. All these bco Immm structures (with [2 + 2] or four-sided rings in plane bindings) have been shown to be metallic by ab initio calculations [44, 45]. Other 3D polymeric structures with different symmetry have also been proposed and investigated by ab initio calculations [46]. More recently, Yamanaka et al. [41] succeeded in isolating a single crystal ($0.3 \times 0.2 \times 0.2$ mm^3) of a 3D polymeric phase synthesized by compressing the 2D orthorhombic polymer at 15 GPa and 600°C. Structural refinement from single crystal x-ray diffraction still assigned a bco Immm space group. The internal structure consists of C_{60} molecules linked via (3 + 3) bonds only between adjacent planes with no in plane bonding (Fig. 6.11). This phase has been experimentally recovered at normal conditions upon decompression. Several modifications of the experimentally assigned structure (Fig. 6.11) have also been investigated from first principles [47–50]. The most stable configuration is obtained by adding additional in plane intercage bonds with respect to the structure proposed experimentally. Namely, by still keeping the Immm symmetry detected experimentally, a [2 + 2] and a [8 + 8] ((4 + 4) bonds in the notation of Ref. [47]) cycloaddition bondings are added along the *a* and *b* directions, respectively (Fig. 6.12).

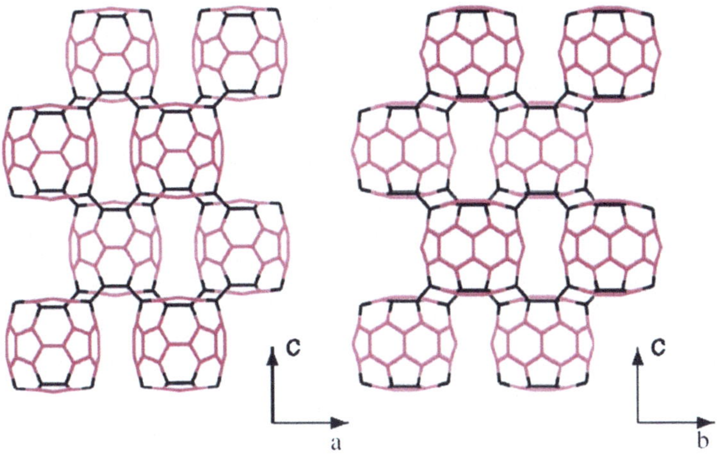

Fig. 6.11 View of the (010) and (100) planes of the 3D polymeric structure proposed in Ref. [41]. A C_{60} molecule is linked by (3 + 3) bonds to eight nearest neighbor cages. There is no covalent bonding in the *ab* plane. The atoms depicted in *gray* are threefold coordinated (44 atoms) and the *black* ones are fourfold coordinated (16 atoms)

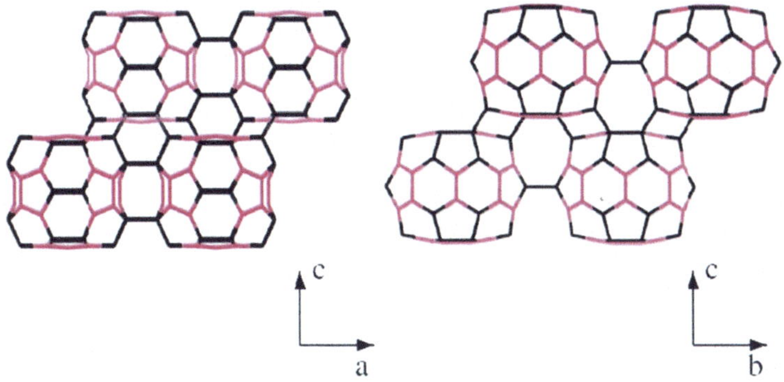

Fig. 6.12 View of the (010) and (100) planes of the 3D polymeric structure proposed as structure B in Ref. [47]. Adjacent planes along *c* are linked by (3+3) bonding. In the *ab* plane C_{60} are linked by [8 + 8] and [2 + 2] cycloadditions in directions *a* and *b*, respectively. The atoms depicted in *gray* are threefold coordinated (32 atoms) while the *black* ones are fourfold coordinated (28 atoms)

However, large misfits with the experimental lattice parameters are found for all the Immm phases investigated so far by first principles [47,49,50]. These discrepancies have been ascribed [49] to the fact that the phase observed experimentally might actually consist of a metastable structure not fully relaxed from the high pressure state. Therefore, at the moment a compelling identification of the detailed structure of the 3D polymer C_{60} decompressed at normal condition is still lacking.

6.3 Mechanical Properties of Carbon Clathrates

6.3.1 Elastic Properties

The theoretical ground-state properties of simple clathrates compared to cubic diamond are listed in Table 6.1. They include the energy per atom E and its difference from the value for cubic diamond E_d, the ratio ρ/ρ_d of the clathrate density to that of cubic diamond, the bulk modulus B and the elastic constants c_{ij} [19]. The calculations, based on the phenomenological Tersoff potential [38], favorably compare with the available ab initio calculations by Adams et al. for fcc-C_{34} and sc-C_{46} [34] and by Bernasconi et al. for hex-C_{40} [37]. The above data allow to derive the anisotropy

Table 6.1 Calculated ground-state properties of cubic diamond and simple clathrates: energy per atom E and its difference from cubic-diamond value E_d, density ratio ρ/ρ_d referred to cubic diamond ($\rho_d = 3.52$ g/cm^3), bulk modulus B and elastic constants. Ab initio calculations by Adams et al. [34] and by Bernasconi et al. [37] are explicitly marked (**a–c**); the other data [19] are obtained from the phenomenological Tersoff potential [38]. The corresponding anisotropy constant and velocities along the [100] direction of the bulk transverse (v_T) and longitudinal (v_L) waves and of the Rayleigh wave for the (001) surface are given in the last four rows

	Diamond	fcc-C_{34}	hex-C_{40}	sc-C_{46}
E [eV/atom]	−7.370	−7.231	−7.156	−7.166
$E - E_d$ [eV/atom]	0.	0.139, 0.113[a],0.134[b]	0.124, 0.17[c]	0.204, 0.144[a]
ρ/ρ_d	1.	0.859, 0.843[a]	0.854, 0.855[c]	0.863, 0.877[a]
B [Mbar]	4.43	3.64	3.72, 3.65[c]	3.64
c_{11} [Mbar]	10.9	10.27	9.80	9.79
c_{12} [Mbar]	1.2	0.33	0.39	0.56
c_{44} [Mbar]	6.4	4.81	2.46	4.96
c_{33} [Mbar]	–	–	9.81	–
c_{13} [Mbar]	–	–	0.59	–
$\eta = 2c_{44}/(c_{11} - c_{12})$	1.32	0.97	–	1.07
$v_{L[100]}$ [km/s]	17.60	18.44	–	17.95
$v_{T[100]}$ [km/s]	13.48	12.62	–	12.77
$v_{R[100]}(001)$ [km/s]	11.05	11.00	–	10.85

[a]Ab initio calculation with a local basis (G. B. Adams et al. [34]).
[b]Ab initio calculation with a plane-wave basis (G. B. Adams et al. [34]).
[c]DFT-LDA calculation (M. Bernasconi et al. [37])

constant $\eta = 2c_{44}/(c_{11} - c_{12})$ and the sound velocities along the [100] direction of the bulk transverse (ν_T) and longitudinal (ν_L) waves and of the Rayleigh wave for the (001) surface (last four rows of Table 6.1). Roughly speaking the bulk modulus is seen to scale with the density, i.e., with the number of bonds per unit volume, consistently with the fact that the bond lengths in clathrates remain very close to that of diamond. On the other hand c_{11} in cubic clathrates is not as much reduced with respect to diamond, so that the speed of longitudinal sound in the [001] direction (Table 6.1) is larger in cubic clathrates than in cubic diamond: it may indeed be considered as the fastest sound ever. On the contrary the transverse sound velocity in clathrates is about 7% smaller than in diamond, whereas the Rayleigh wave velocity travelling along the [100] direction of the (001) surface is practically identical to that of diamond for fcc-C_{34} and slightly slower for sc-C_{46}.

Another interesting aspect of cubic clathrates with respect to diamond is their pronounced elastic isotropy ($\eta \cong 1$), and even in the hexagonal clathrate $c_{33} \cong c_{11}$. This may be viewed as an effect of the many different orientations of the sp^3 bonds in clathrates with respect to diamond where they have only four different orientations.

The bulk modulus of hollow diamonds and of the series of clathrates generated by replacement algorithms can be inferred by assuming that B is proportional to the number of bonds per unit volume. Restricting the argument to hollow diamonds, elastic relaxation may be neglected, which means to further assume that all bonds have the same length as in diamond. In this case the unit cell volume for the n-th member of the series starting from the clathrate with N atoms per unit cell scales as

$$V_{N,n} \cong \frac{1}{27}(1 + 2n)^3 V_{N,1} \tag{6.9}$$

and the corresponding bulk modulus is

$$B_{N,n} \cong \frac{4n(1 + n)}{(1 + 2n)^2} B_d \tag{6.10}$$

with B_d the bulk modulus of diamond. Obviously for $n \to \infty$, $B_{N,n} \to B_d$, whereas for the simple clathrates $B_{N,1} \cong 8/9 B_d$. This value is slightly larger that those reported in Table 6.1 due to neglecting relaxation. A better approximation is given in Ref. [14]. As long as the hollow diamonds represent a reasonable model for icosahedral nano-diamonds, Eq. 6.10 provides a simple way to predict the bulk modulus of these materials as a function of their size.

The phonon density of states at zero wave-vector for the three carbon clathrates have been obtained from classical molecular dynamics simulations [19] based on the Tersoff potential [38] (Fig. 6.13). The largest peak at about 1,250 cm^{-1} is the signature of the sp^3 bond stretching mode; its frequency is however slightly softer than in cubic diamond (1,333 cm^{-1}), its value scaling approximately as $\sqrt{B}$.

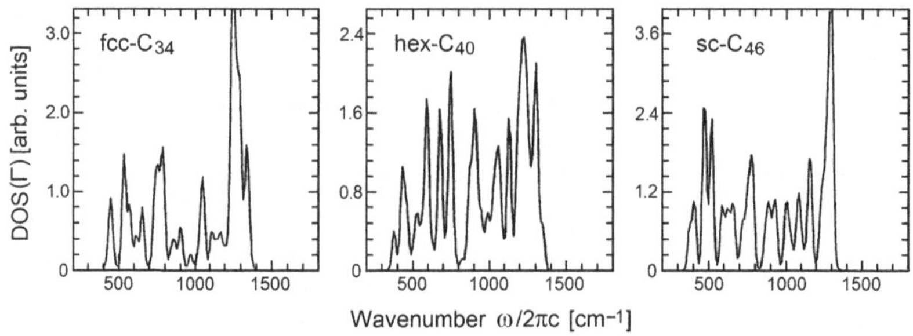

Fig. 6.13 Phonon density of states at zero wave-vector for the carbon clathrates as obtained from classical molecular dynamics simulations based on the Tersoff potential (from Refs. [14, 15]). The largest peak at about 1,250 cm^{-1} is the signature of the sp^3 bond stretching mode; its frequency is however slightly softer than in cubic diamond (1,330 cm^{-1})

6.3.2 Plastic Properties: Harder than Diamond?

In 2003, a team of geophysicists discovered in the so-called Popigai impact crater (Russia) a "new natural, super-hard, transparent polymorph of carbon" [51]. Its polishing hardness was found to be greater than that of lonsdaleite, the hexagonal allotrope of diamond. Micro-beam synchrotron X-ray diffraction revealed a cubic structure with the Pm3m space group and a lattice parameter of a = 14.697 Å. Even though very reminiscent of the structure of carbon clathrates, no further study was performed and the identification of such a phase to carbon clathrates remains elusive. The claim that novel carbon-based structures may scratch diamond is however not isolated since high-pressure polymerized C_{60} or C_{70} phases (fullerites) [40, 42, 52–54] were reported to display such a property as well [52, 55].

The search for ultrahard materials has generated much work in the last few decades, both for its technological implications and the fundamental difficulty in finding relevant criteria for the design of novel structures [56, 57]. In particular, as the full study of real materials with moving dislocations remains a challenge to computer simulations, there have been numerous attempts to relate hardness to the properties of the ideal crystal at equilibrium (bond length, ionicity, band gap, etc.) or close to equilibrium (bulk modulus, elastic constant, etc.) [58].

Recently, several theoretical works [59–61] focused on studying the plastic properties of diamond, the hardest known material. In particular, the stress/strain load curve beyond the elastic regime was studied in order to provide a value for the ideal shear and tensile strengths in various directions. This theoretical strength [62] is the maximum stress that a perfect crystal can sustain before yielding to a plastic deformation. Even though this theoretical strength is an upper limit for the real strength, such calculations were very successful in explaining that the <111> plane is the easiest slip plane in diamond. Further, it was shown [59–61] that under <111> tensile load or shear along the {111} slip systems, the diamond was unstable with respect to the graphite phase, a theoretical confirmation of the Gogotsi et al. [63] experimental results for the indentation of diamond.

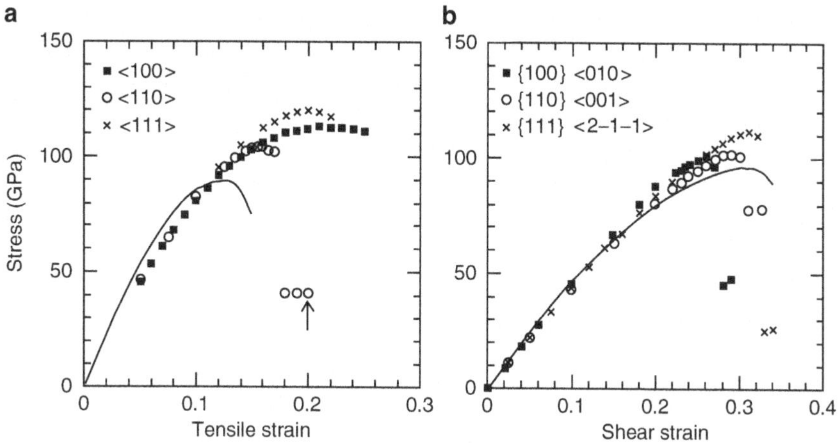

Fig. 6.14 Stress vs strain under (**a**) tensile load and (**b**) shear load, for several directions. The full lines correspond to the diamond phase. In the case of sc-C_{46}, the stress directions and slip planes are indicated next to their corresponding symbols (reproduced from Ref. [64] with permission from the APS)

The topology of clathrates, with their face-sharing cage-like structure, does not allow the presence of easy slip planes as in diamond. In particular, the continuous transformation between diamond and graphite is frustrated, with important consequences on the ideal strength that have been unraveled thanks to DFT calculations [64]. We reproduce in Fig. 6.14 the stress/strain curves for diamond strained along the <111> direction (left, full line) and sheared along the {111} slip-plane (right, full line) showing a ~90 GPa ideal strain in good agreement with the nano-indentation experiments of Gogotsi. This is compared to what we obtained for the sc-C_{46} structure for several tensile or shear stress directions (various symbols). Clearly, *the maximum stress that can be sustained by the clathrate structure before plastic deformation is larger than the 90 GPa obtained for diamond.* Beyond the maximum stress, bonds start to break and the clathrate evolves towards of more opened structure mixing sp^3 and sp^2 atoms (see e.g. arrow in Fig. 6.14a). The resulting material is still fully connected with covalent bonds, so that the stress value remains high (contrary to the graphite structure for which slipping two planes on top of each other requires much less stress).

Besides the hypothetical character of the carbon clathrate structure, the relation between the ideal stress (in the absence of defects) and the actual hardness of a given material is still unclear. In a nano-indentation experiment, where the strained area is sufficiently small so that hardly any dislocation is present, the ideal stress becomes relevant as confirmed by the results of Ref. [63]. For a macroscopic sample, however, the relation between ideal strength and "macroscopic" strength and hardness is certainly less direct.

Concerning the "superhard" fullerite structures cited above, the bulk modulus or elastic constants of various poly-C_{60} phases were calculated [45, 46, 65] to be smaller than their diamond analogs, so that the interpretation of the experimental results and their relation to hardness, the exact nature of the "superhard" phases (sp^2 to sp^3 bond ratio, crystallinity, etc.) remained elusive [66, 67]. We emphasize however that the ideal strength of carbon clathrates have been shown to be larger than that of diamond, even though their bulk or shear moduli are always smaller than their diamond analogs. This suggests that extrapolating hardness from considerations based on the elastic response of the materials may not always lead to significant results.

6.4 Electronic Properties of Clathrates

6.4.1 Electronic Properties of Empty Clathrates

We review in this section the electronic properties of empty carbon and silicon clathrates, that is clathrates without guest atoms within the cages. We emphasize that silicon clathrates are always synthesized with intercalants. As a matter of fact, it is quite difficult to remove all dopant atoms from the most "opened" Na_xSi_{136} (that is Na-doped fcc-Si_{34}) structure [68]. However, understanding the properties of empty clathrates is the starting point for appreciating the physics of intercalated structures.

Since Si and C clathrates are all fourfold coordinated within a slightly distorted sp^3 environment, they satisfy the octet rule and show a band gap as their diamond analog. The most interesting feature however is that the band gap of silicon clathrates is larger by ~0.4–0.8 eV than that of the Si-2 diamond phase (~1.2 eV) and can be nearly direct in the fcc-Si_{34} cubic structure. Such results have been established by several calculations within tight-binding [34], density functional theory (DFT) [35, 36, 69, 70] or more accurate quasiparticle calculations within the GW approximation [71]. These predictions were further confirmed experimentally using optical adsorption and transport measurements on the "nearly emptied" Na_xSi_{136} structure with a measured band gap of 1.9–2.0 eV. [68] The opening of such a band gap was tentatively explained by the presence of pentagons in the network which frustrates the formation of completely bonding states with p character at the top of the valence bands, leading to a contraction of the occupied bands, measured by photoemission, and the subsequent opening of the band gap [35, 72].

Similar theoretical studies have been conducted for guest-free carbon clathrates with hexagonal [37] or cubic [71] structures, showing that as expected carbon clathrates are wide-band gap insulators as their diamond analog. However, and contrary to the silicon case, the band gap of the carbon clathrates is not larger than the ~5.4 eV band gap of the diamond phase. Such results can be established comparing the DFT band gap of all structures (with similar running parameters and exchange-correlation functional) or within a more accurate quasiparticle approach, such as the GW approximation, known to provide for such systems band gaps in excellent agreement with experiments (while DDT-LDA calculations significantly underestimate the band gap).

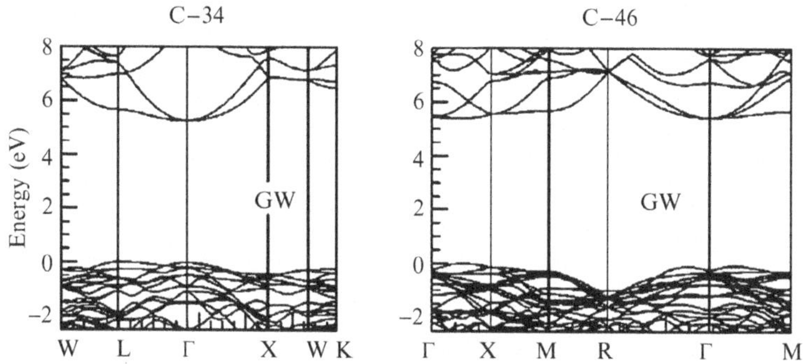

Fig. 6.15 Quasiparticle band structures of fcc-C_{34} and sc-C_{46} along high symmetry directions of the Brillouin zone (reproduced from Ref. [71] with permission from the APS)

We reproduce in Fig. 6.15 the GW band structure along high symmetry directions of cubic fcc-C_{34} and sc-C_{46} [71]. The band gap of fcc-C_{34} is found to be nearly direct at zone-center with a value of ~5.15 eV. Similar results are obtained for sc-C_{46} with a slightly larger 5.35 band gap. This is smaller than the 5.6 eV band gap found for diamond within the same GW approach. The reason why the band gap of guest-free carbon clathrates is not larger than that of diamond, in great contrast with the silicon case, is still to be clarified.

6.4.2 Endohedrally Doped Silicon Clathrates

As emphasized above, clathrates are synthesized "around" the intercalated atoms and guest-free structures can only be obtained in Na_xSi_{136} (that is Na-doped fcc-Si_{34}) by diffusion of the "small" Na atoms through the Si network. Endohedral doping, that is the intercalation of atoms within the cages, certainly offers most of the interesting features of clathrates. While doping of diamond and silicon is limited in standard conditions to a very few percent by the solubility limit [73], the cage-like structure of clathrates allows a very large intercalation, with a guest to host atomic ratio of up to 8/46 in type I clathrates and 24/136 in type II clathrates (all cages filled). For such a large incorporation of impurities, the word "alloying" may sound more appropriate than doping. As shown below however, the electronic properties of intercalated clathrates can often be understood in a close to rigid-band-like model of doping, justifying the use of that terminology. This very large doping of column-IV semiconducting sp^3 structures allows to unravel properties that are difficult to observe in standard semiconductors such as silicon in the diamond phase. We briefly review here below the case of silicon clathrates before turning to endohedrally doped carbon clathrates. We will discuss the superconducting transition under appropriate intercalation in a separate section.

The first synthesis of intercalated silicon clathrates was performed by thermal decomposition of akali-metal silicides MSi in the Zintl phase (M = Na, K, Rb, Cs) [8]. By mixing several silicides, such as $NaSi_2$ and $BaSi_2$, silicon clathrate compounds of the form $(Na, Ba)_xSi_{46}$ could be synthesized [74], the Ba atoms occupying the six larger 14-hedral cages and the Na atoms the two smallest 12-hedral ones mainly. Since then, HPHT techniques led to the intercalation of a much larger variety of atoms from the periodic table including various alkali, alkali-earth, halogens, chalcogens (oxygen column), lanthanides or transition metals (see compilation by Bobev and Sevov in 2000 [75] and san Miguel and Toulemonde in 2005 [76]). These novel guest elements extend in particular the *n*-type doping by alkali and alkali-earth to *p*-type doping in the case e.g. of halogen atoms such as iodine [77]. We emphasize however that in many situations, the guest atom may also substitute in the silicon host network, as it is the case for rubidium, iodine [77], transition metals [78, 79] or tellurium [80]. We focus in this section on clathrates with preserved silicon network.

A first interesting issue related to intercalation is the observation that it may dramatically modify the stability under pressure of the clathrate phases. It was shown [81, 82] that, even though a low density phase of silicon diamond, "nearly emptied" type II Na_xSi_{34} clathrates (x < 0.5) would not turn into diamond at high pressure, but directly into the β-tin phase around 8–10 GPa. This is roughly the same value than for the diamond to β-tin transition pressure [83], but with a much larger 1/3 volume collapse. Thermodynamic considerations based on free-energy calculations under pressure lead to a much smaller 3–4 GPa pressure transition, suggesting a kinetic barrier to overcome related to the breaking of Si bonds. Such a "bond-breaking" barrier can certainly be invoked as well for the frustration of the empty clathrate to diamond phase transition.

The intercalation of type II clathrate by Na just retards the transition to the β-tin phase to about 12 GPa [84]. However, upon intercalation of bigger K, Ba, I or Rb atoms, the transition to the β-tin phase is also frustrated [84–87]. Instead, several *isostructural* transitions occur at intermediate pressure depending on the guest, but the clathrate structure remains stable up to above 30 GPa where a transition to an amorphous phase takes place. These isostructural transitions seem to be related to a motion of the Si atoms around their equilibrium transition and, at higher pressure, to the motion of the intercalated atoms. In summary, intercalation significantly extends the domain of stability in pressure of the clathrate structure. Such an observation, together with the fact that intercalation may lead to a significant increase of the bulk modulus [88], suggest that experiments to test the influence of endohedral doping on hardness (see Section 6.3.2) may be of much interest.

A second important issue concerns the thermoelectric properties of intercalated clathrates, and the interplay between the electronic transport on the crystalline host network and the interaction of the host phonons with the localized "rattling" modes of the guest atoms inside the cages. In that perspective, clathrates may be regarded as "phonon glass/electron crystal" materials [89] with excellent figures of merit. Even though a few theoretical predictions for silicon clathrates are available [90], most experimental work are related to germanium hosts [91–93] with substitution of Ge atoms by Ga atoms in compounds such as $Sr_8Ga_{16}Ge_{20}$ (Sr inside the cages). Both the cage structure and the intercalation of guest atoms have been shown to reduce the heat conductivity by one order of magnitude each with respect to the diamond phase [94]. The latest experimental analysis emphasizes in particular the role of avoided crossing between the rattler modes and the acoustic-phonon branches [95]. Concerning the electronic properties, the donor character of Sr in intercalation with the acceptor one of Ga in substitution of Ge, compete to tune the electronic properties of the compound around the semiconducting nature of the perfectly stoichiometric $Sr_8Ga_{16}Ge_{20}$ system. The tuning of the thermal and electronic transport properties of such systems is still a very active field with e.g. recent variations such as (Si,Ga)-substituted $Ba_8Ga_{16}Si_xGe_{30-x}$ clathrates [96].

Back to the electronic properties of intercalated silicon clathrates, and as mentioned here above in the germanium case, guest atoms act either as donor or acceptors depending on their electronegativity. While alkali and alkali-earth atoms are donor for the Si network, halogen atoms are acceptors. The important issue in the case of clathrates is that the doping is so large that it is beyond the so called metal-insulator (MIT) transition [97]. Namely, the dopants do not form an impurity band in the forbidden gap of the host semiconductor, but the Fermi level enters the valence/conduction bands for acceptors/donors, turning the system into a truly metallic "degenerate semiconductor". This is of much importance for the superconducting transition discussed below. For sake of example, in the case of Ba_8Si_{46} clathrates, the Fermi level is located nearly 1 eV above the bottom of the conduction bands [70, 98]. While in the case of Na intercalation, a rigid-band like model of doping may be invoked, indicating weak hybridization between Na and Si orbitals, a significant hybridization at E_F has been observed between Ba 5*d* and Si orbitals (see below). An interesting observation is the evolution of the value. While *n*-type doping was shown to reduce the band gap [70], *p*-type doping by iodine was shown on the contrary to open the band gap by as much as $0.5-0.6$ eV, leading to a 2.25 eV value. While enhanced screening was invoked to explain the band gap reduction by Na or Ba doping, a clear group symmetry analysis led to the conclusion of a strong repulsion between guest orbitals and the bottom of the conduction band for *p*-type intercalation [98]. Intercalation, besides doping, can therefore also be used to tailor the band gap.

6.4.3 Endohedrally Doped Carbon Clathrates

The potential application of diamond in high-power, high-frequency and high-temperature electronic devices has stimulated a great deal of experimental and theoretical research on the doping properties of diamond. [99] Considerable efforts have been directed mainly to the doping of crystalline diamond, synthesized by chemical vapor deposition technique (CVD) [100–103]. Although it is well established that boron atom can be incorporated in the crystal to form a shallow acceptor level (0.35 eV above the top of the valence band), making diamond an effective p-type semiconductor [100, 103], n-type doping in CVD diamond is still a challenging problem. Incorporation of group V and VI elements in diamond have been studied extensively. Nitrogen is well-known to induce a deep level 1.7 eV below the conduction band [100], while more recently phosphorus [101] and sulfur [102] have been reported to induce shallower acceptor levels in CVD diamond films (0.56 and 0.38 eV below the conduction band for phosphorus and sulfur impurities, respectively). Still, improvements of n-type doping is needed for device applications.

In this perspective, it has been proposed that clathrates might be an alternative form of tetrahedral carbon suitable to be both p-type and n-type doped [37, 98, 104–106].

The n-doping can be realized by inserting lithium atoms in the large cages of the clathrates structure. Ab initio calculations on Li-doped hex-C_{40} [37] and cubic sc-C_{46} [104] clathrates have shown that the guest Li atom fully ionizes by donating its outermost electron to the host frame which becomes metallic. In hex-C_{40} Li can be inserted at the center of the C_{26} and C_{24} cages with small changes in the equilibrium cell parameters (within 1%) [37]. The most favorable site for Li insertion is at the centres of the largest C_{26} cages (cf. Fig. 6.16). However, the other site at the centre of the C_{24} cages is only 0.08 eV higher in energy. The Li ion is confined in the cages by a very high energetic barrier for diffusion across the hexagonal rings. For instance, the energy barriers for the diffusion across the large and small hexagonal rings of the C_{24} cages are 1.97 eV and 3.97 eV, respectively. The large energy barriers prevent Li diffusion up to very high temperature and therefore assure the thermal stability of

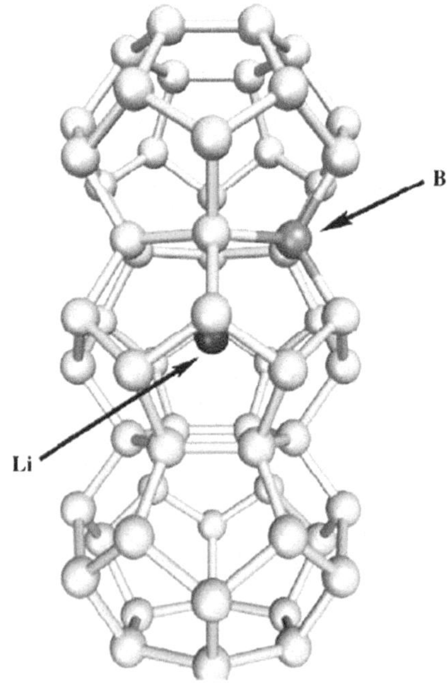

Fig. 6.16 Channel of C_{24} cages running along the *c* direction in the hex-C_{40} clathrate. The position of the Li atom at the centre of one of the C_{24} cages and the most stable site for boron substitution in the largest hexagonal ring of the C_{24} cage are shown [37]

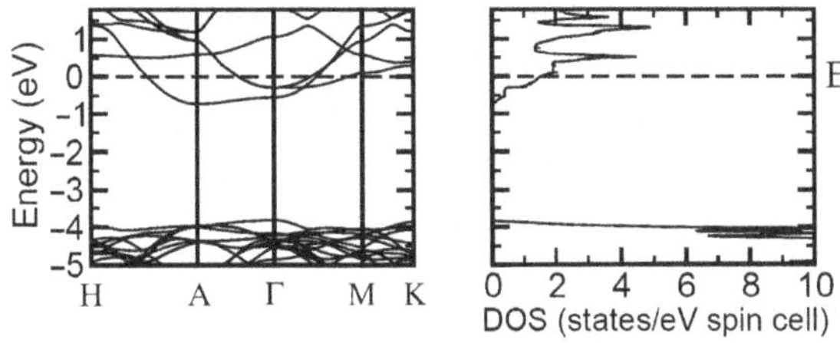

Fig. 6.17 Electronic energy bands (at DFT level of theory) around the Fermi level (E_F) and density of states of LiC_{40} (adapted from Ref. [105])

the Li doping. On the other hand, the presence of high diffusion barriers imply that Li, and *a fortiori* larger alkali ions, can not be inserted by electrochemical means, but must be implanted or introduced in the growth process as done for instance in the synthesis of the Si and Ge clathrates. At low Li content (LiC_{40}), the band structure does not change significantly with respect to the pure compound, but for the shift of the Fermi level inside the conduction bands and a small downward shift in energy of the lowest conduction band at the Γ point [37, 105]. Therefore, the electronic properties of the system at low Li content can be described in the rigid-band approximation where Li is fully ionized and donates its outermost electron to the host without changing the band structure (Fig. 6.17). At higher Li concentration (Li_2C_{40}) deviations from the rigid-band approximation are appreciable [105].

Similar metallic behavior has been predicted for Li_8C_{46} where Li occupies the center of all the fullerenic cages of cubic sc-C_{46} [104].

Concerning p-doping, ab initio calculations have shown that substitutional boron is an effective p-dopant in the hex-C_{40} clathrate [37]. Among the five possible inequivalent sites for substitution, boron is found to lies preferentially in the larger hexagonal ring as shown in Fig. 6.16. The boron insertion induces the formation of an acceptor state in the gap, 0.53 eV above the top of the valence band, mostly localized on the boron atom and on the two first neighbors of boron which do not lie on the hexagonal ring. This occurs even at the relatively heavy boron doping of one Li atom per unit cell (B_1C_{39}). The same is true for the cubic clathrate fcc-C_{34} [106]. Ab initio calculations have shown that insertion of substitutional boron in any of the three positions independent by symmetry in pure fcc-C_{34} (8a, 32e and 96g of space group $Fd3m$) induces three additional bands well above the top of the valence band of the host. These bands are mostly (up to 50%) localized on the boron atoms and its four nearest neighboring carbon atoms (cf. Fig. 6.18). This is at odds with heavy boron doped crystalline diamond synthesized at high pressure and temperature ($P \sim 10$ GPa and $T = 2{,}500 - 2{,}800$ K) [107]. In this system boron (up to 3%) induces a metallization by hole injection into the intrinsic diamond valence bands as demonstrated by ab initio calculations [64, 108–110] and later confirmed by angle resolved photoemission measurements [111].

However, metallization by hole injection in the valence bands can be realized in carbon clathrates as well by inserting halogen atoms at the center of the fullerenic cages [106], as previously demonstrated theoretically and experimentally in silicon clathrates [98]. Ab initio calculations on halogen doping (with iodine and fluorine)

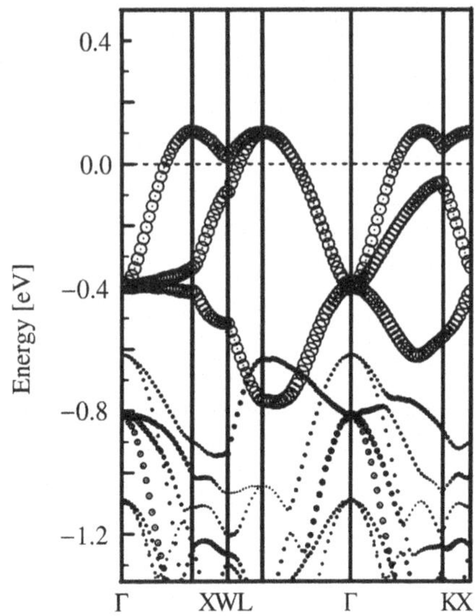

Fig. 6.18 Electronic energy bands (at DFT level of theory) of BC_{33}, i.e. fcc-C_{34} with substitutional boron in position 8a (Adapted from Ref. [106]). The zero of energy is the Fermi level. The size of the empty dots represents the size of the projection of the bands on the orbitals of boron and of the neighboring four carbon atoms which contribute to at most 62 %, with completeness 0.99

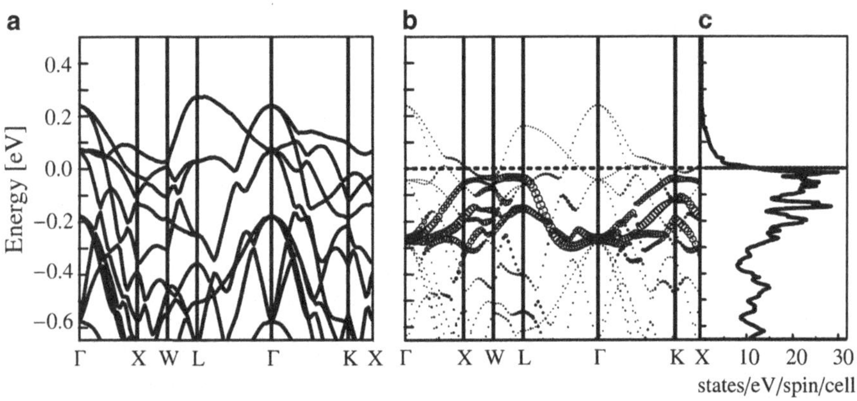

Fig. 6.19 Electronic energy bands of (**a**) pure fcc-C_{34} and (**b**) FC_{34}. The zero of energy is the Fermi level of FC_{34}. The electronic (indirect) gap of pure fcc-C_{34} is 3.9 eV. The top of the valence band at the *Γ* point is aligned for the two systems. The size of the empty dots represents the fluorine-like character, whose maximum value is 77 % (with completeness 0.99). (**c**) Density of states. Adapted from Ref. [106]

have been performed for fcc-C_{34}, sc-C_{46} and hex-C_{40} clathrates [106]. The insertion of the halogen atoms generates additional bands which in the case of FC_{34} (one F atom per unit cell at the center of the C_{28}) lie deep in the valence band (Fig. 6.19). The halogen is fully reduced and holes are introduced in the valence bands of the host. The band structure close to the Fermi level does not show sizeable changes with respect to the pure compound (Fig. 6.19).

On the contrary, in FC_{40} two bands which contribute to the DOS at the Fermi level have a strong contribution from *p* orbitals of the halogen. Fluorine is thus reduced only partially. The situation is even worse for hex-C_{40} doped with iodine. In fact, in the band structure of IC_{40} a set of bands are present in the gap $\sim$1 eV above the top of the valence bands of the host; they correspond to the atomic states of iodine and are partially filled as expected for a neutral, not-reduced halogen atom [106]. For FC_{46} (one F atom per unit cell at the center of the C_{24} cage) the situation is similar to that of FC_{40}. Holes are injected in the conduction bands of the host, but the Fermi level also crosses bands which are mainly atomic-like orbitals of the fluorine guest [106]. Although to a different extent, metallization can be achieved in all the three clathrates by fluorine insertion, but only in FC_{34} the DOS at the Fermi level is mostly due to electronic bands of the host.

6.5 Superconductivity

As discussed above, empty clathrates are "standard" column-IV sp^3 semiconductors such as diamond and silicon (Si-2). Even though the discussion of the superconducting properties of semiconductors may seem rather odd, it turns out that this subject attracted recently much attention with the discovery of a superconducting transition in highly boron doped diamond [107], silicon [112] and silicon-carbide [113] (for a review see [114]). Furthermore, the idea that semiconductors doped beyond the MIT could become superconducting was already discussed in the mid-sixties [115]. Superconductivity was observed subsequently in reduced $SrTiO_3$ perovskite single crystals [116] and in $Ge_{1-x}Te$ alloys [117].

In the limit of sufficient doping, beyond the MIT critical concentration, diamond and silicon become metallic. These are very special metals in the sense that the bonds still have the strength of covalent systems, with a stiffness exceeding considerably that of standard metals. As a matter of facts, superconductivity in highly B-doped diamond was made possible by the developments of HPHT synthesis techniques designed primarily in a community devoted to ultrahard materials, with much focus in particular on B_xC_{1-x} systems [118]. Similarly, a novel BC_5 cubic structure was recently publicized as a "superhard diamondlike" system by the team of experimentalists at the origin of the synthesis [119], while theoreticians predicted for the same material, through ab initio calculations, a transition temperature of the order of that of MgB_2 [120].

The relation between hardness and superconductivity, and the interest of covalent metals as superconducting materials, lie in the strength of the electron–phonon coupling. The so-called deformation potentials, which are a measure of the shift of the electronic energy levels upon distortion of the atomic network along a phonon mode [121], are extremely large. For instance, the deformation potential at the top of the valence bands in diamond for zone-center optical modes is much larger than that associated with the E_{2g} mode responsible for the 40 K superconducting transition in MgB_2 [108, 109]. Even though no precise relation between stiffness and deformation potentials is available to date, one can picture that the localization of electrons in very directional bonds drives both large elastic response and large electron–phonon

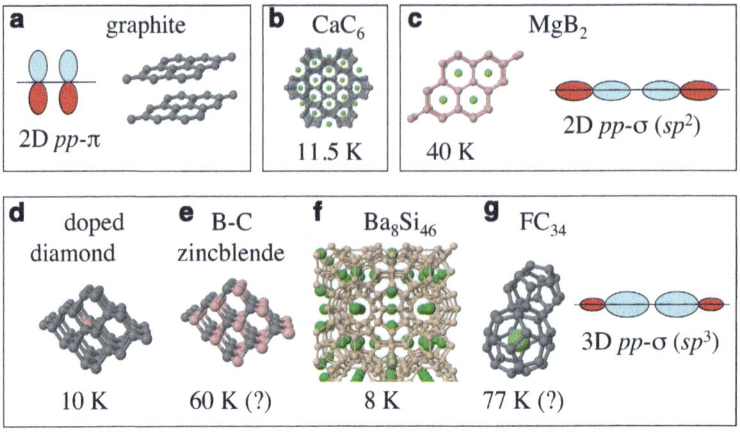

Fig. 6.20 Symbolic representation of several superconducting systems belonging to the family of covalent metals. Superconducting transition temperatures are indicated below. Temperatures in parenthesis indicate theoretical predictions (adapted from Blase et al. in Ref. [114])

coupling matrix elements. The same arguments explain the large T_c in MgB_2 with the activation of in plane σ bonds and the prediction of even larger T_Cs in graphite-like systems such as Li_2B_2 [122], BC_3 [123], LiBC [124] or MgB_2C_2 [144], tailored so as to bring the Fermi level into the σ bands.

Even though the deformation potential responsible for the superconducting properties of doped diamond is much larger than that of MgB_2, the largest T_c measured so far is of the order of 10 K, that is significantly smaller than that of MgB_2. The reason lies in the rather low density of states at the Fermi level related both to the difficulty of doping the diamond phase beyond a few percents, and to the 3D nature of the crystal which forbids any van Hove singularities in the density of states. As discussed below, both large deformation potentials and electronic density of states at E_F are needed to enhance T_c. This is where the clathrates are quite unique, with the sp^3 directional covalent bonds of the diamond phase, but with the possibility of large endohedral doping. Before discussing in more details the superconducting properties of silicon and carbon clathrates, we overview in Fig. 6.20 a few covalent metals that have been extensively discussed in the literature in the last few years for their superconducting properties.

6.5.1 Doped Silicon Clathrates

Nearly one decade before the discovery of the superconducting behavior of heavily doped diamond and silicon (Si-2), a superconducting transition was observed in (Na,Ba)-intercalated Si clathrates with a T_c of 4 K [74]. The transition temperature is increased up to 8 K when all Na atoms are replaced by Ba in the Ba_8Si_{46} structure [125–127]. An important point is that such a T_c is much larger than that observed in doped cubic silicon (~0.35 K), a dramatic increase that can be tentatively rationalized by the two following arguments: (a) the cage structure allows a very large doping concentration, and (b) the hybridation of baryum d-states with the silicon s and p orbitals results in an enhanced density of states at the Fermi level.

Ab initio calculations (DFT) were performed to study the origin of the SC transition in the clathrates, adopting the "standard" phonon-mediated approach to superconductivity within the Eliashberg formalism [128] which treats perturbatively the influence of the electron–phonon interaction on the electron self-energy. Several simplifications lead to the celebrated but approximate McMillan formula:

$$T_c = \frac{\hbar\omega_{ln}}{1.2k_B}\exp\left[\frac{-1.04(1+\lambda_{ep})}{\lambda_{ep} - \mu^*(1+\lambda_{ep})}\right]$$

relating the transition temperature T_c to the electron–phonon coupling constant λ_{ep} and an average phonon frequency ω_{ln}, both calculated ab initio. An important complication in such an approach is related to the screened and retarded Coulomb repulsion term μ^* which is difficult to calculate. μ^* is usually taken as an "adjustable" parameter of which the value is however known to be close to $\sim 0.1 - 0.3$ in standard metals. Such a term reduces the attractive electron–electron interaction mediated by phonons which is responsible for the superconducting pairing (leading to the famous "Cooper" pairs). It was shown however that the measured evolution (reduction) of T_c for Ba_8Si_{46} under applied pressure was very nicely reproduced by ab initio calculations, recalculating λ_{ep} at each pressure (reduced lattice parameter) but without changing the μ^* value from its ambient pressure value [127]. Further, the measured λ_{ep} and μ^* turned out to be in excellent agreement with the calculated ones, with $\lambda_{ep}^{exp} = 0.8 - 1.2$ and $\mu^{*,exp} = 0.23 - 0.31$ at ambient pressure [126], to be compared to the theoretical values, $\lambda_{ep}^{theo} \sim 1$ and $\mu^{*,theo} \sim 0.24$ [127]. Such a good agreement suggests that (a) the transition can be well understood within a phonon-mediated scenario, and (b) ab initio calculations perform rather nicely in predicting and interpreting the superconducting transition.

The λ_{ep} value of ~1 is rather large, significantly larger than the 0.3–0.5 values found in doped diamond or Si-2 silicon. Such an enhancement of the coupling constant in clathrate has been analyzed in terms of a peak in the density of states at E_F originating from the presence of a rather flat Ba 5d band at E_F which hybridizes with the Si Bloch states [35]. Concerning the participating phonon modes, the analysis of the Eliashberg function [98], a phonon density of states weighted by the strength of the coupling with the electrons at E_F, shows that the zone-center Si optical modes contribute significantly to λ_{ep}, as in doped Si-2 diamond. In addition, low optical modes have been shown to provide a non negligible contribution, pointing out to the rattling modes of Ba within the cages as suggested experimentally [129] and theoretically [130, 131].

The much larger value of T_c in Ba_8Si_{46} as compared to B-doped Si-2, the predictions of an increasing T_c in the fullerene family upon cage size reduction, and the much smaller mass of carbon atoms, suggest that turning to carbon clathrates may lead to significant T_c, that is values potentially larger than the 10 K observed in B-doped diamond. This is what we now discuss in the next section.

6.5.2 Doped Carbon Clathrates

Superconductivity in carbon-based systems is a very active area of research. The discovery of high superconducting transition temperature in alkali-doped C_{60} ($T_c = 10 - 40\,K$) has attracted much attention on these materials as representatives of a new class of superconductors [132, 133]. In this respect, the smaller fullerenes (C_{36}, C_{28}, C_{20}) are the most interesting since the electron–phonon coupling (and hopefully T_c) has been calculated to become larger with increasing curvature of the molecule [134–140]. In fact, a strong electron–phonon coupling is expected for states with large σ character as discussed earlier in this section. In fullerites the conduction bands have mainly π character with only a partial σ-character produced by the curvature of the cage-like structure. Ab initio calculations of the electron–phonon coupling have been performed for solid forms of C_{36} [135], C_{28} [140] and C_{20} fullerenes [139] which show that indeed the larger the curvature of the molecular cage, the larger is the electron phonon coupling in the solid state as well. Experimental evidences of the synthesis of solids made by C_{36} [141] or C_{20} [142, 143] clusters covalently interlinked have also been provided.

On the other hand the electron–phonon coupling is expected to be maximum in a carbon material with conduction bands of pure σ-character, i.e. in a diamond-like, tetrahedral form of carbon. As already mentioned (Section 6.5.2) metallization of diamond has been achieved by heavy doping with boron (up to 3 %) at high pressure and temperature ($P \sim 10$ GPa and $T = 2{,}500 - 2{,}800\,K$) [107]. In spite of a large electron–phonon coupling superconductivity has been reported at the relatively low temperature of $T_c \sim 4$ K which is due to the still low electron density of states at the Fermi level. In this system, metallization is produced by hole injection into the intrinsic diamond valence bands [64, 108–111]. The possibility to tune T_c in diamond by changing the carrier concentration in its intrinsic bands, either by p-type or n-type doping is thus intriguing and acquired a new boost from this recent discovery.

Clathrates represent another tetrahedral form of carbon suitable to turn metallic upon doping. As discussed in Section 6.5.2, metallization in carbon clathrates (hex-C_{40}, sc-C_{46} and fcc-C_{34}) can be achieved by n-doping with Li insertion or by p-doping with insertion of halogen atoms. Electron–phonon coupling has then been investigated in these systems by first principles aiming at predicting their superconducting properties [98, 105, 106].

In systems with a small dispersion of phononic and electronic bands around the Fermi level (either because of the large unit cell or of the molecular-like character of the solid) a reliable estimate of the electron phonon coupling constant $\lambda_{ep}/N(0)$ can be obtained in the so-called molecular-like approximation as [133, 134]

$$\frac{\lambda_{ep}}{N(0)} = \sum_{\alpha} \frac{1}{\omega_{\alpha}^2 g^2} \sum_{n,m}^{g} | \langle u_m \mid \mathbf{M}^{-\frac{1}{2}} \varepsilon_{\alpha} \nabla V_{\text{eff}} \mid u_n \rangle |^2 \tag{6.11}$$

where α runs over the phonons at the Γ point with frequency ω_α and polarization vector ε_α. $\mathbf{M}$ is the atomic mass matrix. n, m run over the g electronic states at the Γ point whose corresponding bands cross the Fermi level, eventually also at a different point in the Brillouin Zone. In this manner all the bands contributing to the Fermi surface are included in the calculations, although the change in the electronic states with the wavevector **k** is neglected. ∇V_{eff} is the derivative of the Kohn–Sham effective potential with respect to the atomic displacement and $N(0)$ is the electronic density of states at the Fermi level per spin and per cell. This approximation for the electron–phonon coupling constant has been demonstrated to hold with good accuracy for fullerenic systems [133, 134, 139] and MgB_2-related compounds [144]. A similar approximation for the calculation of electron–phonon coupling constant which involves Γ-point phonons only and an average of the electronic states over the Fermi surface has been used for silicon clathrates in ref. [98]. Phonons at the Γ point and electron–phonon coupling constant can be computed within density functional perturbation theory [145].

This framework has been used to compute the electron–phonon coupling il Li-doped hex-C_{40} in Ref. [105]. Since the rigid band approximation holds to a large extent in Li_xC_{40} (cf. Fig. 6.17) the electron–phonon coupling $\lambda_{ep}/N(0)$ given by Eq. 6.11 has been computed for the pure system with an additional electron in the lowest conduction band, mimicking LiC_{40}. The total electron–phonon coupling constant turned out to be $\lambda_{ep}/N(0) = 0.248\,\text{eV}$ which is substantially larger than theoretical estimates for C_{60} (0.07 eV [133]). From the density of states (DOS) in Fig. 6.17, one obtains $N(0) = 1.55$ or 1.85 states/eV/spin/cell and thus $\lambda_{ep} = 0.38$ and 0.46, for LiC_{40} and Li_2C_{40}, respectively. Although the characteristic phonon frequency and the electron–phonon interaction potential $\lambda_{ep}/N(0)$ are comparable or even larger than in other low-Z superconductors (MgB_2, C_{60}), the electronic DOS at the Fermi level is low in Li_2C_{40} which results into a low value of λ_{ep}.

However, larger values for λ_{ep} have been envisaged for hole-doped clathrates in Ref. [106]. As a matter of fact the electronic DOS is much larger for the top valence bands than for the lowest conduction bands (cf. Fig. 6.17) while the electron–phonon coupling constants are similar for valence and conduction bands. As discussed in Section 6.5.2 metallization of clathrates by hole doping can be realized via insertion of halogen atoms in the fullerenic cages. For instance, the electron–phonon coupling constant for FC_{34} has been calculated to be $\lambda_{ep}/N(0) = 96$–$100\,\text{meV}$. The coupling is produced by stretching modes of the C–C bonds mostly due to two phonons at 957 cm^{-1} and 1,162 cm^{-1} which contribute for ~60% to the total electron phonon coupling constant. By properly averaging the density of states over a narrow energy range of the order of ω_{ln} around the Fermi level [128, 146] (where ω_{ln} is the weighted logarithmic average of the phonon frequencies equal to 1,023 cm^{-1} for FC_{34}) one obtains $\lambda_{ep} = 0.85$. The critical temperature T_c can then be estimated by making use of the McMillan's solution [128, 146] of the Eliashberg equations given in Section 6.6.1 [128, 146]. For typical values of the renormalized Coulomb pseudopotential μ^* in the range 0.1–0.2, and for $\lambda_{ep} = 0.85$ and $\omega_{ln} = 1{,}023\,\text{cm}^{-1}$ we obtain T_c in the range 77–35 K for FC_{34}, which is within the weak-coupling regime ($T_c/\omega_{ln} \ll 1$) for the McMillan solution to be applicable. The hole concentration can be increased in C_{34} by adding a second fluorine atom per unit cell at the centre of the second C_{28} cage. However, the three additional p bands produced by the second fluorine atom now cut the Fermi level indicating a partial reduction of the second fluorine. The increased DOS at the Fermi level is thus expected not to contribute to λ_{ep}.

Analogously, by inserting fluorine in the largest fullerenic cages of the other two carbon clathrates hex-C_{40} and sc-C_{46}, the guest is only partially reduced; a sizeable contribution to the electronic density of states at the Fermi level comes from atomic-like states of fluorine (cf. Section 6.5.2). The latter clathrates (FC_{40} and FC_{46}) are thus expected to display a sizeably lower electron–phonon coupling constant.

As a final remark we mention that also 3D polymeric forms of C_{60} have been proposed as possible superconducting materials. Most of the phases proposed for the 3D polymer are metallic from the analysis of DFT electronic band structure [44, 45, 47, 49, 50]. However, some of these phases including the most stable one (phase B in Ref. [47]) display an antiferromagnetic ordering due to the presence of unpaired electrons on p_z-like orbitals which turns the metallic non-magnetic phase into a semiconductor with a small band gap (of the order of 0.5 eV) [47]. The energy gain due to antiferromagnetic ordering is, however, overall small for all the structures suggesting a Néel temperature far below room temperature. Whether the existence of an insulating antiferromagnetic ground state might be responsible for transport properties different from those envisageable for the metallic non-magnetic phases, it remains to be seen. Measurements of the electrical conductivity σ show no supercondutive transition down to 4 K [147]. Instead a behavior $ln\sigma \propto T^{-1/4}$ typical of variable range hopping among localized states has been recorded [41]. A variable range hopping conductivity might be consistent with an intrinsic metallicity of the perfect crystal which could turn into an Anderson insulator due to disorder. However, one might not exclude that the perfect crystal is a band insulator with defect states in the gap responsible for the variable range hopping conductivity.

6.6 Conclusion

There may be still a long way to the controlled synthesis of carbon clathrates. Previous examples of potentially interesting compounds – on the basis of computer simulations – but seemingly extremely difficult to synthesize, such as the much debated ultra-hard cubic BC_2N structure [56, 57], suggest that care must be taken in guessing on the future discovery of carbon clathrates. While ab initio simulations are quickly improving in their ability to predict the properties of complex but well defined structures, the study of growth mechanisms and phase diagrams remains a computational challenge. The recent direct synthesis of guest-free clathrates [148] made of germanium shows that novel routes of synthesis can still be opened. As a matter of fact, the small size of C_{20} cages may stand as a barrier to the synthesis of intercalated carbon systems, and the geological Popigai phase suggest the possible synthesis of empty carbon clathrates in adequate HPHT conditions. The potential properties of such systems, as described above in the present chapter, are a great incentive for keeping the search of sp^3 polymerized carbon cage-like systems.

Acknowledgements X.B. is indebted to D. Connétable, E. Bourgeois and V. Timoshevskii, for important contributions to part of the work presented in this chapter. G.B and M.B gratefully acknowledge the contributions to the work presented here by J. Cariboni, L. Colombo, E. Galvani, S. Gaito, A. Gambirasio, G. Onida, S. Sanguinetti, S. Serra, I. Spagnolatti and F. Zipoli. One of us (GB) acknowledges the support of the Ikerbasque Foundation (project ABSIDES).

References

1. Powell HM (1948) J Chem Soc 1:61
2. Pauling L, Marsh RE (1952) Proc Natl Acad Sci 36:112
3. Davy H (1811) Philos Trans R Soc Lond 101:30
4. Faraday M (1823) Quart J Sci Lit Arts 15:71
5. Sloan ED Jr (1998) Clathrate hydrates of natural gases, 2nd edn. Marcel Dekker, New York
6. Dickens GR, Paull CK, Wallace P (1997) Nature 385:426428
7. Cros C, Pouchard M, Hagenmuller P (1965) C R Acad Sci 260:4764
8. Kasper JS, Hagenmuller P, Pouchard M, Cros C (1965) Science 150:1713
9. McMillan PF (2002) Nat Mater 1:19
10. Kaczmarski M, Bedoya-Martinez ON, Hernández ER (2005) Phys Rev Lett 94:095701
11. Benedek G, Bernasconi M, Gambirasio A (2003) Phys Stat Sol (b) 237:296
12. Wells AF (1977) Three-dimensional nets and polyhedra. Wiley, New York
13. Williams R (1979) The geometrical foundation of natural structure. Dover, New York
14. Benedek G, Colombo L (1996) Mater Sci Forum 232:247
15. Benedek G, Colombo L, Gaito S, Serra S (1997) In: Paoletti A, Tucciarone A (eds) The physics of diamond. IOS Press, Amsterdam, pp 575–598
16. Benedek G, Bernasconi M, Donadio D, Colombo L (2001) In: Benedek G, Milani P, Ralchenko VG (eds) Nanostructured carbon for advanced applications. Kluver, Dordrecht p 89
17. Alexandrov PS (1998) Combinatorial topology. Dover, New York
18. Ceulemans A, Fowler PW (1991) Nature 353:52
19. Benedek G, Galvani E, Sanguinetti S, Serra S (1995) Chem Phys Lett 244:339
20. Breza J, Kadlecikova M, Vojs M, Michalka M, Vesely M, Daniś T (2004) Microelectr J 35:709
21. Narayan J, Srivatsa AR, Peters M, Yokota S, Ravi Al KV (1988) Phys Lett 53:1823
22. Shevchenko V Ya, Madison AE (2006) Glass Phys Chem 32:118
23. Enyashin AN Ivanovskii AL (2007) Phys Sol St 49:392 [Russian original: (2007) Fiz Tverd Tela 49:378]
24. Pokropinny VV, Ivanovskii AL (2008) Rus Chem Rev 77:837
25. Hu S, Sun J, Du X, Tian F, Jiang L (2008) Diamond Rel Mater 17:142
26. Raty J-Y, Galli G, Bostedt C, van Buuren TW, Terminello LJ (2003) Phys Rev Lett 90:037401
27. Strong RT, Pickard CJ, Milman V, Thimm G, Winkler B (2004) Phys Rev B 70:045101
28. Aste T, Weaire D (1996) The pursuit of perfect packing. IoPP
29. Kelvin L (Sir William Thomson) (1887) Phil Mag 24:503
30. Weaire D, Phelan R (1994) Phil Mag Lett 69:107
31. Weaire D, Hutzler S (2001) The physics of foams. Oxford University Press; see also the review by de Gennes P-G (2001) Phys Today 54:54
32. Grason GM (2006) Phys Reports 433:1
33. Nester R, Vogel K, Blöchl PE (1993) Angew Chem 32:701
34. Adams GB, O'Keefe M, Demkov AA, Sankey OF, Huang Y (1994) Phys Rev B 49:8048
35. Saito S, Oshiyama A (1995) Phys Rev B 51:2628
36. Galvani E, Onida G, Serra S, Benedek G (1996) Phys Rev Lett 77:3573
37. Bernasconi M, Gaito S, Benedek G (2000) Phys Rev B 61:12689
38. Tersoff J (1988) Phys Rev B 37:6991; (1989) Phys Rev Lett 61:2879

39. Eklund PC, Rao AM (1999) Fullerene polymers and fullerene polymer composites. Springer, Berlin
40. Marques L, Mezouar M, Hodeau J-L, Nunez-Regueiro M, Serebryanaya NR, Ivdenko VA, Blank VD, Dubitsky GA (1999) Science 203:1720
41. Yamanaka S, Kubo A, Inumaru K, Kini NS, Inoue T, Irifune T (2006) Phys Rev Lett 96:76602
42. Chernotatonskii LA, Serebryanaya NR, Marvrin BN (2000) Chem Phys Lett 316:199
43. Blank VD, Buga SG, Dubistky GA, Serebryanaya NR, Popov MY, Sundqvist B (1998) Carbon 36:319
44. Okada S, Saito S, Oshiyama A (1999) Phys Rev Lett 83:1986
45. Berber S, Osawa E, Tománek D (2004) Phys Rev B 70:85417
46. Burgos E, Halac E, Weht R, Bonadeo H, Artacho E, Ordejón P (2000) Phys Rev Lett 85:2328
47. Zipoli F, Bernasconi M (2008) Phys Rev B 77:115432
48. Yang J, Tse JS, Tao Y, Iitaka T (2007) Angew Chem Int Ed 46:6275
49. Yang J, Tse JS, Tao Y, Iitaka T (2007) J Chem Phys 127:134906
50. Yamagami Y, Saito S (2009) Phys Rev B 79:045425
51. El Goresy A et al (2003) C R Geosci 335
52. Blank VD et al (1994) Phys Lett A 188:281
53. Brazhkin VV et al (1995) JETP Lett 62:350
54. Núnez-Regueiro M et al (1995) Phys Rev Lett 74:278
55. Blank VD et al (1995) Phys Lett A 205:208; Blank VD et al (1998) Diamond Related Mater 7:427
56. Liu AY, Cohen ML (1989) Science 245:841; Cohen ML (1985) Phys Rev B 32:7988
57. Teter DM (1998) MRS Bull 23:22
58. Gao F et al (2003) Phys Rev Lett 91:015502
59. Chacham H, Kleinman L (2000) Phys Rev Lett 85:4904
60. Telling RH, Pickard CJ, Payne MC, Field JE (2000) Phys Rev Lett 84:5160
61. Roundy D, Cohen ML (2001) Phys Rev B 64:212103
62. Frenkel J (1926) Z Phys 37:572; Kelly A, MacMillan NH (1986) Strong solids. Clarendon, Oxford
63. Gogotsi YG, Kailer A, Nickel KG (1999) Nature (London) 401:663
64. Blase X, Gillet P, San Miguel A, Mélinon P (2004) Phys Rev Lett 92:215505
65. Perottoni CA, da Jornada AH (2002) Phys Rev B 65:224208
66. Brazhkin VV, Lyapin AG, Popova SV (1998) J Appl Phys 84:219; Brazhkin VV, Lyapin AG, Hemley RJ (2002) Phil Mag A 82:231
67. Takashi H, Kaichi S, Michihiro K, Ryuichiro O (2005) Jpn J Appl Phys 44:3141-3146
68. Gryko J, McMillan PF, Marzke RF, Ramachandran GK, Patton D, Deb SK, Sankey OF (2000) Phys Rev B 62:R7707
69. Smelyanski VI, Tse JS (1997) Chem Phys Lett 264:459
70. Moriguchi K, Munetoh S, Shintani A (2000) Phys Rev B 62:7138
71. Blase X (2003) Phys Rev B 67:035211
72. Mélinon P, Kéghélian P, Blase X, Le Brusc J, Perez A, Reny E, Cros C, Pouchard M (1998) Phys Rev B 58:12590
73. Zhang SB, Wei S-H, Zunger A (2000) Phys Rev Lett 84:1232
74. Kawaji H, Horie H-O, Yamanaka S, Ishikawa M (1995) Phys Rev Lett 74:1427
75. Bobev S, Sevov SC (2000) J Solid State Chem 92:153
76. San Miguel A, Toulemonde P (2005) High Press Res 25:159
77. Reny E, Yamanaka S, Cros C, Pouchard M (2000) Chem Commun 2505–2506
78. Imai M, Nishida K, Kimura T, Yamada K (2002) J Alloy Compd 335:270
79. Herrmann RFW, Tanigaki K, Kuroshima S, Suematsu H (1998) Chem Phys Lett 283:29
80. Jaussaud N et al (2005) Inorg Chem 44:2210-2214
81. San-Miguel A et al (1999) Phys Rev Lett 83:5290
82. Ramachandran GK et al (2000) J Phys Condens Matter 12:4013
83. McMahon MI, Nelmes RJ (1993) Phys Rev B 47:8337; McMahon MI, Nelmes RJ, Wright NG, Allan DR (1994) Phys Rev B 50:739; Hu JZ, Merkle LD, Menoni CS, Spain IL (1986) Phys Rev B 34:4679

84. san Miguel A et al (2002) Phys Rev B 65:054109
85. Tse JS et al (2002) Phys Rev Lett 89:195507
86. Kume T et al (2003) Phys Rev Lett 90:155503
87. Machon D et al (2009) Phys Rev B 79:184101
88. Connétable D, Timoshevskii V, Artacho E, Blase X (2001) Phys Rev Lett 87:206405
89. Slack GA (1995) In: Rowe DM (ed) CRC handbook of thermoelectrics. CRC Press, Boca Raton, FL, p 401
90. Tse JS et al (2000) Phys Rev Lett 85:114
91. Nolas GS et al (1998) Appl Phys Lett 73:178
92. Slack G et al (1999) Phys Rev Lett 82:779
93. Sales BC, Chakoumakos BC, Jin R et al. (2001) Phys Rev B 63:245113
94. Dong JJ, Sankey OF, Myles CW (2001) Phys Rev Lett 86:2361
95. Christensen M et al. (2008) Nat Mater 7:811
96. Martin J et al (2006) J Appl Phys 99:044903; Nenghabi EN, Myles CW (2008) Phys Rev B 77:205203
97. Mott NF (1974) Metal-insulator transitions. Taylor and Francis, London
98. Connétable D, Timoshevskii V, Masenelli B, Beille J, Marcus J, Barbara B, Saitta AM, Rignanese G-M, Mélinon P, Yamanaka S, Blase X (2003) Phys Rev Lett 91:247001
99. Nesladek M (2005) Semicond Sci Technol 20:R19-R27
100. Collins AT (1993) In: Davies G (ed) Properties and growth of diamond. INSPEC, the Institution of Electrical Engineers, London, p 263
101. Fujimori N et al (1990) Mat Res Soc Symp Proc 162:23; Alexenko AE, Spitsyn BV (1992) Diamond Relat Mater 1:705; Koizumi S et al (1997) Appl Phys Lett 71:1065; Saito T et al. (1998) Jpn J Appl Phys Part 2, 37:L543; Nesládek N et al (1999) Phys Rev B 59:14852
102. Sakaguchi I et al (1999) Phys Rev B 60:2139
103. Fox BA et al (1995) Diamond Relat Mater 4:622; Yamanaka S et al (1998) Jpn J Appl Phys Part 2 37:L1129
104. Rey N, Munoz A, Rodriguez-Hernandez P, San-Miguel A (2008) J Phys Condes Matter 61:215218
105. Spagnolatti I, Bernasconi M, Benedek G (2003) Eur Phys J B 34:63
106. Zipoli F, Bernasconi M, Benedek G (2006) Phys Rev B 74:205408
107. Ekimov EA, Sidorov VA, Bauer ED, Melńik NN, Curro NJ, Thompson JD, Stishov SM (2004) Nature 428:542
108. Boeri L, Kortus J, Andersen OK (2004) Phys Rev Lett 93:237002
109. Lee K-W, Pickett WE (2004) Phys Rev Lett 93:237003
110. Lee K-W, Pickett WE (2006) Phys Rev B 73:75105
111. Yokoya T, Nakamura T, Matsushita T, Muro T, Takano Y, Nagao M, Takenouchi T, Kawarada H, Oguchi T (2005) Nature 438:647
112. Bustarret E et al (2006) Nature 444:465468
113. Ren ZA et al (2007) J Phys Soc Jap 76:103710
114. Blase X, Bustarret E, Chapelier C, Marcenat C (2009) Nat Mat 8:375
115. Cohen ML (1964) Phys Rev 134:A511-A521; ibid (1964) Rev Mod Phys 36:240243
116. Schooley JF, Hosler WR, Cohen ML (1964) Phys Rev Lett 12:474475; Schooley JF et al. (1965) Phys Rev Lett 14:305307
117. Hein RA, Gibson JW, Mazelsky R, Miller RC, Hulm JK (1964) Phys Rev Lett 12:320322
118. Solozhenko VL, Dubrovinskaia NA, Dubrovinsky LS (2004) Appl Phys Lett 85:1508-1510; Dubitskiy GA et al (2005) JETP Lett 81:260263
119. Solozhenko VL et al (2009) Phys Rev Lett 102:015506
120. Calandra M, Mauri F (2008) Phys Rev Lett 101:016401
121. Khan FS, Allen PB (1984) Phys Rev B 29:3341; Resta R (1991) Phys Rev B 44:11035
122. Liu AY, Mazin II (2007) Phys Rev B 75:064510; Calandra M, Kolmogorov AN, Curtarolo S (2007) Phys Rev B 75:144506
123. Ribeiro FJ, Cohen ML (2004) Phys Rev B 69:212507
124. Rosner H, Kitaigorodsky A, Pickett WE (2002) Phys Rev Lett 88:127001
125. Yokoya T et al (2001) Phys Rev B 64:172504

126. Tanigaki K, Shimizu T, Itoh KM, Teraoka J, Moritomo Y, Yamanaka S (2003) Nat Mater 2:653
127. Connétable D et al (2003) Phys Rev Let 91:247001
128. Carbotte JP (1990) Rev Mod Phys 62:10271157
129. Lortz R et al (2008) Phys Rev B 77:224507
130. Toulemonde P et al (2005) Phys Rev B 71:094504
131. Tse J et al (2005) Phys Rev B 72:155441
132. Hebard AF, Rosseinsky MJ, Haddon RC, Murphy DW, Glarum SH, Palstra TTM, Ramirez AP, Kortan AR (1991) Nature 350:600
133. Gunnarsson O (1997) Rev Mod Phys 69:575
134. Schlüter M, Lannoo M, Needels M, Baraff GA, Tománek D (1992) Phys Rev Lett 68:526
135. Coté M, Grossman JC, Cohen ML, Louie SG, Phys Rev Lett 81:697 (1998); Grossman JC, Louie SG, Cohen ML (1999) Phys Rev B 60:R6941
136. Devos A, Lannoo M (1998) Phys Rev B 58:8236
137. Adams GB, Sankey OF, Page JB, O'Keeffe M (1993) Chem Phys 176:61
138. Breda N, Broglia RA, Colò G, Onida G, Provasi D, Vigezzi E (2000) Phys Rev B 62:130
139. Spagnolatti I, Bernasconi M, Benedek G (2002) Europhys Lett 59:572
140. Romero NA, Kim J, Martin RM (2007) Phys Rev B 76:205405
141. Piskoti C, Yarger J, Zettl A (1998) Nature 393:711
142. Iqbal Z, Zhang Y, Grebel H, Vijayalakshmi S, Lahamer A, Benedek G, Bernasconi M, Cariboni J, Spagnolatti I, Sharma R, Owens FJ, Kozlov ME, Rao KV, Muhammed M (2003) Eur Phys JB 31:509
143. Wang ZX, Ke XZ, Zhu ZY, Zhu FY, Ruan ML, Chen H, Huang RB, Zheng LS (2001) Phys Lett A 280:351
144. Spanò E, Bernasconi M, Kopnin E (2005) Phys Rev B 72:14530
145. Baroni S, de Gironcoli S, Dal Corso A, Giannozzi P (2001) Rev Mod Phys 73:515
146. Allen PB, Mitrović B (1982) Solid State Phys 37:1
147. Yamanaka S, Kubo A, Kini NS, Inumaru K (2006) Phys B 389:59
148. Guloy AM, Ramlau R, Tang Z, Schnelle W, Baitinger M, Grin Y (2006) Nature 443:320

Chapter 7
Exotic Carbon Phases: Structure and Properties

M.G. Ganchenkova[1], T.T. Vehviläinen[1], and R.M. Nieminen[1]

Abstract The present chapter describes a number of "exotic" carbon nanostructures as related to the well-known graphene and carbon nanotube structures, which are characterized by the developed internal and external surfaces. Structural, electronic and magnetic properties as well as some insights into synthesis of these structures are considered and discussed.

7.1 Introduction

In recent years, nanomaterials have become a core focus of materials research and development due to their unique structural and electronic properties that underlie their important uses in diverse fields, such as ion exchange and separation, catalysis, biomolecular isolation and purification as well as in chemical sensing [53]. One of the ways to enhance their functional properties is to increase their specific surface area. This can be achieved either by the creation of a large number of nanostructured elements or by the synthesis of a highly porous material. Both recipes have actively been developed during the last decades. In this chapter we concentrate our attention to the general class of porous carbon-based materials, exhibiting large and functionalised external and internal surface areas.

Classically, porous matter is seen as material containing three-dimensional voids, exhibiting translational repetition, while no regularity is necessary for a material to be termed "porous". In general, the pores can be classified into two types: open pores which connect to the surface of the material, and closed pores which are isolated from the outside [53]. This defines the area of applications that the porous material can be effectively used in. Indeed, if the material exhibits mainly open pores, which can be easily penetrated, then one can consider its use in functional applications such as adsorption, catalysis and sensing. The closed pores, in turn, find their use

[1]COMP/Department of Applied Physics, School of Science and Technology, Aalto University, P.O. Box 11000, FI-0076 Aalto, Espoo, Finland
e-mail: mganchen@cc.hut.fi; tve@fyslab.hut.fi; rni@tkk.fi

L. Colombo and A. Fasolino (eds.), *Computer-Based Modeling of Novel Carbon Systems and Their Properties*, Carbon Materials: Chemistry and Physics 3,
DOI 10.1007/978-1-4020-9718-8_7,

in sonic and thermal insulation, or lightweight structural applications. The use of porous materials offers also new opportunities in such areas as inclusion chemistry, guest–host synthesis and molecular manipulations and reactions for manufacture of nanoparticles, nanowires and other quantum nanostructures.

The International Union of Pure and Applied Chemistry (IUPAC) defines porosity scales as follows:

- **Microporous** materials 0–2-nm pores
- **Mesoporous** materials 2–50-nm pores
- **Macroporous** materials >50-nm pores

This definition, it should be noted, is somewhat in conflict with the definition of nanoscale objects, which typically have large relative porosities (>0.4), and pore diameters between 1 and 100 nm. In order to classify porous materials according to the size of their pores the sorption analysis is one of the tools often used. This tool is based on the fact that pores of different sizes lead to totally different characteristics in sorption isotherms (see Fig. 7.1). The correlation between the vapor pressure and the pore size is given by the Kelvin equation:

$$r_p\left(\frac{p}{p_0}\right) = \frac{2\gamma V_L}{RT\,\mathbf{ln}(p/p_0)} + t\left(\frac{p}{p_0}\right)$$

where $r_p \approx$ pore radius, $\gamma \approx$ surface tension, $t \approx$ thickness of the adsorbate film, $V_L \approx$ molecular volume of the condensate. Therefore, the isotherms of microporous materials show a steep increase at very low pressures (relative pressures near zero) and reach a plateau quickly. Mesoporous materials are characterized by a so-called capillary condensation step and a hysteresis (a difference between adsorption and desorption). Macroporous materials show a single or multiple adsorption steps near the pressure of the standard bulk condensed state (relative pressure approaches one). An example of these three different pore systems is given in Fig. 7.2, where the ordered pore systems are shown. Although for porous materials the order is no premise, the ordered porous materials are found to be more attractive due to possibility to control over pore sizes and pore shapes. Moreover, the ordered porous materials seem to be much more homogeneous.

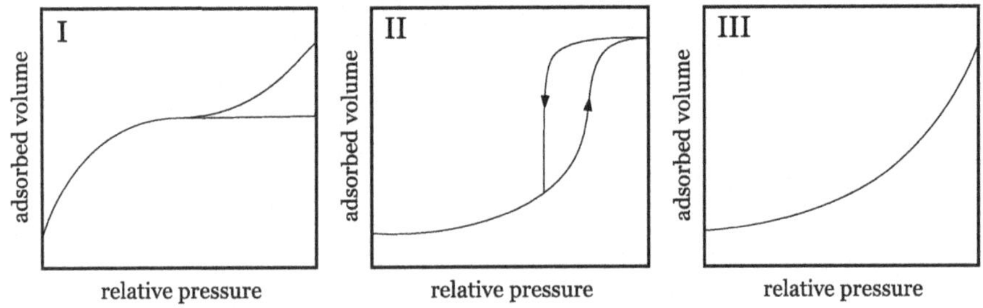

Fig. 7.1 Sorption isotherms of microporous materials (I), mesoporous materials (II), and macroporous materials (III)

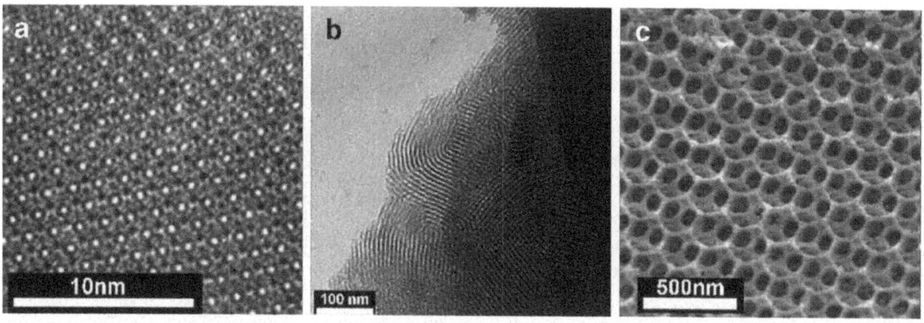

Fig. 7.2 Three different types of nanoporous silica materials. (**a**) A microporous material: zeolite Y. (**b**) A mesoporous materials. (**c**) A macroporous material [78]

Nanoporous materials abound in nature, both in biological systems and in natural minerals. Some nanoporous materials have been used industrially for a long time. For example, petroleum industry has been using naturally nanoporous materials called zeolites as catalysts for decades, much before the first synthetic structure had been synthesized. Recent progress in characterization and manipulation on the nanoscale has led to significant improvement in understanding and making a variety of nanoporous materials: from the merely opportunistic to directed design. This is most strikingly the case in the creation of a wide variety of membranes where control over pore size is increasing dramatically, often to atomic levels of perfection, as is the ability to modify physical and chemical characteristics of the materials that make up the pores.

Nanoporous solids have been out of a wide variety of substances, including carbon, silicon, silicates, various polymers, ceramics, and various metals or organic materials.

The traditional porous carbon materials exhibit rather disordered pore structures. However, the recent implementation of template techniques for the fabrication of nanostructured materials [45, 87, 94] allowed to synthesize different nanoporous carbons with periodic pore structure [5, 34, 35, 45, 46, 54, 87, 109, 110]. They have revealed themselves as very good candidates for electrochemical double-layer capacitors (EDLCs), hydrogen storage and electro-catalytic devices, being thus appealing materials for digital communication devices and fuel-cell technologies. Researchers in Korea, for example, have developed a template technique using silica nanoparticles which can create activated carbon with uniform 8 and 12 nm pore sizes. The resulting material showed adsorption greater than 10 times that of commercial activated carbon.

Porous inorganic solids such as zeolites (crystalline aluminosilicate materials with pore sizes ranging from 3 to 13 Å, see some examples in Fig. 7.3) and mesoporous silicates (amorphous solids with ordered pore structures) are extensively used as nanotemplates for the preparation of nanoporous carbon and silicon materials. Surface areas of these nanoporous solids are generally in the hundreds of square meters per gram. Porous carbons are obtained via the template carbonization

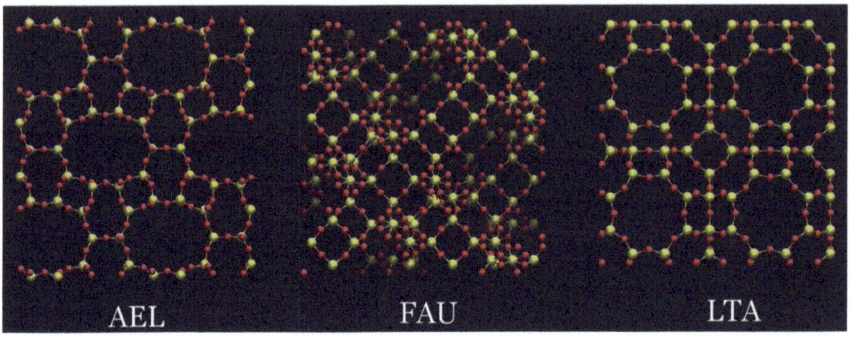

Fig. 7.3 Examples of some zeolite structures

method, which involves the introduction of suitable carbon precursors into the pores of a hard template, followed by carbonization and finally removal of the template. After removal of the template material, two different behaviours have been observed: some materials retain the same structure as before sieve elimination, and others develop structural transformations. In every case, the final result is a thermally and mechanically stable periodic carbon material with ordered pores. In general, the use of zeolites and mesoporous silicas as hard templates results in microporous or mesoporous carbons, respectively. However, while the structural ordering in mesoporous carbons is usually a faithful replica of the pore channel ordering of the silica templates, it is much more difficult to replicate the zeolite pore structure. This poor replication has been attributed to several reasons, one of the most important being the effect of the zeolite nanotemplate type. For instance, in the work of Terasaki et al. [54], the very small size of the zeolite pores, the lack of evidence for graphitic planes after thermal treatment up to 900 K, and the strong evidence of sp^2 bond hybridization given by the solid state NMR spectra support the idea of ordered aromatic rings conforming curved surfaces as structural blocks of the resulting nanoporous carbon.

One should understand, however, that the number of zeolite structures, which could be used as a template, is enormously large. Hence, experimental "sorting out" of nanotemplates, which would provide an appropriate structure and properties of the replica, is too challenging task. A computational approach that could allow to predict the structure and properties of a replica to be synthesized within a predefined nanotemplate would be of an immense practical value. The theoretical space of possible permutations for such materials is vast and will present many opportunities for the control of substances at the molecular and atomic levels for years to come.

Nanoporous carbons with new geometries can be created also by other methods, exploiting arrangements of positively curved surfaces, such as carbon fullerene crystals [41] and silicon clathrates [22]. Although a few decades ago carbon chemistry was thought to be completely understood in terms of the sp^2 graphene hybridization and sp^3 diamond hybridization, the discovery of nanostructured carbon forms (fullerenes [41], nanotubes [28], nano-onions [101], etc.) brought attention to the rich variety of chemical and physical properties related more strongly to the geometry, and particularly to the curvature of carbon structures, which imply intermediate degrees of hybridization. The specific structures of these carbon structure arrangements, which affects the electronic and vibrational properties, as well as the vast amount of void space in fullerite crystals along with their cage-like structure, open a wide range of possibilities for their potential applications.

One of the exotic examples of these structures is highly nanoporous carbon with a fractional internal geometry (fractals are patterns that show similar structures as different scales, such as coastlines or branches of trees), which is believed to have potential as methane (natural gas) storage for vehicles.

In late 2002, a group of Italian and British researchers created another form of carbon that had been hypothesized a decade earlier [56, 96, 97]. The material uses carbon rings containing more than six carbon atoms (which create the flat hexagons in graphite) to create an internal structure with negative curvature that is highly porous, with pore sizes in the 50–600 nm range, much larger than in activated carbon. These periodic arrangements [56, 71, 76, 96, 97, 103], called *schwarzites*, divide space into two regions possessing interconnected channels with pores similar to those found in zeolites. Relationships between zeolites and carbon networks have also been explored in the literature in the context of fullerite crystals [25].

Summing up, the above mentioned carbon structures, which are sometimes called "exotic" as related to the well-known graphene and carbon nanotubes, open a new research field with interesting prospects for creation of a novel class of materials for practical applications. This chapter is devoted to describe structural, electronic and magnetic properties of these materials.

7.2 Fullerene-Based Solids

Kroto, together with his colleagues Curl, Smalley, Heath and O'Brian, in their experiments on production of cold carbon clusters, managed to produce long carbon chains [41], which were characterized as those consisting of 60 carbon atoms. This structure was concluded later to be that of a truncated icosahedron with I_h symmetry (60 vertices) and was called *buckminsterfullerene* or C_{60}. This discovery gave a new strong signal for scientific community of the possibilities of carbon materials.

In these experiments, the researchers also found an evidence for the existence of a 70-atom based structure, which later was identified as another carbon cage molecule consisting of 12 pentagons and 10 additional carbon atoms around the waist of C_{60} (Fig. 7.4). Later Kroto introduced the idea that graphite could produce a whole family of nested cages, called giant fullerenes (Fig. 7.5). Despite of extensive research investigations, however, a number of different size carbon cages confirmed experimentally is still very limited. There is experimental evidence about some carbon

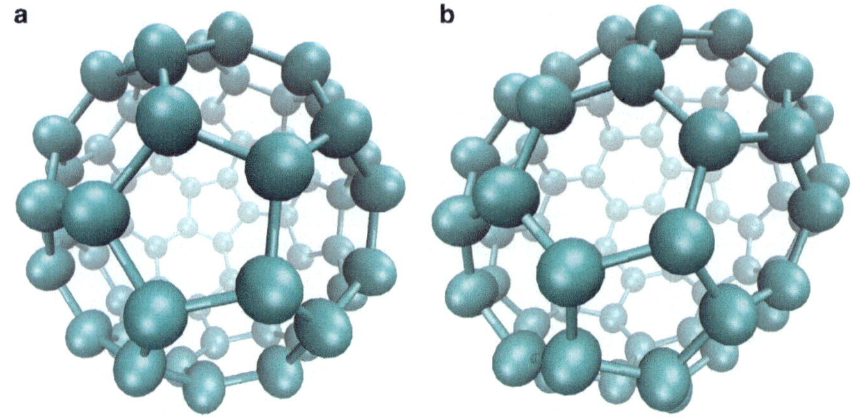

Fig. 7.4 (**a**) Buckminsterfullerene C_{60} (I_h) composed of 60 carbon atoms. (**b**) C_{70} (D_{5h}) composed of 70 carbon atoms

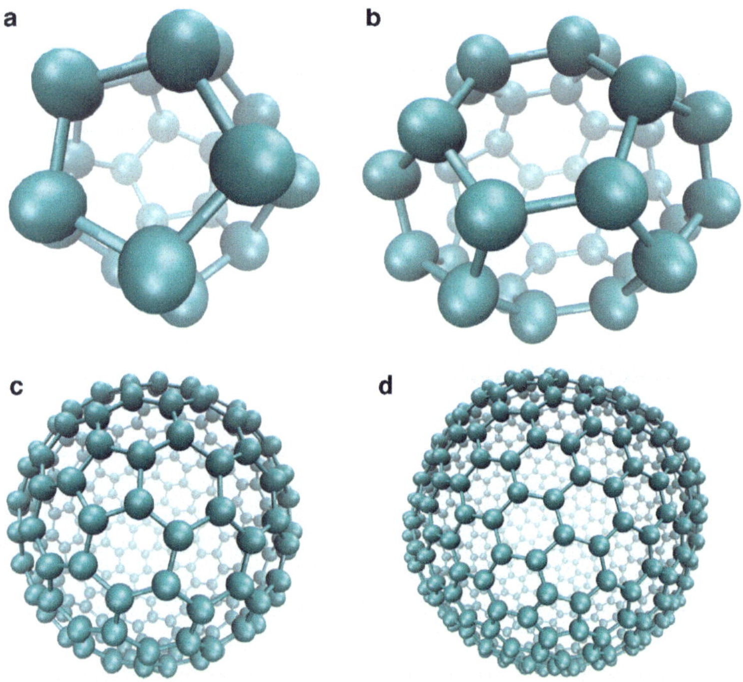

Fig. 7.5 Family of fullerenes. (**a**) C_{20}, a dodecahedron; (**b**) C_{36}; (**c**) C_{240}; (**d**) C_{540}

cages larger than C_{60}: C_{70}, C_{76} [13] and C_{84} [42], which are nowadays produced industrially. Recently, a number of small fullerenes have also been synthesizes. In particular, in 1998 Piskoti et al. [77] claimed to have synthesized crystals of C_{36} and in 2000 Prinzbach et al. [79] reported a successful C_{20} synthesis.

After the synthesis of buckminsterfullerene C_{60}, there has been growing interest in creating fullerene-based polymers as possible candidates for superconductors [63], strong and light materials, hydrogen-storage materials [39, 83, 115], quantum dots [3,36] and biological applications [65,93]. C_{60} based polymers have been studied intensively both experimentally and theoretically [6, 70]. To the date, several polymerized phases of C_{60} have been synthesized by using photo-induced [80], alkaline-atom-induced [7], and pressure-induced [32, 72, 94] polymerization procedures. Among these procedures, pressure-induced polymerization is known to be able to control network geometries and dimensionality by varying the pressure and temperature [33, 60]. Three different types of the polymerized phases have been experimentally synthesized under moderate hydrostatic pressure at elevated temperature, which can be classified as another form of crystalline carbon consisting of both sp^2-like (threefold coordinated) and sp^3-like (fourfold coordinated) C atoms. The x-ray diffraction experiments have clarified structures of one-dimensionally polymerized orthorhombic and two-dimensionally polymerized tetragonal and also rhombohedral phases. First-principles electronic structure calculations performed by Okada and Saito ([67, 68] for the structure as shown in Fig. 7.6) for the experimentally determined atomic structures of these polymerized phases have revealed that they are semiconducting with a narrow fundamental energy gap. Furthermore, the electronic structures of these phases are considerably different from that of the face-centered cubic phase of C_{60} fullerite [89], due to the presence of the sp^3-like carbon atoms which modulate network topologies of the p-electron system [20, 58, 81]. It has been shown that if these fullerene crystals are doped with alkali metals, such as potassium, caesium or rubidium, they can manifest superconducting properties at temperatures as high as 33 K [14, 24]. The insertion of ^{13}C isotopes in C_{60} might shed some light on the vibrational properties of fullerenes [90], and therefore on the superconductivity mechanism.

The recent synthesis of other fullerene cages has opened perspectives to creation of new kinds of polymer structures with different fullerenes as building blocks. Among different fullerenes, C_{20} is considered as a perspective polymer building

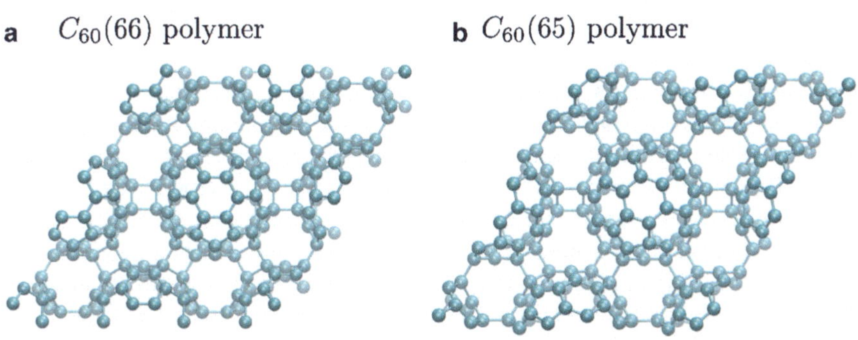

Fig. 7.6 Optimized geometries of each polymerized layer of (**a**) $C_{60}(66)$ polymer and (**b**) $C_{60}(65)$ polymer (based on [70])

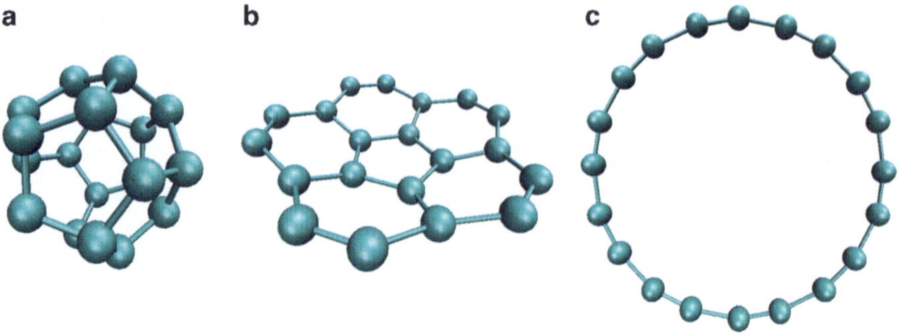

Fig. 7.7 Isomers of C_{20} (**a**) cage, (**b**) bowl, and (**c**) ring

block. C_{20} molecule has been predicted to have three different isomers, namely cage, bowl and ring, see Fig. 7.7. The question about the ground state isomer of the C_{20} molecule has been controversial for a long time. Different theoretical calculations give different predictions. For example, density-functional theory (DFT-BLYP) based calculations predicted that the ring isomer has the lowest energy, while diffusion Monte Carlo calculations (DMC) show that bowl is the most energetically favourable one [21]. However, recently Prinzbach et al. managed to chemically synthesize both cage and bowl isomers [79].

The C_{20} cage is the smallest fullerene and consists of 12 pentagons and no hexagons. The reason why C_{20} seems to be highly promising for polymer creation is related to the fact that though carbon atoms in the C_{20} molecule exhibit sp^2 hybridization, the extreme curvature of the cage surface makes the dihedral angles between bonds (108°) more appropriate to sp^3 hybridization (109.5°). One can expect that for C_{20} the change of the hybridization type, which is a prerequisite for chemical bonding between fullerenes during polymerization, occurs especially easy and hence C_{20} should manifest extreme reactivity and easily form compounds. However, so far only one three-dimensional C_{20}-based polymer has been experimentally produced [29, 91]. Based on the comparison of computational and experimental Raman and IR spectra, Iqbal et al. [29] have shown that the best fit to the experimental data is demonstrated by the face-centered cubic C_{22} structure, where the C_{20} molecules are located on face-centered cubic lattice sites and are bound by two additional C atoms (Fig. 7.8). At the same time, according to computational studies other stable 3D polymers structure might exist, including simple cubic (sc) [8, 63], face-centered cubic (fcc) [8] and body-centered cubic (bcc) [8, 69] phases. Recently a number of additional C_{20} structures to those reported earlier have been proposed by T. Vehviläinen et al. [104]. First of all, in this work [104] there are two computationally generated structures of the sc type. The first one, which is referred to as sc_{open}, reproduces the sc structure proposed earlier by Okada [69] and Chen [8] (Fig. 7.9) with the lattice parameter of 5.45 Å and distorted down to −2% of the bond lengths (Table 7.1). During the polymerization process six bonds per

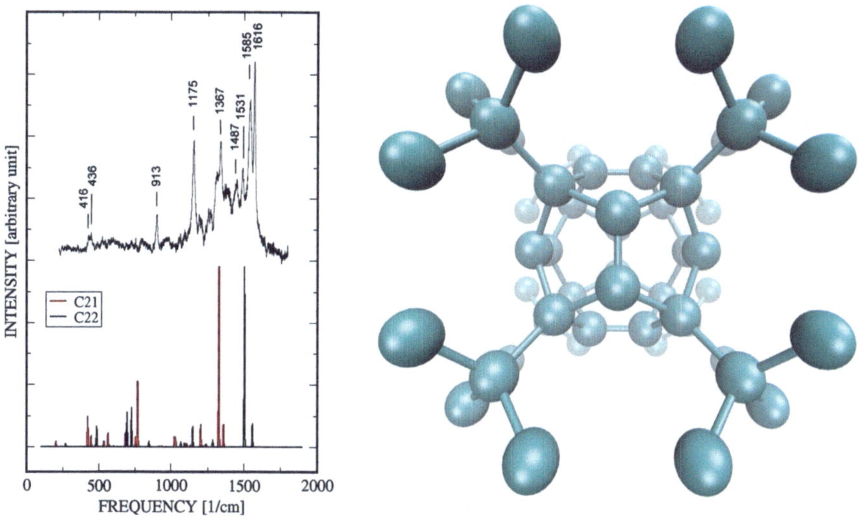

Fig. 7.8 Representation of C22 structure and its Raman spectrum compared to the experimental one for the synthesized structure as shown in [29]

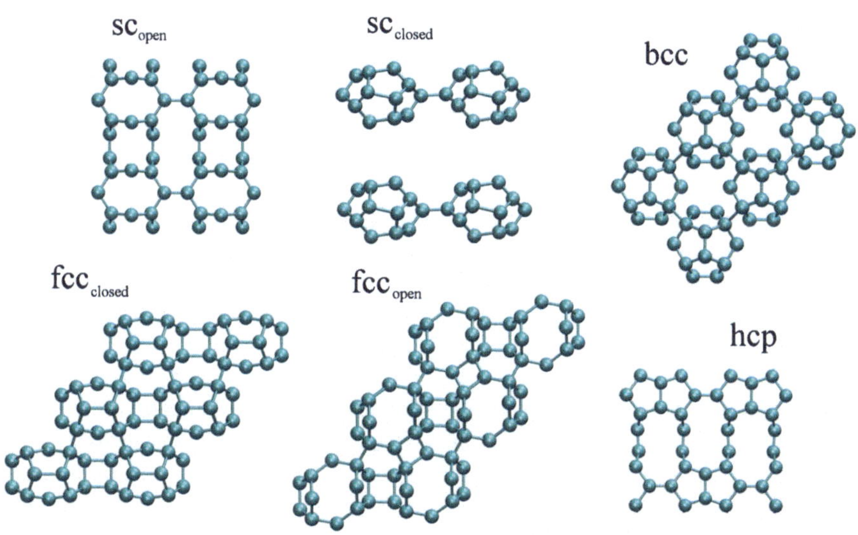

Fig. 7.9 Structures for most energetically favourable polymers considered in [104]. The sc$_{closed}$ and hcp structures are layered materials. The in-plane projection of hcp is shown

Table 7.1 Energies and lattice parameters for studied polymers

Structure	E/atom[a] (eV)	E/ato[b] (eV)	Lattice vectors (Å)	Bond lengths (Å)
C-atom	−1.27	–	–	–
Diamond	−9.10	−9.22	a = 3.573	1.54
Graphite	−9.25	−9.24	a = b = 2.46, b = 6.2	1.42
C_{20} molecule	−8.07	−8.01	–	1.45
C_{20} sc_{open}	−8.49	−8.48	a = 5.45	1.42, 1.46
C_{20} sc_{closed}	−8.36	–	a = 6.06, b = 5.86, c = 6.48	1.38, 1.42, 1.33, 1.48, 1.52, 1.53, 1.55, 1.66
C_{20} fcc_{closed}	−8.56	−8.56	a_1 = [4.28, 3.81, −0.07] a_2 = [−0.07, 3.81, 4.28] a_3 = [4.21, 0.00, 4.22]	1.29, 1.44, 1.50, 1.54, 1.61, 1.62, 1.64, 1.69
C_{20} fcc_{open}	−8.79	–	a_1 = [4.31, 4.54, 0.32] a_2 = [0.32, 4.54, 4.31] a_3 = [3.98, 0.65, 3.98]	1.34, 1.50, 1.53, 1.54, 1.56, 1.57
C_{20} bcc	−8.51	–	a_1 = [−3.37, 3.37, 3.37] a_2 = [3.37, −3.37, 3.37] a_3 = [3.37, 3.37, −3.37]	1.34, 1.54
C_{20} hcp	−8.49	–	a_1 = [2.86, 5.57, 0.00] a_2 = [2.86, −5.57, 0.00] a_3 = [0, 00, 0.00, 6.2]	1.34, 1.39, 1.41, 1.47, 1.50

[a] Vehviläinen et al. 2009 [104].
[b] Chen et al. 2004 [8].

fullerene molecule become broken, opening the cages, but at the same time 12 new bonds between neighbouring molecules are created, leading to the overall energy gain of about 0.4 eV/atom (Table 7.1). About the same energy is found for the considered hcp structure, that is −8.49 eV/atom (Table 7.1).

The second type of the sc polymer structure is referred to as sc_{closed}, because there are no broken bonds in it (Fig. 7.9). This structure is not exactly cubic and looks rather like a set of planar layers with square C_{20} lattice, stacked together in a graphite-like manner. It is called sc type only because of the way it was formed. In order to create this type of a polymer structure, the fullerenes were initially packed in a simple cubic lattice, facing each other with parallel edges, and then isotropically compressed [108]. In this way, the fullerene cage remains closed but the hybridization of six carbon atoms changes from sp^2 to sp^3. The geometry optimization of this structure was performed taking into account the van der Waals interactions as described in [104]. This structure was found to be energetically less favorable compared to the sc_{open} (Table 7.1), but can be used to form even more favourable polymers.

The next structure is bcc structure (Fig. 7.9). In this case, when molecules are bound together, the hybridization of eight carbon atoms changes from sp^2 to sp^3, leading to the configuration with binding energy of 8.51 eV/atom, which is energetically more favourable than either sc or hcp (Table 7.1).

Among the considered typical lattice symmetries, the fcc-type structures are found to be energetically the most favourable ones (Table 7.1). Different fcc-based structures can be created using initial fcc-type lattice symmetry by variation of mutual orientation of C_{20} molecules constituting the lattice and applying different strains to the structure. In [104] two fcc-type polymer structures are considered (Fig. 7.9), which, following the analogy with the sc structures, are called here fcc_{open} and fcc_{closed}. The fcc_{closed} structure has been proposed earlier by Okada [69] with the same energy as calculated in [104] (−8.56 eV/atom). However, another fcc-type structure, predicted in [104], turned out to be pronouncedly more energetically favourable, being characterized by the energy of −8.79 eV/atom. Such a significant difference in energy between the structures is the result of notably different atomic geometries. Indeed, in the fcc_{closed} structure almost all carbon atoms are sp^3 hybridized while only four atoms remain sp^2 hybridized. All the bonds of the fullerene cage remain unbroken and the cage is closed. However this is not the case for fcc_{open} structure, where two bonds of the fullerene cage are broken, which leads to the formation of four new bonds between neighbouring molecules and to cage opening. Each carbon atom of the fullerene cage takes part in the bonding between neigbouring fullerenes. Carbon atoms with dangling bonds remain sp^2 hybridized, while the others become sp^3 hybridized during the polymerization process. Also, as can be seen in Fig. 7.9, the cage constituting the fcc_{open} polymer structure, is slightly inclined compared to the fcc_{closed} structure.

It should be also noted that even though dangling bonds exist in some of the considered structures, no magnetic moment was observed in the calculations for any of the structures [104].

Finally, under the isotropic compression a phase transformation in sc_{closed} polymer structure was found to take place when the pressure reached the value of 30 GPa. This led to the creation of a new high-pressure phase that is called the "quasi-graphite phase" (qgp) [17]. This will be considered in more details in the next section.

Fingerprints by which the studied structures can be identified experimentally are their vibrational and Raman spectra. They have been calculated by [105] as shown in Fig. 7.10.

The elastic constants calculated for all considered C_{20} based structures are summarized in Table 7.2. The results show, in particular, that elastic constants for the fcc structures are large and the structures are therefore very rigid, which correlates with the large number of sp^3 hybridized carbon atoms. The bulk modulus of the fcc structures is also large, though not as large as that for diamond. The calculated bulk moduli for the layered structures are quite small compared to three-dimensional polymers, which results from the weak van der Waals interaction between the planes.

The results of the electronic-structure calculations for all the found equilibrium C_{20} polymer structures are summarized in Table 7.3 and Fig. 7.11, where the Kohn–Sham density-of-states (DOS) plots are presented. As can be seen in the table and the figure, all the considered three-dimensional structures manifest semiconducting properties and have band gaps varying from 1.4 eV (for fcc_{closed}) to 2.32 eV (for bcc) except sc_{open} which is metallic. The results shown in [104] are in good agreement with calculations by Chen [8] for fcc_{closed} structure and by Miyamoto [63] for sc_{open}

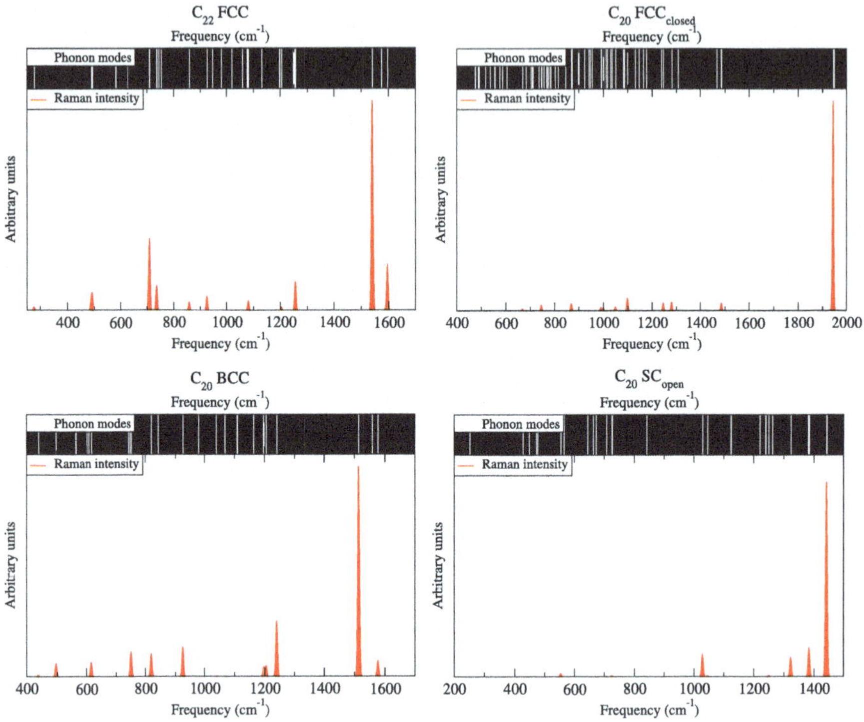

Fig. 7.10 Calculated Raman spectra for some C_{20} based polymer structures [105]

Table 7.2 Elastic properties of studied polymers ([104])

Structure	c_{11} (GPa)	c_{22} (GPa)	c_{33} (GPa)	c_{44} (GPa)	c_{66} (GPa)	c_{12} (GPa)	c_{23} (GPa)	B[a] (GPa)
Diamond	1,085	–	–	569	–	140	–	447
Diamond[a]$_{exp}$	1,079	–	–	575	–	124	–	–
Graphite	1,058	–	49	–	–	–	–	30
Graphite[b]$_{exp}$	1,060	–	36	–	–	–	–	–
Nanotube[c]	1,060	–	40	–	–	–	–	–
C_{20} sc$_{open}$	468	–	–	41	–	134	–	247
C_{20} sc$_{closed}$	221	607	20–35	–	67	110	–	15–20
C_{20} fcc$_{closed}$	723	–	581	207	304	108	179	286
C_{20} fcc$_{open}$	753	–	863	324	431	111	211	345
C_{20} bcc	375	–	–	165	–	106	–	190
C_{20} hcp	633	300	20–35		17	49	–	20–30

[a] McSkimi 1972 [62].
[b] Ahmadieh 1973 [2].
[c] Lu 1997 [52].

Table 7.3 Electronic properties of studied polymers

Structure	[104]	[8]	[69]
Diamond	I, $E_{gap} = 5.6\,eV$	–	–
$Diamond_{exp.}$	I, $E_{gap} = 5.46–5.6\,eV$	–	–
Graphite	M	–	–
C_{20} sc_{open}	M	M	M
C_{20} sc_{closed}	S, $E_{gap} = 1.0\,eV$	–	–
C_{20} fcc_{closed}	S, $E_{gap} = 1.4\,eV$	S, $E_{gap} = 1.4\,eV$	S, $E_{gap} = 3.5\,eV$
C_{20} fcc_{open}	S, $E_{gap} = 1.5\,eV$	–	–
C_{20} bcc	I, $E_{gap} = 2.32\,eV$	–	–
C_{20} hcp	M	–	–

S, semiconducting; I, insulator; M, metallic.

structure. Two-dimensional structures (qgp and hcp) have metallic conductivity in planar directions (Fig. 7.11) and a set of discrete levels in the interplane direction similar to graphite.

The performed molecular-dynamics study of the polymer structure stability showed that all the structures can sustain pressures up to 20 GPa at temperatures below 1,000 K [104]. However, at higher pressures phase transformations can occur, especially when the pressure is rapidly increased. Structures that are energetically less favourable, such as the sc structures, undergo phase transformation much easier than fcc or qgp structures.

Other small fullerenes such as C_{28} observed experimentally by [9], C_{32} synthesized by Kietzmann et al. [37], C_{36} obtained by Kratschmer et al. [40], and C_{50} synthesized by Xie et al. [108] have also attracted much attention from the scientific community. Many theoretical investigations have been devoted to the study of formation and stability of possible fullerenic solid structures. Unfortunately, among the all considered structures only few have been synthesized experimentally to date, namely those consisting C_{20} [29] and C_{36} [37] fullerene cages.

7.3 Quasi-Graphite Structure

Recently, a new carbon structure called quasi-graphite has been proposed by Ganchenkova et al. [17] based on density-functional calculations. It is very stable against external strain and high temperatures as will be shown later. The structure resembles carbon nanotubes welded into planes, which are arranged similar to layers in graphite (Fig. 7.12).

One of the possible schemes of qgp generation is shown in Fig. 7.13. According to the general scheme, the fullerene cages are initially packed in a simple cubic lattice, facing each other with parallel edges (the rightmost inset in Fig. 7.13). Gradual isotropic compression of this system eventually results in the equilibrium layered sc polymer structure with the lattice parameter of 6 Å, as shown in the middle inset in

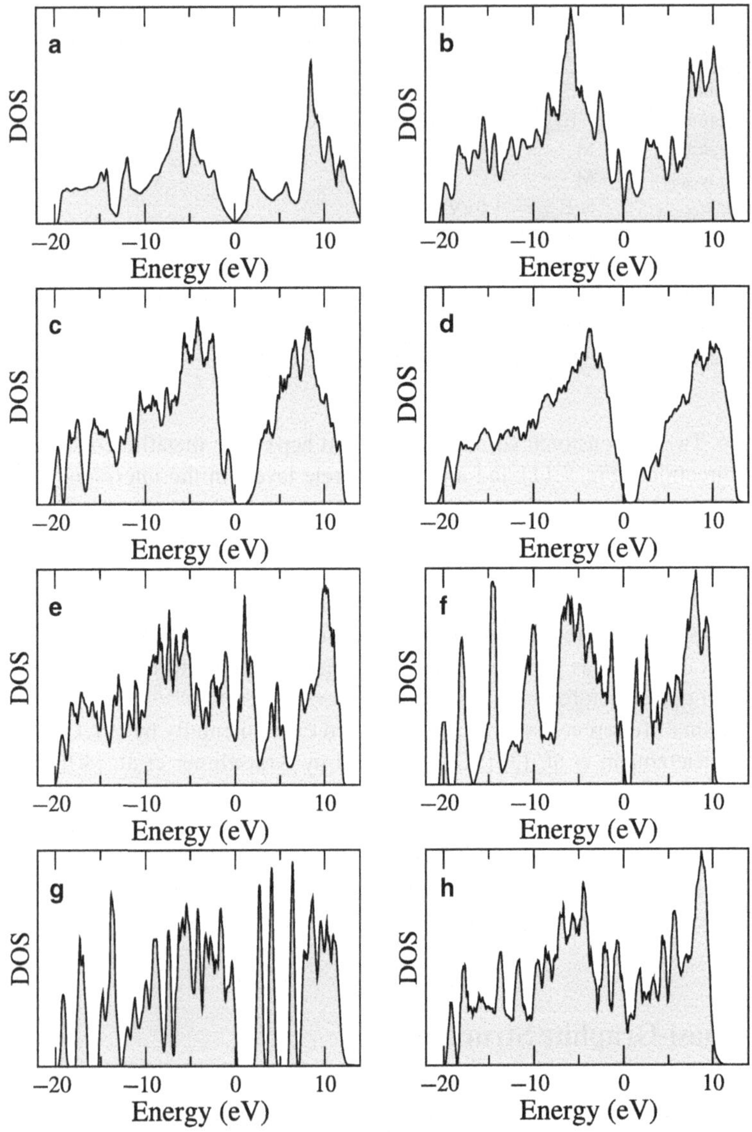

Fig. 7.11 Electronic density of states for (**a**) graphite, (**b**) qgp, (**c**) fcc_{closed}, (**d**) fcc_{open}, (**e**) sc_{open}, (**f**) sc_{closed}, (**g**) bcc, and (**h**) hcp [104]

Fig. 7.13. If, however, the hydrostatic compression is continued, another phase transformation takes place when the pressure reaches the value of 30 GPa (see Fig. 7.13) and a new high-pressure phase, that is qgp, is formed. The activation barrier associated with the transition is about 3 eV/fullerene.

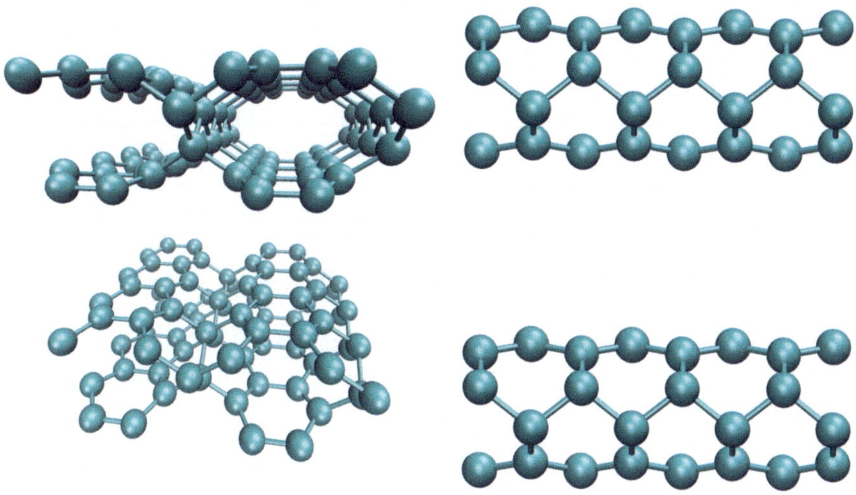

Fig. 7.12 In the quasi-graphite structure small tubes are joined together forming layers. The atoms are sp^2 hybridized except those atoms which join tubes together

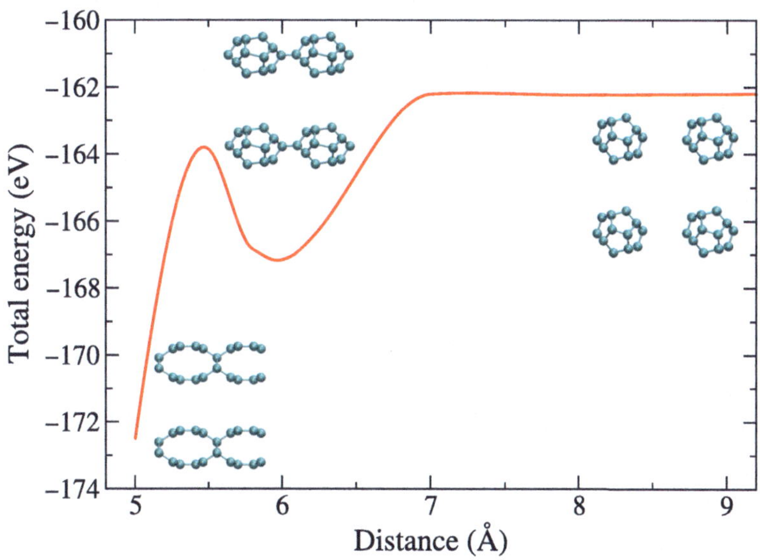

Fig. 7.13 The pressure-induced phase transformation of the C_{20}-based layered cubic structure. Here, the distance means the size of the supercell, which contains only one C_{20}

The resulting structure, which is characterized by the cohesive energy of 8.91 eV/atom, is not only more energetically favourable than the intermediate sc structure, but turned out to have the largest cohesive energy among all the equilibrium 3D polymers based on C_{20} (see Table 7.1 and [104]).

Note that the transition is induced exclusively by application of pressure, without any assistance of temperature (the latter is always zero for the molecular static relaxation used here). The structure of quasi-graphite phase can be described as an arrangement of nanotubes welded together in planar layers, which are, in turn, stacked in a graphite-like manner (Fig. 7.12). The atoms that provide nanotube "welding" manifest the sp^3 hybridization, while the others constituting the nanotubes have sp^2 hybridization. Correspondingly, the quasi-graphite phase is characterized by two different sets of bond lengths, those corresponding to the diamond-like sp^3-type bonds (with the bond length of about 1.54 Å), and those corresponding to graphite-like sp^2 hybridized bonds (1.42 Å).

Thus the quasi-graphite phase consists of two dimensional layers of packed nanotubes, while the layer arrangement and interlayer separations are similar to those observed for individual carbon sheets in graphite. The latter indicates that the interaction between layers is, like in graphite, due mostly to van der Waals (vdW) forces. Correspondingly, the interlayer separation predicted by DFT should be taken with caution, because the vdW interaction is known to be treated improperly by standard exchange-correlation functionals. In order to improve the description of vdW interaction, the obtained equilibrium qgp structure was reoptimized using CPMD, which allows to estimate the additive vdW contribution to the standard DFT energies using an empirical damped dispersion force model. (See the detailed descriptions in [17] and references therein.)

This model ensures that bonded interactions are not included since the effect of vdW correction drops to zero when the distance between atoms is about 3 Å. The application of vdW corrections resulted in the optimized distance between qgp planes of 3.0 Å, see Fig. 7.14. For comparison, the same procedure repeated for

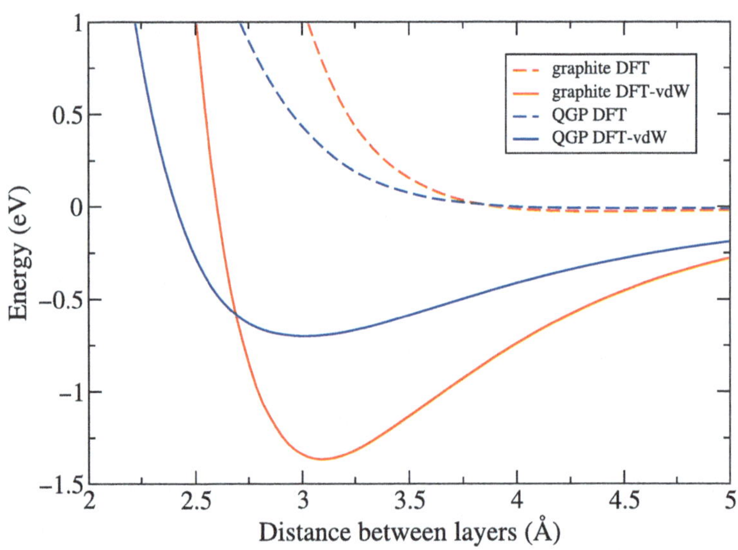

Fig. 7.14 The dependence of the binding energy between layers for 20 atom supercell for quasi-graphite phase and graphite vs. the distance between layers when van der Waals correction is taken into account

Table 7.4 Elastic properties of carbon phases (in GPa)

Structure	c_{11}	c_{22}	c_{33}	c_{44}	c_{66}	c_{12}	c_{23}	B (GPa)
Diamond	1,085	–	–	569	–	140	–	447
Diamond[a]$_{exp}$	1,079	–	–	575	–	124	–	442
Graphite	1,058	–	49	–	–	–	–	39
Graphite[b]$_{exp}$	1,060	–	36	–	–	–	–	34
Nanotube[c]	1,060	–	40	–	–	–	–	19
qgp	1,050	636	48	–	340	90	–	50

[a] McSkimi 1972 [62].
[b] Ahmadieh 1973 [2].
[c] Lu 1997 [52].

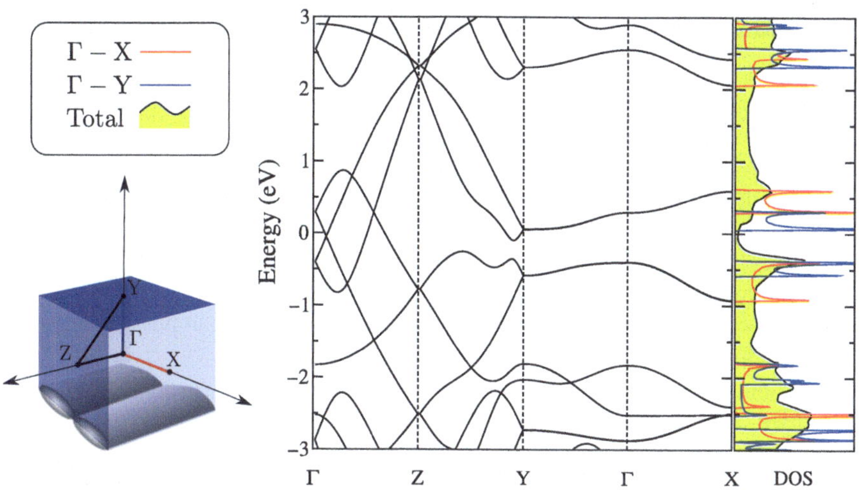

Fig. 7.15 Band structure and DOS calculated for quasi-grahite phase. In the DOS plot filled (*gray*) curve is the total DOS for QGP, red line corresponds to the DOS for the Γ-X section and blue line corresponds to the DOS for the Y-Γ section. For clarity DOS curves are scaled

graphite, predicted the graphite interlayer distance of 3.1 Å. This value differs from the experimental one, 3.35 Å, by less than 10%, giving a rough estimate of the accuracy of interplane separation determination for qgp as well.

The calculated elastic constants and bulk modulus for qgp are presented in Table 7.4. The elastic constants c_{11} and c_{33} for qgp, graphite and carbon nanotubes are found to be very similar, as could be expected since their geometries are similar in the strain direction. The difference between calculated values of the bulk modulus and experimental values is small, less than 3% for solids and 10% for layered materials.

Electronic structure calculations show that the qgp structure manifests metallic properties along the axis of the tube and no metallic conductivity in in-plane (Γ-X) and perpendicular (Y-Γ) directions (see the band structure and electronic DOS shown in Fig. 7.15). This behaviour is to some extent similar to that for graphite.

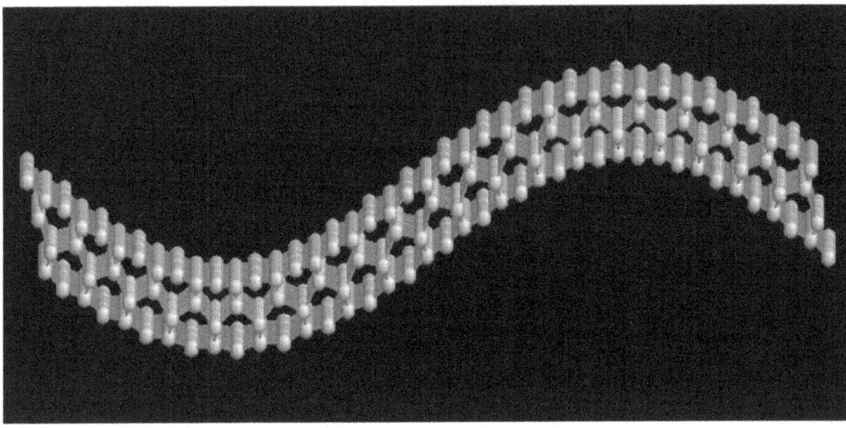

Fig. 7.16 Quasi-graphite layer at temperature of 2,000 K and pressure of 5 GPa molecular dynamics simulations (courtesy of V.A. Borodin, RRC Kurchatov Institute, Russia)

However, in contrast to the graphite electronic structure, for QGP there are two well localized states in (Γ-X) and (Y-Γ) directions seen in the band gap. Moreover, in order to excite an electron from the lowest of these localized states, which is fully occupied, to the next empty level one needs only 0.69 and 0.40 eV in (Γ-X) and (Y-Γ) directions, respectively. This electronic band property can possibly be used in applications which require photons with high frequency.

To examine the stability of the qgp structure the molecular-dynamics simulations were used. Both isotropic and anisotropic strains were considered. It is found that the qgp structure endures pressures up to 50 GPa and 2,000 K without any phase transformations (Fig. 7.16). At higher pressures, however, a phase transformation may occur. In particular, a shock induced phase transformation takes place when the pressure is rapidly increased to values higher than 50 GPa. In this case bonds between quasi-graphite sheets are formed and the structure transforms to the phase which can be considered as amorphous diamond.

The stability of the qgp structure is also confirmed by the absence of the zero- or negative frequency modes in the calculated phonon dispersion spectrum (Fig. 7.17). The phonon dispersion spectrum shown in Fig. 7.17 was calculated using a finite-difference approach implemented in the ABINIT code within the local-density approximation (LDA). In order to estimate the error in calculated frequencies introduced by the choice of the exchange-correlation approximation and the code used, the phonon modes were also calculated using both LDA and semilocal (GGA) approximations and the VASP code. In all tested cases GGA gives approximately 50 cm^{-1} lower frequencies than LDA, since LDA overestimates binding while GGA underestimates it. With LDA, VASP gives identical results compared to ABINIT. Since vibrational information is very specific for the chemical bonds in the structure, Raman spectroscopy provides a fingerprint by which the studied structure can be identified. Therefore Raman spectra for qgp have been calculated. They can be

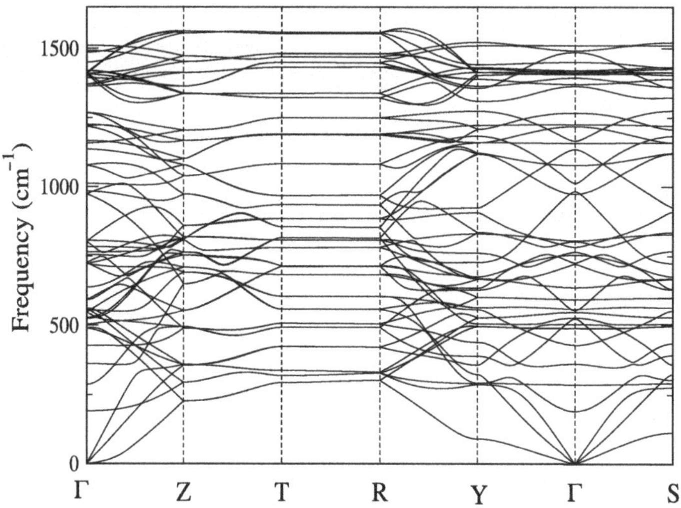

Fig. 7.17 Phonon dispersion relations for qgp

used by experimentalists to identify synthetized materials. Raman intensities are calculated within the LDA approximation using ABINIT.

The Raman spectrum for qgp is presented in Fig. 7.18. Raman active modes with significant intensity are located at 379, 454, 748, 964, 1,008, 1,203, 1,372, 1408 and 1,510 cm^{-1}. In experimental Raman spectra for graphite there is an intense peak at 1,580 cm^{-1} (the G band) and a band at 1,355 cm^{-1} (the D band) which is associated to defects in the carbon aromatic structure [82]. In the qgp Raman spectra the peak at 1,510 cm^{-1} corresponds to the G band of graphite, but the peaks at 748 and 964 cm^{-1} have larger intensity compared to peak at 1,510 cm^{-1}. Moreover, the character of this vibration mode for qgp is similar to that for carbon nanotubes (Fig. 7.19). These peaks are distinctive features and are not seen in other C_{20} polymer structures. There is also a small peaks at 1,372 and 1,408 cm^{-1} close to the D band in graphite.

Summing up, following the similarities between the quasi-graphite phase and both carbon nanotubes and graphene one could expect a number of promising properties of the new structure.

7.4 Negatively Curved Graphite 'Schwarzites'

In 1991, Mackay and Terrones [56] proposed that the inclusion of carbon rings with more than six atoms in a graphite hexagonal mesh could produce stable periodic graphitic arrangements with negative curvature, analogous to zeolites (Fig. 7.20). These new hypothetical arrangements, called 'schwarzites', exhibit topologies similar to the triply periodic minimal surfaces (TPMSs) [56, 57, 96, 98, 99]. To date, several periodic graphitic structures with negative Gaussian curvature have been

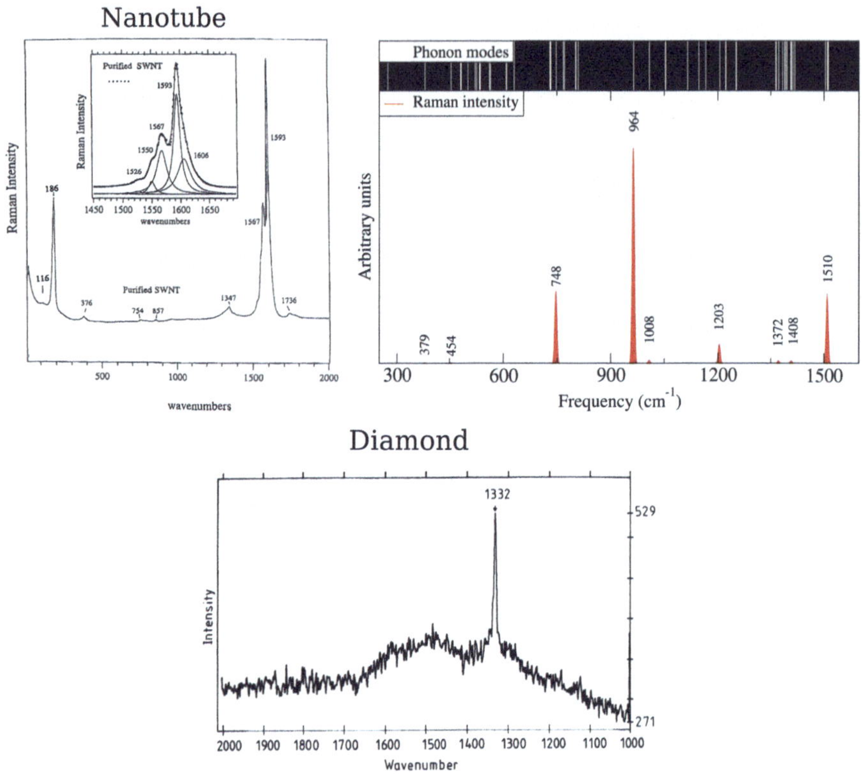

Fig. 7.18 Phonon modes and Raman spectra for qgp calculated with ABINIT and the LDA approximation [17], compared to the Raman spectra for carbon nanotubes and diamond

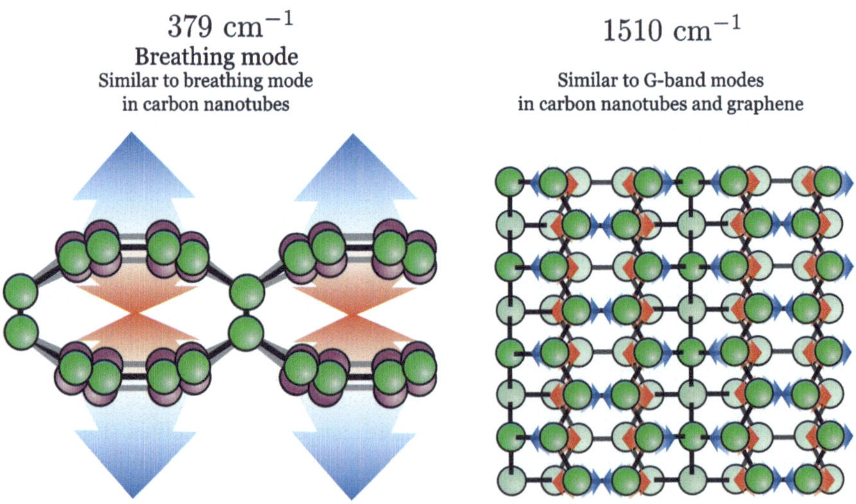

Fig. 7.19 The schematic representation of vibrational modes of QGP

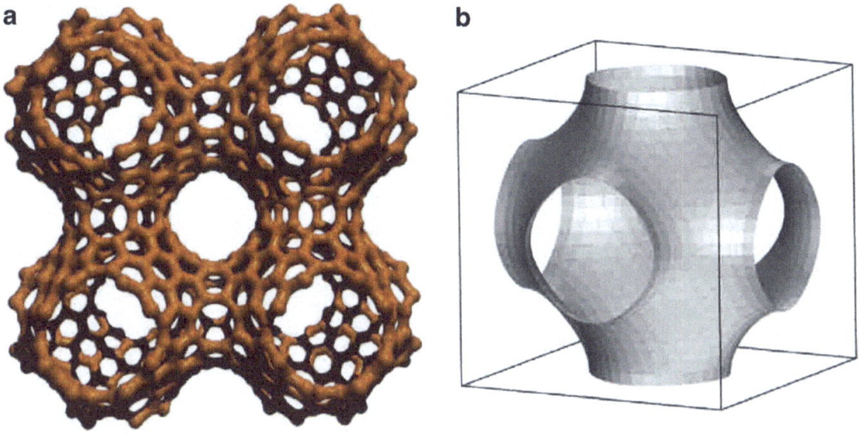

Fig. 7.20 (**a**) Four cubic cells of the P (primitive) TPMS decorated with graphite using hexagonal and octagonal rings of carbon (192 atoms per cubic cell). (**b**) The P TPMS [100]

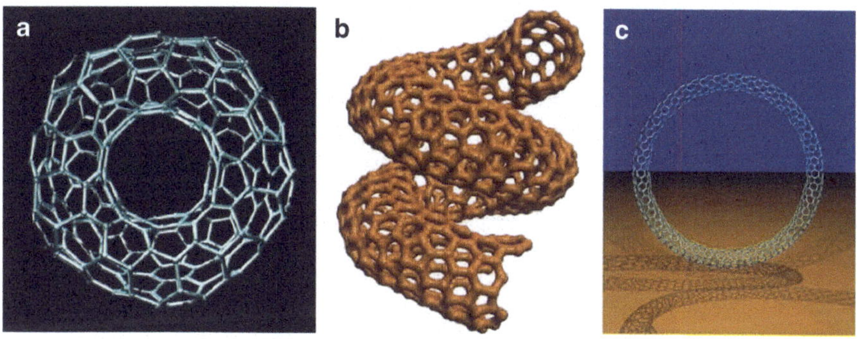

Fig. 7.21 (**a**) Torus decorated with graphite with heptagons, pentagons and hexagons. (**b**) Helicoidal graphite with heptagons, pentagons and hexagons. (**c**) Torus decorated with graphite formed by joining the opposite ends of a (8, 0) nanotube, possessing just hexagons

proposed in the literature [27,56,71,76,96,97,103]. Some of them have been shown to be stable under normal experimental conditions, making them good candidates for crystalline carbon materials with porosity at the mesoscopic or nanoscopic level [5]. Moreover, according to [23] carbon schwarzite phases have already been identified and may be present in ordered nanoporous materials such as those obtained by Ma and co-workers [54], and also in non-graphitizing meso- and nanoporous carbons, such as those reported in [1].

Itoh, Ihara and Kitakami showed the possibility of existence of a closed complex structure without periodicity, namely the toroidal graphite [30, 31] (see Fig. 7.21). Moreover, helicoidal graphitic tubes have been found experimentally [4].

Other proposed graphitic structures with negative Gaussian curvature [27, 56, 71, 76, 96, 97, 103], are classified as diamond-like (D), gyroid (G), and primitive (P), according to the symmetry of the network formed by the interconnected

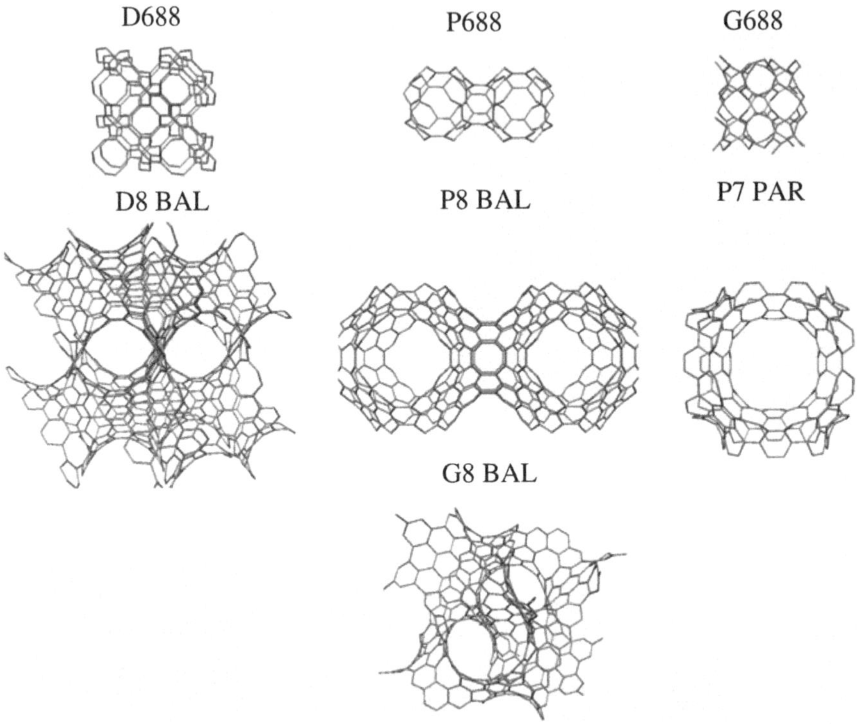

Fig. 7.22 Negative curvature surfaces are joined in space, forming porous structures. From top to bottom and left to right eight cubic cells of D688, four cells of P688, one cell of G688, eight cells of D8BAL, two cells of the P8BAL, one cell of P7PAR, and one cell of G8BAL [102]

channels [56]. The D, G and P (TPMSs) divide the space in two equivalent subspaces, and for this reason they are also referred as balanced (BAL) structures. These TPMS can be decorated with carbon atoms, thus obtaining the D8BAL, G8BAL, and P8BAL structures. On the other hand, there are periodic graphitic structures in which each carbon atom is shared by two octagons and one hexagon (type 688 vertices), thus generating the arrangements D688, G688, and P688 (as defined in [102]). The representations of these structures are shown in Fig. 7.22.

In [102] the tight-binding model was used to optimize the bonds and angles in the carbon schwarzite structures, which are found to be rather close to those of a graphene layer (Table 7.5). The DFT energy difference between the P688 unrelaxed and relaxed structures is about 0.35 eV/atom – both in the LDA and GGA calculations, quite similar to the tight binding result of 0.33 eV/atom. For G688 the figures are about 0.11 eV in DFT results and 0.15 eV in the tight-binding method.

As the total-energy calculations show, the schwarzite structures lie between 0.23 and 0.53 eV above graphene or diamond (Table 7.6). The energies calculated for the P688 and polybenzene (D688) structures compare well with those reported in the work of O'Keeffe et al. [71].

Table 7.5 Geometries of the negatively curved graphitic structures

Structure	Number of atoms/cm^3	Bond lengths (Å)	Angles (deg.)
D688	24	1.399	119.99 ± 0.010
		1.494	
P688	48	1.348	113.89
		1.489	122.71
G688	96	1.409	114.31
		1.492	119.96
			121.99
D8BAL	96	1.427 ± 0.029	119.97 ± 5.86
P8BAL	192	1.427 ± 0.021	119.68 ± 2.69
P7PAR	216	1.428 ± 0.025	119.6 ± 3.440
G8BAL	384	1.428 ± 0.016	119.77 ± 4.16

Table 7.6 Energies per atom relative to the value of a graphene layer [102]

Structure	D688	P688	G688	D8BAL	P8BAL	P7PAR	G8BAL	Diamond
Energy (eV)	0.259	0.523	0.479	0.276	0.233	0.260	0.266	0.032

Table 7.7 Lattice constants and elastic moduli for the schwarzites. Values computed for diamond are shown for comparison (based on data reported in [102])

Structure	Lattice parameter a (Å)	B (GPa)	C_{11}–C_{12} (GPa)	C_4 (GPa)	G (GPa)	E_y (GPa)	v
Diamond	3.57	484	938	607	547.1–547.5	1,193	0.09
D688	6.07	325	161	186	131.7–135.4	325	0.32
P688	7.83	273	101	79	66.1–66.3	184	0.39
G688	9.53	293	123	124	93.3–94.8	255	0.36
D8BAL	11.69	179	148	76	75.5	198	0.32
P8BAL	14.87	149	61	82	54.3–56.8	148	0.33
P7PAR	16.11	126	69	54	45.1–45.3	121	0.34
G8BAL	18.37	159	115	78	69.0–69.1	182	0.31

B, Bulk modulus; G, Isometric shear modulus; E_y, Young's modulus; v, Poisson's ratio.

Among the considered structures, P8BAL exhibits the lowest energy, 0.236 eV/atom, although the difference with polybenzene (D688) is only about 0.026 eV/atom. The P688 structure exhibits the larger energy difference with respect to graphene (0.52 eV/atom) and this and the G688 structures are the only ones with energies above that of the C_{60} structure [71].

The calculated [102] elastic constants of the schwarzites are summarized in Table 7.7. As follows from the table, compared to diamond all the schwarzites are softer: the bulk moduli are larger than, for example, the value for pure silicon (98 GPa and, typically, between 80 and 100 GPa in LDA calculations). All the shear moduli are larger than zero, but still rather small, meaning the structures are

Table 7.8 Band gap values for schwarzites. Values computed for diamond are shown for comparison (based on data reported in [102])

Structure	Diamond	D688	P688	G688	D8BAL	P8BAL	P7PAR	G8BAL
Gap (eV)	7.8	2.9	0.04	1.5	0.2	1.4	1.3	0.2

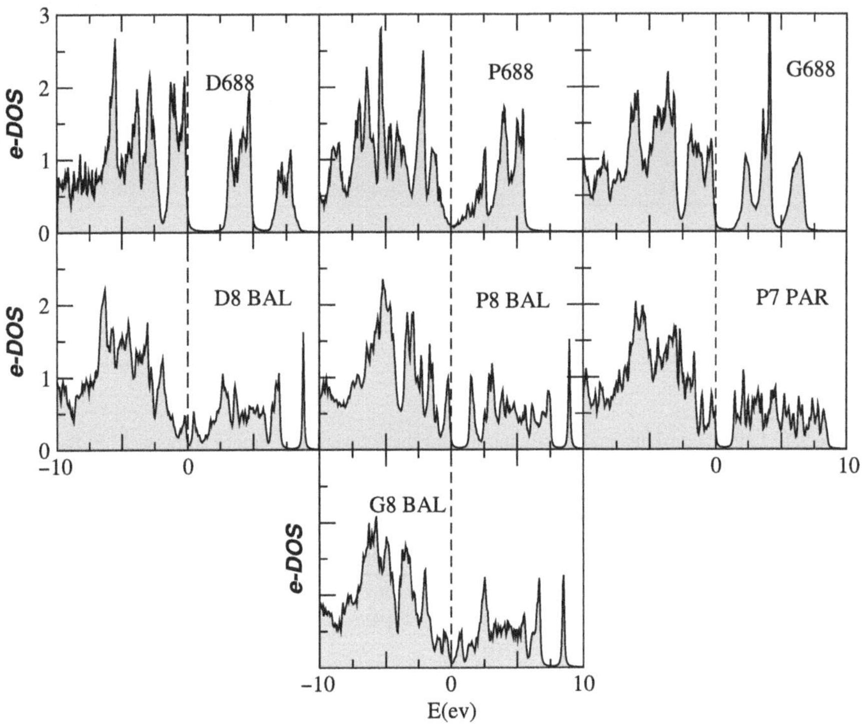

Fig. 7.23 Electronic DOS, normalized to unity. The reference level is the Fermi level (or the highest occupied energy level) [102]

mechanically stable and that these materials would be much softer under shear than under compression.

In Table 7.8 the calculated band gaps are shown. The band gap for diamond is remarkably overestimated in the tight-binding calculations (about 42% over the actual value of 5.47 eV) as could be expected. Therefore, the same trend is to be expected for the other values. Electronic DOSs calculated in [102] are presented in Fig. 7.23, revealing four semiconducting structures (D688, G688, P8BAL, and P7PAR), and three "gapless systems" structures (P688, G8BAL, and D8BAL).

The vibrational DOSs for schwarzites as computed in [102] are shown in Fig. 7.24. The most remarkable feature in these v-DOS spectra is the peaks-and-gaps structure beyond 1,500 cm^{-1} in the smaller cells, although the 10 cm^{-1} Lorentzian broadening used tends to somewhat obscure this behaviour. For D688 there is a

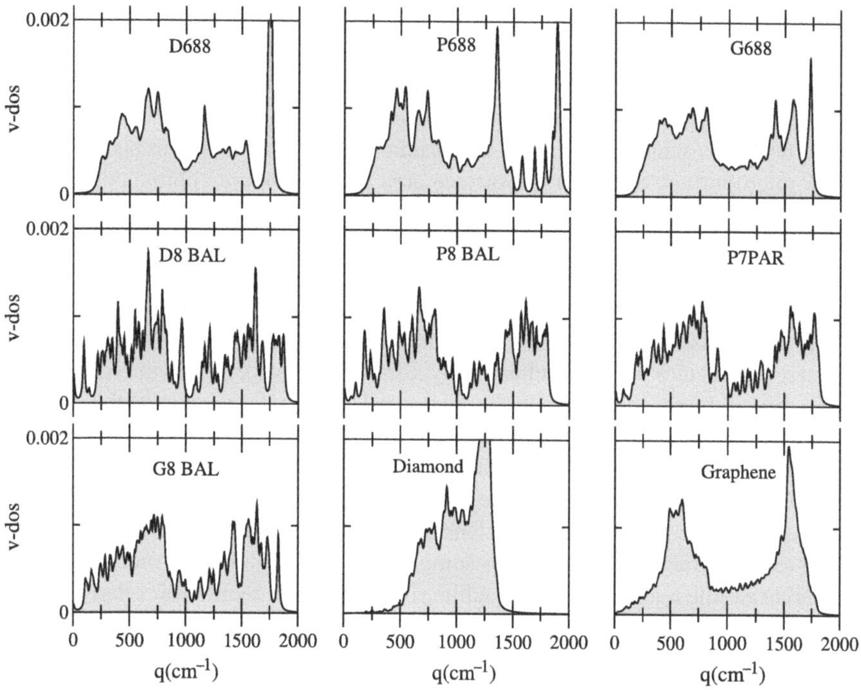

Fig. 7.24 Vibrational DOSs for the seven schwarzite models, normalized to unity. The spectra shown for P8BAL and larger structures actually correspond only to Γ-point vibrations. A Lorentzian broadening of $10\,cm^{-1}$ has been used [102]

clear gap centered at 1,670 and of $85\,cm^{-1}$ width. For P688 more gaps appear: the first one centred at 1,535.5 and $50\,cm^{-1}$ wide, another one at 1,642.5 and $88\,cm^{-1}$ wide, a third one centered at 1,767 and $141\,cm^{-1}$ wide, the fourth is centered at 1,817 and $41\,cm^{-1}$ of width, and finally a fifth, very narrow, gap appearing centered at $1{,}863\,cm^{-1}$ and only $3\,cm^{-1}$ wide. For G688 there are no longer gaps in the investigated region. For the larger structures it is difficult to establish the existence of gaps as long as we are limited to the vibrations at the gamma point. These vibrations, nevertheless, seem to fill rather densely the range of allowed wave vectors and these structures are unlikely to develop vibrational gaps.

7.5 Carbon Replica of Zeolite Nanotemplates

Porous carbons are multipurpose materials that have been widely used in many fields such as air and water purification, catalyst supports, electrodes for supercapacitors [10, 15, 47], etc. The template approach for synthesis of new materials has been a research focus in materials science for many years. Many different porous materials have been prepared using various types of templates. Highly ordered mesoporous carbons were first synthesized using cubic mesoporous silica MCM-48 as the template in 1999 [87]. Since then, ordered porous carbons with different structure have been synthesized using HMS [46], MCM-48 [109] and SBA-15 [88] as templates. Kyotani et al. succeeded to produce high surface area carbon by use of zeolite Y as the template, which has a three-dimensional nano-array structure and is nearly 100% microporous, of which the surface area and micropore volume could be as high as 5,100 m^2/g and 2.1 cm^3/g, respectively [43,44,55]. The synthesis procedure consists of several steps such as pore filling of the parent zeolite with the carbon precursor, carbonization by thermal treatment, and subsequent liberation of the carbon replica from the zeolite framework by dissolving in HF. During the pore filling different processes such as diffusion, adsorption, desorption, and polymerization of precursor molecules occur simultaneously. Since these processes compete partially, the quality of the resulting carbon should depend on the relative rates of the individual processes. In case of incomplete pore filling the carbon framework will be less stable and the pore structure of the carbon material will be less well defined [19].

In 2005 A. Garsuch et al. [18] have reported about the synthesis of ordered carbon replicas of HY-zeolite by using ethylene as carbon precursor. The obtained carbon preserves the structural regularity to some extent and it shows the morphology of the parent zeolite template. The crystallite size of the used zeolite was relatively low (about 0.5 μm). In 2006 A. Garsuch and co-workers reported another investigations by using a zeolite with a considerably larger particle size of about 3 μm [19]. The carbon preparation can be briefly described as follows [18], the zeolite (protonic form, obtained by deammoniation at 623 K) was loaded with ethylene in a batch reactor at a pressure of 10 bar followed by heating up to 523 K. Afterwards, the polymer/template composite was calcined at 723 K for 3 h in nitrogen flow. The composite was loaded in a second step and finally calcined at 1,073 K in nitrogen. The zeolite template was removed by dissolving in HF. The resulted carbons were characterized by means of nitrogen adsorption at 77 K, x-ray diffraction (XRD) and scanning electron microscopy (SEM). The analysis of experimental results was actively supported by the density-functional theory (DFT) calculations used for the pore size distribution determination.

Two different samples of carbons have been compared. The material obtained using large zeolite crystallites was denoted CL. Sample CS has been synthesized from small zeolite crystallites. In Fig. 7.25a–c scanning electron microscope (SEM) images of the zeolite, the zeolite/carbon composite, and the resulting carbon CL as reported in [19] are shown. For comparison purpose, the SEM image of carbon CS obtained from Y-zeolite with smaller crystallite size of about 0.5 μm has been depicted in Fig. 7.25d. The images confirm the morphological similarity between the zeolite template, the composite, and the carbon material. There was no significant carbon deposition on the external surface of the zeolite as shown by the smooth surface of the carbon and the composite particles [26]. The particle size of the resulting carbons corresponds to the parent zeolite template. For sample CL an average size of about 3 μm (Fig. 7.25c) was determined. Both materials show mainly pore diameters between 0.8 and 2 nm indicating micropores.

Fig. 7.25 SEM images of large zeolite Y crystallites (**a**), the corresponding zeolite/carbon composite (**b**), the resulting carbon (**c**), and carbon obtained from small zeolite crystallites [19]

Lei et al. [50] exploited chemical vapor deposition of methanol using zeolite LTA as a template to prepare nanocarbons. Figure 7.26 demonstrates the SEM images of zeolite template and synthesized carbon materials as obtained by Lei et al. [50]. The zeolite (LTA in this case) has the cubic structure as can be seen in Fig. 7.26a. Figure 7.26b,c show that synthesized carbon materials have cubic and plate forms that are similar to the template structure, suggesting that the resultant carbon copies the template structure.

XRD patterns of zeolite LTA and porous carbons synthesized of Z-973-CT shown in Fig. 7.26d exhibits many sharp peaks for the zeolite LTA as due to the inherent crystal structure. As can be seen, some of sharp peaks are also observable in the patterns of all the resultant carbons, suggesting that the zeolite framework of the template is present. This also takes place in the zeolite LTA of template, which is an indication of long-range structural ordering, and suggest that synthesized carbon partly reflect template structure [19, 38, 92, 107]. When the heat-treatment temperature is elevated from 1,073 to 1,273 K, the XRD peaks become stronger and sharper. Consequently the carbon grown in the pores of zeolite should be more crystallized, leading to more stable and regular structure.

It has been noticed also that the higher the decomposition temperature, the more graphitic and ordered the structure [50]. The analysis of the XRD peaks nanocarbon indicates the formation of three-dimensional regular structure of synthesized

Fig. 7.26 SEM images of LTA zeolite template and carbon synthesized. (**a**) zeolite template; (**b**), (**c**) synthesized carbon materials; (**d**) XRD patterns of (**a**) zeolite LTA, (**b**), (**c**), and (**d**) carbon replica at 1073, 1123, and 1273 K, respectively [50]

carbon irrespective of the growth in highly narrow pore spaces with the periodicity of 1.2 nm. The supporting experimental study grand-canonical Monte Carlo simulation of N_2 adsorption isotherm indicates that synthesized nanoporous carbon has a hollow hemispherical structure of which diameter is less than 0.7 nm.

Yang et al. [113] reported the synthesis of zeolite-like carbon materials that exhibit well-resolved powder XRD patterns and very high surface area. In this work, the zeolite-like carbons were prepared via chemical vapor deposition (CVD) at 800 or 850C using zeolite β as solid template and acetonitrile as carbon precursor. SEM images of the zeolite β template and the resulted sample (Fig. 7.27) indicate that

Fig. 7.27 Representative SEM images of (**a**) zeolite β template (ZTA) and (**b, c**) ZTA-templated carbon prepared at a CVD temperature of 850°C [113]

the morphology (shape and size) of the zeolite particles is replicated in the carbon material. Carbon particles with solid cores and smooth surfaces (Fig. 7.3c), and hardly any irregular stand-alone particulates are formed. The high level of zeolite-type structural ordering in the present carbon materials is confirmed by powder XRD patterns with at least two well resolved diffraction peaks and TEM image analysis that reveal well-ordered micropore channels [113]. The resulted carbons possess surface area of up to 3,200 m^2/g and pore volume of up to 2.41 cm^3/g. The material exhibits a significant proportion of the porosity in the carbons (up to 76% and 56% for surface area and pore volume, respectively). Both TEM and nitrogen sorption data indicate that porosity is dominated by pores of size 0.6–0.8 nm.

The hydrogen uptake of the carbon materials was measured via two methods, gravimetric and volumetric. The carbon replica of zeolite β nanotemplate was found exhibit enhanced (and reversible) hydrogen storage capacity, with measured uptake of up to 6.9 wt% and estimated maximum of 8.33 wt% at -196 °C and 20 bar. At 1 bar, hydrogen uptake capacity as high as 2.6 wt% is achieved. Isosteric heat of adsorption of 8.2 kJ/mol indicates a favorable interaction between hydrogen and the surface of the carbons. Hirscher and Panella, in particular, examined hydrogen storage for a wide range of high surface area carbon nanostructures and reported a maximum capacity of 4.5 wt% at 70 bar. Yang et al. [111, 112] have also observed uptake of up to 4.5 wt% at 20 bar for zeolite-templated carbons with low levels of zeolite-type structural ordering. The storage capacity of the discussing zeolite-like carbons is generally higher than that of metal organic frameworks (MOFs) [11, 12, 16, 48, 51, 73, 74, 84–86, 106, 114], mesoporous carbons [75, 95], zeolites [64, 116], or organic microporous polymers [61]. Thus, the hydrogen uptake capacity observed for the zeolite-like carbon materials is among the highest ever reported for carbon (activated carbon, mesoporous carbon, CNTs) or any other (MOFs, zeolites) porous material.

As was discussed in the Introduction, the number of zeolite structures, which could be used as a template is enormously large. Therefore, a computational approach that might allow us to predict the structure and properties of a replica to be synthesized within a predefined nanotemplate would be of an immense practical value. However, due to the complexity of the systems the simulation of the replica formation is an extremely challenging task and to date the authors can not refer to any.

7.6 Local Moments and Dangling-Bond Magnetism of Carbon Structures

The existence of ferromagnetic order in elemental carbon has attracted considerable interest in recent years, summarized in [59]. There is an obvious technological driving force to create light, non-metallic magnetic materials with Curie temperatures well above room temperature. Obviously it is also important to understand the origin of magnetic order in systems traditionally thought to exhibit diamagnetic behavior only.

In particular, C_{60} based polymers have attracted a lot of attention since the magnetic phase of these materials has been discovered. This magnetic phase is characterized by a high Curie temperature, measured to be 500–820 K, and small saturation magnetization values [59]. This discovery was highly unexpected, since the nature of the magnetism of a carbon based structure was unusual. These materials contain no d- or f-electrons, providing only s- and p-state derived moments, which are to be considered as candidates for the source of the magnetism. While much research has been carried out on the origin, the picture of carbon magnetism is still far from being complete.

A much-debated and controversial question of experimental nature is whether the observed magnetic behavior is an intrinsic property or due to the presence of unintentional magnetic impurities. Ferromagnetic transition metals, such as Fe and Ni, are often used as catalysts to produce carbon nanostructures in reactor conditions.

Leaving the experimental discussion aside, one can approach the issue from the theoretical point of view by focusing on two basic questions:

(i) Under what conditions can stable local magnetic moments appear in all-carbon systems, and how large can the moments be?
(ii) What is the nature and range of the interactions between the moments (if any), and can they lead to long-range magnetic order at finite temperatures?

In order to explain the magnetism, first the explanation of the existence of unpaired electrons, and second, the understanding of origin of the ferromagnetic (FM) coupling mechanism is needed. So far, most of the theoretical explanations have focussed on justifying the existence of the local moments, leaving the origin of the FM coupling mechanism still unclear.

Based on a review of C_{60}-based and related materials ([59] and references therein) the viewpoint emerges that these materials belong to a sub-class of FM materials which follow a common formation of ferromagnetism and is classified as a magnetism induced by defects (structural and/or topological). The same conclusion can be obviously extended to the whole class of nanoporous carbon materials.

There is a large body of work addressing in particular this question. Much of this work is reviewed in the recent article by Lehtinen et al. [49] and will not be repeated here. There are many mechanisms that can give rise to the appearance of localized spin moments. In particular, defects in the carbon networks such as under-coordinated atoms (dangling bonds) readily produce large paramagnetic moments arising from strongly localized spin densities. The magnitude of the moment is, however, sensitive to the specific defect geometry, as is the extent of the spin density. The defect-mediated mechanism has been explored especially in graphite, fullerenes and various types of nanotubes, focusing particularly on vacancies, adatoms (interstitials), and their complexes. It appears that nonmagnetic impurities, such as hydrogen and nitrogen, common in carbon, can act to stabilize and enhance the moments considerably. Finite systems, such as grapheme ribbons and flakes, have plenty of under-coordinated edge atoms which readily form spin-polarised configurations.

Electron and particle irradiation have been used to produce controlled amounts of defects in carbon materials in order to correlate the magnetism to defect concentrations. This indeed produces the most direct evidence for defect-induced magnetism. Hydrogen from proton irradiation is particularly efficient in stabilizing defect-associated moments. It also has a role in immobilizing such defects that otherwise would move rapidly already at low temperature, such as a C adatom on graphite, and disappear from the system through recombination and/or clustering.

Thus it is fully conceivable that local moments of the order of 1–2 μ_B can readily appear in defected all-carbon systems with undercoordinated atoms, also in conjunction with reactive non-magnetic impurities such as hydrogen, oxygen and nitrogen. However, to explain possible magnetism one needs (i) large enough defect/moment densities to produce macroscopic magnetization and (ii) ferromagnetic interactions between the moments. These issues are much discussed presently. The reported experimental observations usually give very low saturation magnetizations and can conceivably be reconciled with expected defect densities, such as arising from particle irradiation and other nanostructure manipulation. The inter-moment interactions, when calculated using spin-density-functional theory and defect–defect distances in accord with the defect densities, tend to be oscillatory and usually weak. The spin density around the local moment is usually well localized, and mechanisms such as superexchange or double exchange predict weak coupling. Given also the random structural positions of the defects, it is difficult to explain the high Curie temperatures reported in some experiments in terms of defect-induced magnetic order only. The situation is different for edge-associated magnetism in grapheme nanoribbons and flakes, where spin-polarised extended states can readily form.

In summary, it is quite possible that magnetism in various exotic carbon systems may be associated with defects, albeit that the origin a magnitude of the interactions between moments is still somewhat of a mystery. Nonmagnetic impurities, in particular hydrogen, play a role, as they can both suppress defect annealing and enhance the stable moment and its interaction with other moments. Irradiation appears to be the best tool to control defect densities and manipulate carbon magnetism.

References

1. Acharaya M, Mathews J, Strano M, Billinge SJL, Petkov V, Subramoney S, Foley HC (1999) Phil Mag B 79:1499
2. Ahmadieh AA, Rafizadeh HA (1973) Phys Rev B 7:4527
3. Alexandrov AS, Bratkovsky AM, Williams RS (2003) Phys Rev B 67:075301

4. Amelinckx S, Zhang XB, Bernaerts D, Zhang XF, Ivanov V, Nagy JB (1994) Science 265:63
5. Benedek G, Vahedi-Tafreshi H, Barborini E, Piseri P, Milani P, Ducati C, Robertson J (2003) Diamond Relat Mater 12:768
6. Beu TA, Onoe J, Hida A (2005) Phys Rev B 72:155416
7. Chauvet O, Oszlanyi G, Forro L, Stephens PW, Tegze M, Faigel G, Janossy A (1994) Phys Rev Lett 72:2721
8. Chen Z, Heine T, Jiao J, Hirsch A, Thiel W, Schleyer P (2004) Chem Eur J 10:963
9. Cox DM, Reichmann KC, Kaldor A (1988) J Chem Phys 88:1588
10. Chai GS, Yoon SB, Yu JS, Choi JH, Sung YE (2004) J Phys Chem B 108:7074
11. Chen B, Ockwig NW, Millward AR, Contreras DS, Yaghi OM (2005) Angew Chem Int Ed 44:4745
12. Chun H, Dybtsev DN, Kim H, Kim K (2005) Chem Eur J 11:3521
13. Dorset DL, Fryer JR (2001) J Phys Chem B 105:2356
14. Fleming RM, Ramirez AP, Rosseinsky MJ, Murphy DW, Haddon RC, Zahurak SM, Makhija AV (1991) Nature 352:787
15. Fodriguez-Reinoso F (1998) Carbon 36:159
16. Frost H, Duren T, Snurr RA (2006) J Phys Chem B 110:9565
17. Ganchenkova M, Vehviläinen T, Nieminen RM (2008) Phys Rev B 78:195421
18. Garsuch A, Klepel O (2005) Carbon 43:2330–2337
19. Garsuch A, Klepel O, Sattler RR, Berger C, Gläser R, Weitkamp J (2006) Carbon 44:593
20. Goze C, Rachidi F, Núñez-Regueiro M, Marques L, Hodeau J-L, Mehring M (1996) Phys Rev B 54:R3676
21. Grossman JC, Mitas L, Raghavachari K (1995) Phys Rev Lett 75:3870
22. Gryko J, MacMillan PF, Marzke RF, Ramachandran GK, Patton D, Deb SK, Sankey OT (2000) Phys Rev B 62:7707
23. Harris PJF (1997) Int Mater Rev 42:206
24. Hebard AF, Rosseinsky MJ, Haddon RC, Murphy DW, Glarum SH, Palstra TTM, Ramirez AP, Kortan AR (1991) Nature 350:600
25. Heine T, Fowler PW, Seifert G (1999) Solid State Commun 111:19
26. Hou PX, Yamazaki T, Orikasa H, Kyotani T (2005) Carbon 43:2624
27. Huang MZ, Ching WY (1993) Phys Rev B 47:1593
28. Iijima S (1991) Nature 354:56
29. Iqbal Z, Zand Y, Grebel H, Vijayalakshmi S, Lahamer A, Benedek G, Cariboni MJ, Spagnolatti I, Sharma R, Owens FJ, Kozlov ME, Rao KV, Muhammed M (2003) Eur Phys J B 31:509
30. Ihara S, Itoh S, Kitakami J (1993) Phys Rev B 47:12908
31. Itoh S, Ihara S, Kitakami J (1993) Phys Rev B 47:1703
32. Iwasa Y, Arima T, Fleming RM, Siegrist T, Zhou O, Haddon RC, Rothberg LJ, Lyons KB, Carter HL, Hebard AF Jr, Tycko R, Debbagh G, Krajewski JJ, Thomas GA, Yagi T (1995) Science 264:1570
33. Iwasa Y (1996) Solid State Phys (Kotai Buturi) 31:909 (AGNE, Tokyo)
34. Joo SH, Jun S, Ryoo R (2001) Microporous Mesoporous Mater 44/45:153
35. Jun S, Joo SH, Ryoo R, Kruk M, Jaroniec M, Liu Z, Ohsuna T, Terasaki O (2000) J Am Chem Soc 122:10712
36. Ke S-H, Baranger HU, Yang W (2003) Phys Rev Lett 91:116803
37. Kietzmann H, Rochow R, Gantefor G, Eberhardt W, Vietze K, Seifert G, Fowler PW (1998) Phys Rev Lett 81:5378
38. Kim TW, Park IS, Ryoo R (2003) Angew Chem Int Ed 42:4375
39. Kim Y-H, Zhao Y, Williamson A, Heben MJ, Zhang SB (2006) Phys Rev Lett 96:016102
40. Kratschmer W, Lamb LD, Fostiropoulos K, Huffman DR (1990) Nature 347:354
41. Kroto HW, Heath JR, O'Brien SC, Curl RF, Smalley RE (1985) Nature 318:162
42. Kuzuo R, Terauchi M, Tanaka M (1994) Phys Rev B 49:5054
43. Kyotani T, Liu Z, Terasaki O, Tomita A (2001) Chem Mater 13:4413
44. Kyotani T, Ma ZX, Tomita A (2003) Carbon 41:1451
45. Lee J, Yoon S, Hyeon T, Oh SM, Bum Kim K (1999) Chem Commun 21:2177

46. Lee J, Yoon S, Oh SM, Shin SH, Hyeon T (2000) Adv Mater 12:359
47. Lee J, Kim J, Lee Y, Yoon S, Oh SM, Hyeon T (2004) Chem Mater 16:3323
48. Lee JY, Pan L, Kelly SP, Jagiello J, Emge TJ, Li J (2005) Adv Mater 17:2703
49. Lehtinen PO, Krasheninnikov AV, Foster AS, Nieminen RM (2006) In: Makarova T, Palacio F (eds) Carbon-based magnetism. Elsevier, Amsterdam, pp 371–396
50. Lei S, Miyamoto J, Ohba T, Kanoh H, Kaneko K (2007) J Phys Chem C 111:2459
51. Lin X, Jia J, Zhao XB, Thomas KM, Blake AJ, Walker GS, Champness NR, Hubberstey P, Schroder M (2006) Angew Chem Int Ed 45:7358
52. Lu JP (1997) Phys Rev Lett 79:1297
53. Lu GQ, Zhao XS (2004) Nanoporous Materials: Science and Engineering. Nanoporous Materials – An Overview. World Scientific, Imperial College Press, 1–14
54. Ma Z, Kyotani T, Liu Z, Terasaki O, Tomita A (2001) Chem Mater 13:4413
55. Ma Z, Kyotani T, Tomita A (2002) Carbon 40:2367
56. Mackay AL, Terrones H (1991) Nature 352:762
57. Mackay AL, Terrones H (1993) Phil Trans R Soc A 343:113
58. Maniwa Y, Sato M, Kume K, Kozolv ME, Tokumoto M (1996) Carbon 34:1287
59. Makarova T, Palacio F (2006) Carbon-based magnetism. Elsevier, Amsterdam
60. Marques L, Hodeau J-L, Núñez-Regueiro M, Perroux M (1996) Phys Rev B 54:R12 633
61. McKeown NB, Ghanem B, Msayib KJ, Budd PM, Tattershall CE, Mahmood K, Tan S, Book D, Langmi HW, Walton A (2006) Angew Chem Int Ed 45:1804
62. McSkimi H, Andreatch P (1972) J Appl Phys 43:2944
63. Miyamoto Y, Saito M (2001) Phys Rev B 63:161401
64. Nijkamp MG, Raaymakers JEMJ, van Dillen AJ, de Jong KP (2001) Appl Phys A 72:619
65. Noon W, Kong Y, Ma J (2002) PNAS 99:6466
66. Núñez-Regueiro M, Marques L, Hodeau J-L, Bethoux O, Perroux M (1995) Phys Rev Lett 74:278
67. Okada S, Saito S (1997) Phys Rev B 55:4039
68. Okada S, Saito S (1999) Phys Rev B 59:1930
69. Okada S, Miyamoto Y, Saito M (2001) Phys Rev B 64:245405
70. Okada S, Oshiyama A (2003) Phys Rev B 68:235402
71. O'Keeffe M, Adams G, Sankey OT (1992) Phys Rev Lett 68:2325
72. Oszlanyi G, Forro L (1995) Solid State Commun 93:265
73. Panella B, Hirscher M (2005) Adv Mater 17:538
74. Panella B, Hirscher M, Putter H, Muller U (2006) Adv Funct Mater 16:520
75. Pang J, Hampsey JE, Wu Z, Hu W, Lu Y (2004) Appl Phys Lett 85:4887
76. Philips R, Drabold DR, Lenosky L, Adams GB, Sankey OT (1992) Phys Rev B 46:1941
77. Piskoti C, Yarger J, Zettl A (1998) Nature 393:771
78. Polarz S, Smarsly B (2002) J Nanosci Nanotechnol 2:581
79. Prinzbach H, Weiler A, Landenberger P, Wahl F, Worth J, Scott LT, Gelmont M, Olevano D, Issendorff B (2000) Nature 407:60
80. Rao AM, Zhou P, Wang K, Hager GT, Holden JM, Wang Y, Lee W, Bi X, Eklund PC, Cornett DS, Duncan MA, Amster IJ (1993) Science 259:955
81. Rao AM, Eklund PC, Hodeau J-L, Marques L, Núñez-Regueiro M (1997) Phys Rev B 55:4766
82. Reich S, Thomsen C (2004) Phil Trans Soc Lond A 362:2271
83. Ren Y, Ng T, Liew K (2006) Carbon 44:397
84. Rosi NL, Eckert J, Eddaoudi M, Vodak DT, Kim J, O'Keeffe M, Yaghi OM (2003) Science 300:1127
85. Rowsell JLC, Millward AR, Park KS, Yaghi OM (2004) J Am Chem Soc 126:5666
86. Rowsell JLC, Yaghi OM (2005) Angew Chem Int Ed 44:4670
87. Ryoo R, Joo SH, Jun SJ (1999) Phys Chem B 103:7743
88. Ryoo R, Joo SH, Kruk M et al (2001) Adv Mater 13(9):677
89. Saito S, Oshiyama A (1991) Phys Rev Lett 66:2637
90. Soto JR, Calles A, Moran-Lopez JL (2000) J Chem Phys 113:1055
91. Spagnolatti I, Bernasconi M, Benedek G (2002) Europhys Lett 59:572

92. Su FB, Zeng JH, Bao XY, Yu YS, Lee JY, Zhao XS (2005) Chem Mater 17:3960
93. Tabata Y, Ikada Y (1999) Pure Appl Chem 71:2047
94. Tang ZK, Sun HD, Wang J, Chen J, Li G (1998) Appl Phys Lett 73:2287
95. Terres E, Panella B, Hayashi T, Kim YA, Endo M, Dominguez JM, Hirscher M, Terrones H, Terrones M (2005) Chem Phys Lett 403:363
96. Terrones H, Mackay AL (1992) Carbon 30:1251
97. Terrones H, Mackay AL (1993) Chem Phys Lett 207:45
98. Terrones H, Terrones M, Hsu WK (1995) Chem Soc Rev 24:341
99. Terrones H, Terrones M, Moran-Lopez JL (2001) Curr Sci 81:1011
100. Terrones H, Terrones M (2003) New J Phys 5:126
101. Ugarte D (1992) Nature 359:707
102. Valencia F, Romero AH, Hernández E, Terrones M, Terrones H (2003) New J Phys 5:123
103. Vanderbilt D, Tersoff J (1992) Phys Rev Lett 68:511
104. Vehviläinen T, Ganchenkova M, Nieminen RM (2009) J Nanosci Nanotechnol 9:4367
105. Vehviläinen T, Ganchenkova M, Nieminen RM (2009) (unpublished)
106. Wong-Foy AG, Matzger AJ, Yaghi OM (2006) J Am Chem Soc 128:3494
107. Xia YD, Mokaya R (2004) Adv Mater 16(17):1553
108. Xie SY, Gao F, Lu X, Huang RB, Wang CR, Zhang X, Liu ML, Deng SL, Zheng LS (2004) Science 304:699–699
109. Yoon SB, Kim JY, Yu J-S (2001) Chem Commun 6:559
110. Yoon SB, Kim JY, Yu J-S (2002) Chem Commun 14:1536
111. Yang Z, Xia Y, Mokaya R (2005) Microporous Mesoporous Mater 86:69
112. Yang Z, Xia Y, Mokaya R (2005) Stud Surf Sci Catal 156:573
113. Yang Z, Xia Y, Mokaya R (2007) J Am Chem Soc 129:1673
114. Zhao X, Xiao B, Fletcher AJ, Thomas KM, Bradshaw D, Rosseinsky M (2004) Science 306:1012
115. Zhao Y, Kim Y-H, Dillon AC, Heben MJ, Zhang SB (2005) Phys Rev Lett 94:155504
116. Zecchina A, Bordiga S, Vitillo JG, Ricchiardi G, Lamberti C, Spoto G, Bjorgen M, Lillerud KP (2005) J Am Chem Soc 127:6361

Index

M

N

O

P

Q

R

S

The manufacturer's authorised representative in the EU is Springer Nature Customer Service Centre GmbH, Europaplatz 3, 69115 Heidelberg, Germany. If you have any concerns regarding our products, please contact ProductSafety@springernature.com

Printed and bound by CPI Group (UK) Ltd, Croydon, CR0 4YY
29/07/2026
02180636-0003